Gappar Rahmanberdiev

WATER-SOLUBLE CELLULOSE ETHERS

Gappar Rahmanberdiev

WATER-SOLUBLE CELLULOSE ETHERS

ScienciaScripts

Imprint
Any brand names and product names mentioned in this book are subject to trademark, brand or patent protection and are trademarks or registered trademarks of their respective holders. The use of brand names, product names, common names, trade names, product descriptions etc. even without a particular marking in this work is in no way to be construed to mean that such names may be regarded as unrestricted in respect of trademark and brand protection legislation and could thus be used by anyone.

Cover image: www.ingimage.com

This book is a translation from the original published under ISBN 978-620-6-15805-9.

Publisher:
Sciencia Scripts
is a trademark of
Dodo Books Indian Ocean Ltd. and OmniScriptum S.R.L publishing group

120 High Road, East Finchley, London, N2 9ED, United Kingdom
Str. Armeneasca 28/1, office 1, Chisinau MD-2012, Republic of Moldova, Europe
Printed at: see last page
ISBN: 978-620-5-98645-5

RAKHMANBERDIEV GAPPAR RAKHMANBERDIEVICH

WATER-SOLUBLE CELLULOSE ESTERS.
(TASHKENT INSTITUTE OF CHEMICAL TECHNOLOGY)

The monograph deals with water-soluble cellulose ethers. Their physical and chemical properties, solubility and properties of diluted and concentrated solutions, properties of films and fibers are described. Methods for the preparation and properties of water-soluble acetyl cellulose and its mixed ethers are studied in more detail, since data on them and the world literature are lacking. Data on physiologically active polymers derived from water-soluble cellulose esters are also presented.

The book is intended for specialists, employees of research institutes, as well as teachers, masters and doctoral students of chemical universities in the study, production and application of cellulose ethers and materials based on them.

CONTENTS

PREDICTION

Recently in the chemical industry a great attention is paid to the development of production of high-molecular weight compounds. Among various types of these substances used for production of chemical fibers, films and plastic masses cellulose and its derivatives (esters and ethers) occupy a special place. This is due to the enormous raw material resources of cellulose and its widest use.

It is interesting to note in this connection that our Republic annually collects about 3.5 million tons of cotton and is the world's leading producer of this most valuable raw material. Thus, our country has inexhaustible sources of raw materials for pulp production.

In recent years there have been major advances in the study of cellulose. On the basis of all scientific achievements in the field of structure and properties of cellulose and its derivatives, new directions for the use of these compounds are currently outlined. Thus, recently, much attention has been paid to obtaining, studying the properties and application of water-soluble cellulose derivatives.

At present, the application of water-soluble derivatives, more precisely cellulose ethers, is extremely diverse. Their solutions in water are biochemically stable, neutral, allow obtaining stable emulsions and suspensions, have high quenching ability and a number of other valuable properties. This allows the use of this cellulose derivative by such industries as oil, textile, ceramic, food and many others. In a number of cases, water-soluble ethers are used as useful additives in food products.

By their significance and properties, water-soluble cellulose ethers are, at present, an independent specific group of cellulose derivatives. It should be assumed that these cellulose derivatives will get further distribution not only in traditional areas of their application, but certainly in other new directions as well.

Medicine is one of the most interesting and, perhaps, noblest applications of water-soluble cellulose ethers.

The main focus of this book is on the general physical and chemical properties of water-soluble cellulose ethers. To this end, the book discusses the issues of obtaining, properties and applications of the most common water-soluble cellulose ethers and water-soluble cellulose ethers. A significant place in the book is occupied by chapters on water-soluble acetyl cellulose and mixed ethers based on it, since they are extremely understudied. Much attention is also paid to obtaining physiologically active polymers based on simple and complex water-soluble cellulose ethers.

All of the above determines, in our opinion, the need for the publication of this book.

The book is written on the basis of research materials performed by the author at the former Research Institute of Chemistry and Technology of Cotton Pulp, Tashkent Chemical Engineering Institute at the Department of "Technology of pulp and wood processing" with his staff, to whom he expresses his deep gratitude.

Section 3.1 in Chapter III and Section 6.3 in Chapter 6 about cellulose sulfates and physiologically active polymers on its basis were written by A.S. Turaev, Doctor of Chemical Sciences, and Section 6.4 in Chapter 6, "Physiologically Active Polymers Based on Carboxymethylcellulose" was written by A.A. Sarimsakov, Professor.

CHAPTER I. CELLULOSE. MAIN FEATURES OF CELLULOSE STRUCTURE AFFECTING THE PRODUCTION OF ITS WATER-SOLUBLE ETHERS (STIFFNESS OF MOLECULES, HYDROGEN BONDS, MORPHOLOGICAL AND SUPRAMOLECULAR STRUCTURE, ETC.).

It is known that cellulose is insoluble in water and in order to make it soluble, it is necessary to chemically modify it. The directions of these changes result from the structure of natural cellulose.

It is known that cellulose macromolecules are linear unbranched chains constructed of a large number of D-glucopyranose residues (anhydroglucose units) connected by 1,4-β-glucoside (acetal) bonds:

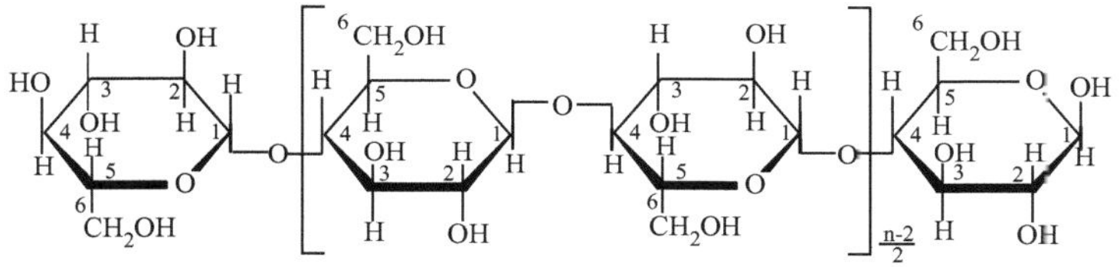

The process of cellulose biosynthesis in nature, which determines its morphological supramolecular organization, is very complex and interesting. Back in 1886 Brown [1,p.6] discovered that bacteria *Acetobacter xylinum* are capable of forming cellulose membranes not only from D-glucose, but also from D-fructose and multi-atomic alcohols D-mannitol and glycerol. Later Schram and coworkers [2] investigated the possibility of cellulose biosynthesis from hexose phosphate. The data obtained confirm that the exogenous substrate is transformed into cellulose and carbon dioxide as a result of the activity of *Axylinum* cells, and the initial substance for cellulose synthesis is probably hexose phosphate.

In [3,4] cellulose biosynthesis was described by a cell-free enzyme system isolated from *Acetobacter xylinum by* milling and subsequent precipitation in an ultracentrifuge. UDFG-(uridine diphosphate-D-glucose) labeled with isotope-14C was used as a donor and cellodextrins and products of deep cellulose hydrolysis destructed to water-soluble state were used as an acceptor.

Barber and Hassid [5] found the ability of enzymes isolated from pea and mung bean to form guanosine triphosphate-D-glucose containing isotope 14C from guanosine triphosphate (GTP) and α-D-glucose-1-

phosphate. Another enzyme [6,7] capable of forming a GDF-D-glucose-14 substrate$_C$ polysaccharide identical in properties to natural cellulose was found in rapidly growing root tissues of mungbean, pea, wheat, in fruit pulp, and in seeds and fibers of immature cotton bolls /8/. The role of cellulose-synthesizing enzyme is the transfer of activated D-glucose residues to the growing polysaccharide chain.

Based on the data obtained in experiments with preparations isolated from plants, the following mechanism of cellulose synthesis has been proposed [1, p.9].

$$GTP +- 2\text{-}D\text{-glucose- I-phosphate} \rightleftarrows$$
$$\rightleftarrows GDF - D - glucose +PF$$
$$n(GDF\text{-}D\text{-glucose}) +Acceptor \rightarrow$$
$$\rightarrow Acceptor \text{ /- } (I \rightarrow 4) \text{ -}D\text{-glucose/}_{II} + pGDF$$

Barber and coworkers [7] showed that GDF-D-glucose is a direct donor of D-glucose during cellulose biosynthesis in plants, but [9] obtained data suggesting that plant cells contain a glucolipid which is converted into cellulose outside the cells. The transfer of the D-glucose residue from nucleoside-diphosphate-D-glucose to glucolipid occurs near the membrane or on the cytoplasmic membrane, then the D-glucose combined with the lipid component migrates outside the cell. In this process, the lipid component performs a transport function, making it possible for the glucose to penetrate from the cell into the outer sphere, where the glucose polymerizes to form cellulose microfibrils.

At present, it can be considered indisputable that all cellulose molecules are combined into microfibrils and that no other forms of molecule aggregation exist.

The processes of formation of cellulose microfibrils can take place inside or outside cells. Intracellular processes are fully biochemical and involve a series of complex transformations, as a result of which cellular products are converted into the activated form of glucose, after which glucose can enter the composition of inert, insoluble, stable microfibrils. Long chains of anhydroglucose links are almost without exception connected with each other by β-(1-4)-glucoside bonds and are considered the primary elements of cellulose structure.

Dennis and Preston [10] suggested that cellulose macromolecules in the center of a microfibril form a highly ordered crystalline core. Around this nucleus there is a region with lower ordering of the structure. According to the authors, regularity disorder may occur due to the presence of non-glucose links (as well as side chains) in macromolecules.

The association of elementary fibrils may consist of the following: four crystalline long primary fibers (elementary fibrils), with a cross-section of 70-30 A^0 , are connected by a paracrystalline phase and thus form a microfibril, whose width is 200 A^0 and thickness is 100 A^0 . It was then assumed [11] that each of the elementary fibrils of 36 macromolecules and behaves as a brittle needle-like crystal, which proves the high degree of regularity of the crystal lattice.

According to these ideas, cellulose microfibrils have a layered structure. It is formed by long, continuous, highly crystalline fibers, bound together by a substance, which has a much less ordered structure than these fibers. As was found in [12], most cellulose microfibrils are formed near the cytoplasm membrane.

The average microfibril velocity is slightly more than 0.1 $\mu m/min$, which corresponds to polymerization of 1.5-10^7 glucose residues per hour per cell, i.e. 10^3 glucopyranose links will join the growing ends of insoluble microfibrils per second. Microfibrils grow in both directions at the same rate and from 20 to 40 new microfibrils can be formed in a cell during one minute [1, p.28-29]).

When studying cotton, Usmanov and Sushkevich [13] found that there are three maxima on the molecular weight distribution curve of young-age cotton pulp. The main cellulose fraction isolated from 5-day-old cotton fiber had a SP close to 200. However, as the age of the fiber increased, the relative content of the low molecular weight fraction decreased and there was only a small peak on the molecular weight distribution curve of 50-day-old cotton pulp corresponding to a SP greater than 4000. It was assumed that low molecular weight cellulose was formed in the primary cell wall and high molecular weight cellulose in the secondary cell wall. They found that initially formed low molecular weight cellulose (i.e., cellulose of the primary cell wall) disappears over time. They and a number of other authors [14] make the assumption from this that originally low molecular weight cellulose has transformed over time into higher molecular weight cellulose or there is so little of it that it could not be quantified.

The degree of polymerization of natural plant cellulose is 1400-12500 (pulp, extracted from mature but not opened capsules of cotton, contains up to 90% of the fraction with a degree of polymerization of 14000), for bacterial cellulose -3750-6000, for technical cellulose - 800-2000.

According to Grahlen and Swedberg [15], the maximum experimentally determined molecular weight of cellulose is about 6,000,000 (for flax cellulose). The molecular weights of the same samples may vary slightly depending on the method of determination. As the molecular weight of the cellulose increases, the characteristic viscosity $/\eta/$ changes and the corresponding relationship is usually written in the form of Mark-Kuhn-Hauvinck expressions:

$$/\eta/ = K_\eta * M^\alpha$$

where, K_η and α are constants characteristic of the given polymer-solvent system;

K is a constant that depends on the properties of the molecules of the dissolved substance and the solvent.

α- is determined by the shape of the dissolved polymer molecule. In the case of rigid rod-shaped molecules, the index of degree α should be equal to 2. In the other extreme case, when the molecules are so tightly coiled that they form balls that are completely impermeable to the solvent, the α-value is 0.5.

Cellulose is a relatively rigid-chain polymer. Due to the cyclic structure of the elementary link, the presence of strongly polar hydroxyl groups in the cellulose macromolecule and the intensive interaction of these groups with the oxygen atom of the glucoside bond, the degree of asymmetry of cellulose macromolecules is much higher than that of polymers without strongly polar groups. Therefore, its α is close to 1. One can think that this is due to increased stiffness, as a result of which both the known permeability and stability of asymmetric shape can be ascribed to tubules [16, p.105].

The degree of asymmetry and conformation of cellulose macromolecules are not constant and can vary significantly depending on various factors (degree of polymerization, temperature, solvent, etc.).

Flexibility of cellulose macromolecule and its esters and the possibility to change their shape are also proved by the presence of highly elastic deformation in fibers and films produced from cellulose and its esters [17]. It was the assumption of bendability of macromolecules that allowed Manley to put forward the hypothesis of folded structure of cellulose macromolecule [18].

However, despite numerous works in the field of supramolecular and fine structure of cellulose, there is still no unified point of view on the micromorphological structure of cellulose materials. For example, a few years later (1978) Manley, at the XXV International Symposium on

Macromolecular Chemistry (Tashkent), rejected the model of elementary fibrils in the form of a ribbon coiled into a helicoid, which he had been developing for more than ten years. Taking into account recent achievements in this direction, Manley suggested that cellulose fibrils in longitudinal direction have "bead-like" structure, i.e. periodicity, and, in his opinion, several levels of periodicity are possible: 100-150 and 30-35 A^0 . The combination of new data suggests that although cellulose chain folding is possible in principle, it does not determine the physical structure of cellulose, because the intense intramolecular bonds tend to maintain the elongated conformation [19].

The question about the phase state of cellulose preparations has been widely discussed during the last decade. Kargin V.A. and his collaborators have paid much attention to this question in their studies [20-21]. Recently, a number of works have appeared where the presence of long-range order in various cellulose materials was shown, i.e. the crystalline state of cellulose. Thus, in works [22,23] the possibility of obtaining cellulose preparations in the form of macro crystals was shown. Hess [23. p.284] isolated cellulose triethers (trimethyl-, ethylcellulose and triacetylcellulose) as monocrystals. Monocrystals were also isolated from low-substituted cellulose esters [24]. The existence of long-range order in cellulose is evidenced by the data obtained by Nelson and O'Connor on the abrupt change in the main structural and thermodynamic parameters of cellulose during the transition from an extremely disordered state to a highly ordered one, as well as spontaneous change in the degree of ordering [25-27], accelerated by plasticizing cellulose or derivatives by water action or heating in polar liquids at elevated temperatures, etc.

Based on the available works in this direction, Nikitin N.I. [23,p.62] draws the following conclusion. Cellulose, like many other polymers, is partially in a crystalline state. This state is sufficiently stable under normal conditions, which is explained by the rigidity of its chains and strong intermolecular interaction due to hydrogen bonds. Existing crystal formations have long-range order, and the molecular chains are not stacked in perfect three-dimensional order (defective crystals), practically this conclusion is reached by Smith, Kitchen, and Matchon /29/. Based on the results of deuterium exchange, they propose to consider cellulose as a set of three phases.

1. The crystalline phase-that gives distinct diffraction maxima on the X-ray image, i.e., the region of three-dimensional far-order.

2.	Non-crystalline but ordered phase-region in which the ordering is lower than in the first one and no clear diffraction pattern is observed.

3.	The amorphous phase-region is disordered.

In [30,31], these three phases were quantified for different types of cellulose.

As we have already noted, cellulose is characterized by strong intermolecular interactions due to hydrogen bonds. It is the formation of hydrogen bonds that is one of the reasons that probably limits phase transformations of cellulose, fixing any of its structure and largely determines its properties, in particular insolubility in water and in other organic liquids.

Therefore, it seems very important to obtain information about the state of various intermolecular bonds existing in the structure of cellulose.

Intermolecular bonds in the structure of cellulose can be of two types:

a) chemical bonds and b) bonds carried out by forces of intermolecular interaction.

Chemical bonds between macromolecules can be formed in the process of cellulose extraction from natural cellulose-containing materials, as well as in the treatment of cellulose with various polyfunctional compounds.

The bonds carried out by intermolecular interaction forces can be divided into two groups: a) intermolecular physical forces with low interaction energy (van der Waals forces) and b) hydrogen bonds.

As is known, hydrogen bonding is formed between cellulose hydroxyls when they approach each other to a distance of 2.7-3 A^0. It is a donor-acceptor interaction in its nature.

The electron pair of the OH bond is somewhat closer to the oxygen atom, the proton is as if "bare" and has the ability to bind to the electron pair of another electronegative atom. When a hydrogen bond is formed, the proton is pulled away from the oxygen atom to which it is chemically bonded (proton donor) to the oxygen atom of the neighboring molecule (acceptor-proton), and thus the oxygen atoms are bonded according to the scheme:

$$O- N.................O \ / \ 32,33/.$$

In addition to the intermolecular hydrogen bond, under well-defined conditions, an intramolecular hydrogen bond is formed, i.e. a bond between hydrogen and oxygen that is part of the same molecule.

Mann and Marrinan [34] give the following possible formation of hydrogen bonds in the structure of cellulose:

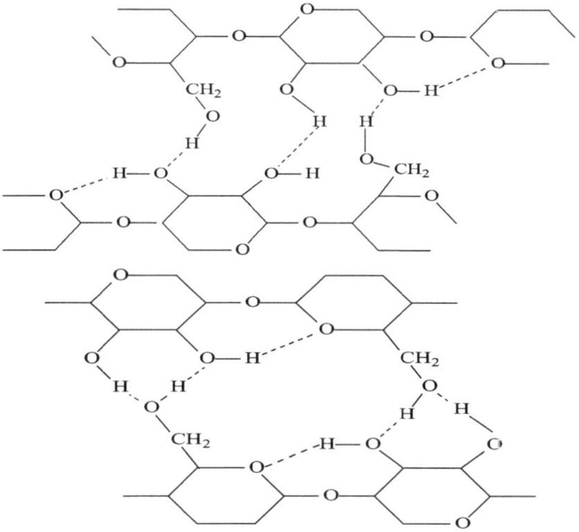

The above schemes are conditional. They show the possibility of formation of only different types of hydrogen bonds. The energy of hydrogen bonding (20.9-33.4 kJ/mol) significantly exceeds the energy of the usual intermolecular interaction, but is 13-17 times less than the chemical bonding. Ellis and Bass [24, p. 58] first studied hydrogen bonding in cellulose using infrared absorption spectra in 1940. These authors, examining the absorption spectrum of natural and mercerized fiber in the region of 1-2.5 microns came to the conclusion that in the first of them hydroxyl groups are included in the hydrogen bond. The presence in cellulose preparations of bonds between neighboring macromolecules, the possibility of their breaking and reformation, and their uniform distribution, is apparently one of the main factors determining a number of important properties of cellulose materials, mainly their solubility.

A more complete and systematic study of hydrogen bonds in cellulose and its derivatives by infrared spectroscopy was carried out by O'Connor, Mann and Marrinan, Nikitin, Zhbankov and a number of other researchers [35-40]. Nikitin showed that hydroxyl groups are included in hydrogen bonding not only in natural fibers, but also in hydrate cellulose. The same studies clearly showed that the process of cellulose esterification is accompanied by hydrogen bond breaking in unsubstituted hydroxyls. One could assume that when hydrogen atoms of hydroxyl groups are partially replaced by more or less voluminous radicals (etherification reaction), hydrogen bonds acting between cellulose chains will be broken, since the distance between oxygen atoms in OH---O will exceed the distances at which hydrogen bonding can exist ($\sim 3A^0$). When studying the absorption spectra of differently nitrated fiber samples, it was found that with increasing degree of substitution there is a gradual appearance of bands 1.44 and 2.07 μm. which characterize the vibrations of free hydroxyls. Thus, the introduction of bulk nitro groups into the fiber leads to the separation of chains and the breaking of hydrogen bonds. It was also found /44/ that when a small amount of acetyl groups which are much more hydrophobic than hydroxyl groups are introduced into the cellulose molecule, hygroscopicity of cellulose material not only decreases but even increases. This fact is explained by the fact that in natural cellulose the majority of hydroxyl groups are included in hydrogen bonds which sharply reduce the possibility of their hydration. When a small number of acetyl groups are introduced into the cellulose macromolecule, the average distance between macromolecules increases, hydrogen bonds are broken and the number of hydroxyl groups capable of binding water increases, which leads to increase of hygroscopicity of cellulose.

The existence of hydrogen bonding in cellulose and its absence in the incompletely substituted ether can be illustrated by the spectrograms shown in Fig. 1.

Curve 1 represents the OH absorption band observed in the dinitrate film. This sharp band-1.44 μm- characterizes the oscillation of hydroxyls not perturbed by hydrogen bonding. Curve 2 is the OH absorption band, where hydroxyls are included in the hydrogen bond. It can be assumed that hydrogen bonding in cellulose to a large extent, determines not only the mutual arrangement of molecular chains in this polymer, but also the physicochemical properties of cellulose.

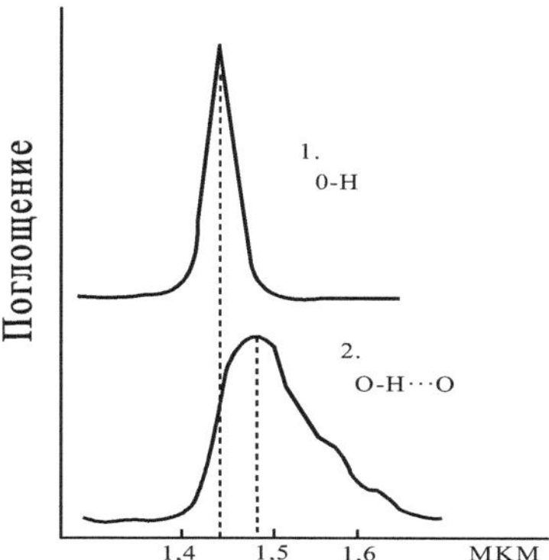

Fig.1. Spectrograms of dinitrate (1) and regenerated cellulose (2).

Thus, the main characteristics of cellulose include: crystallinity or amorphous nature of its structure, i.e. supramolecular structure, length of its main chains (SP) and intensity of intermolecular interaction between them. However, the homogeneity of all these indicators of cellulose is of great importance for the dissolution of cellulose and, in general, in obtaining any of its esters. When obtaining partially substituted cellulose ethers, the main importance is the same accessibility of all cellulose fibers and all macromolecules in the fiber to the reaction.

The issue of cellulose availability in obtaining partially substituted cellulose ethers was given much attention in the works of G.A. Petropavlovskii and co-workers [42-44]. To determine the availability of cellulose fibers they have developed methods based on determining the amount of soluble fractions formed in the process of partial esterification or alkylation reactions under certain conditions. It was shown that any cellulose has fractions that react more or less readily and that the bulk of the cellulose consists of reactive elements. It has been suggested that the reactivity of cellulose fibers is related to the original cells. To prove the relation of cellulose fiber reactivity to initial cells, the authors [44] isolated spring and summer tracheids separately from coniferous wood. It was found that with very similar chemical composition, cellulose fibers from spring tracheids differed significantly higher reactivity than those obtained from fall tracheids.

As a result of the issues considered in Chapter 1 concerning the structure of natural cellulose, we can conclude that natural cellulose is a sum of fibers differing in their availability, and the basis of these fibers is composed of highly reactive microfibrils, which probably have the structure of defective crystallites. In natural cellulose, these crystallites are bound to each other by a system of hydrogen bonds (of different strength) in the lateral direction and by chains of main valence in the longitudinal direction, forming larger aggregates (fibrils, microfibrils, fiber), which leads to a significant decrease in the reaction surface and, therefore, in the reactivity of fiber as compared to its constituent elements.

The different packing options of microfibrils in different fibers and other features of fiber structure are related to the biological origin of fibers by the fact that they are genetically linked to different plant cells.

Without taking into account all structural features, the problem of the reactivity of cellulose, which is inevitably related to the accessibility of the polymer structure to the penetration of solvent molecules or esterifying agents, the possibility of breaking various types of existing intermolecular bonds, the presence of the most favorable conformations of the cellulose macromolecule for the approach of reacting molecules to reaction centers and, finally, the activity of cellulose reaction centers, related to the specific structure and its changes at all stages

The nature in the process of biosynthesis predetermines the fact that cellulose is a relatively inert, virtually insoluble substance, soluble only in a very limited number of aggressive chemicals, since during biosynthesis, as shown above, a polymer is obtained, which has a very high packing density, high crystallinity. In addition, the chemical structure of the cellulose macromolecule - cyclic structure of monomeric links, polar functional groups capable of forming strong hydrogen bonds and, at the same time, the presence of "hinge" bonds between the links (β -1-4-glucoside -C-O-C bonds) which allow conformational changes of macromolecules - all this leads to the fact that cellulose in the solid aggregate state is a rigid polymer with strong intermolecular interactions, high mechanical strength and resistance to external influences.

It is clear that the solubility of such a polymer is limited and, therefore, to increase the solubility, it is necessary to first of all influence all the above-mentioned factors of natural cellulose structure.

Principles of obtaining soluble cellulose preparations.

The ability of polymeric substances to change to a plasticized state or completely dissolve in various liquids has long been used in engineering. Most polymers (in the form of fibers and films) can swell or dissolve in specially selected liquids. Thus, cellulose dissolves in cadoxene, copper-ammonia solution, acetate fibers in acetone or methylene chloride, viscose fibers in diluted alkali solutions, etc.

The use of organic solvents and inorganic complexes, as a rule, is difficult because of their harmfulness, the need for sealing equipment, solvent regeneration and for a number of other reasons.

Therefore, obtaining water-soluble cellulose ethers is of great theoretical and practical importance.

Water-soluble cellulose ethers, possessing a complex of very interesting specific properties from the technical point of view, have won numerous applications in various industries. The possibility of replacing by them such products as starch, gelatin, tragacanth, etc. aroused special interest to them already in the pre-war and wartime years.

In the following years and at present, despite the rapid development of the synthetic polymer industry, water-soluble cellulose ethers have not lost their importance and, on the contrary, their production increases every year. This is explained by the fact that the technology of obtaining water-soluble cellulose ethers is not complicated and there is a real possibility of wide variation of their properties, which allows to obtain products with certain specific properties.

Water-soluble cellulose ethers are a broad class of compounds. It includes simple and compound ethers with different radicals:

Methyl cellulose, carboxymethyl cellulose, oxyethyl cellulose, sulfate cellulose, etc. As a rule, water-soluble ethers, cellulose are partially substituted derivatives of cellulose, in which only part of the hydroxyl is substituted for other radicals.

The solubility of cellulose ethers is in direct correlation with the degree of substitution (SZ).

The values of the degree of substitution (SZ) for some water-soluble cellulose derivatives are given in Table 1.

Table 1.

Solubility of cellulose ethers in water depending on C3.

Cellulose ether	NW
Carboxymethylcellulose (Na-salt)	0,5-1,2
Methylcellulose	1,5-2,4
Oxyethylcellulose	1,3-3,0
Ethyloxyethylcellulose	1,4-1,6
Ethylmethylcellulose	1,0-1,3
Cellulose acetate	0,5-0,6

| Cellulose sulfate | 0,15-0,20 |
| Carboxymethyloxyethylcellulose | $0.3^a : 0.7^6$ |

(a) NW by carboxymethyl groups

b) NW by oxyethyl groups

Water-soluble cellulose ethers are obtained by O-alkylation method and alkyl sulfates, alkyl halides and aromatic sulfonic acid ethers can be used. The peculiarity of the O-alkylation method, in contrast to the esterification methods, is the course of side reactions leading to a significant increase in the specific consumption of reagents, which usually exceeds the consumption of the main O-alkylation reaction by 3-4 times.

LITERATURE

1. Cellulose and its derivatives. Ed. by N. Bakles and L. Segal. M.: Mir, 1974, Vol.1. - 499 c.

2.Schromm M., Gromet Z., Hestrin s. Syntheses1s of cellulose by bakter xilinum. 3. Substrates and inhibitors. Biochem J., 1957, vol. 67, N° 4, pp. 669-679.

3. Koch G., Gordon M. The enzymic synthesis of cellulose by Aceto bakter xylinum. Biochem., Biophys. Acta, 1957, vol. 25 N° 2, pp.436-437.

4. Greathouse G.A. Solution of a cell-free Enzyme System from Acetobacter xylinum Capable of Cellulose Synthes1s. J. Of Amer. Chem. Soc., 1957, N° 16, vol. 79, pp. 4503-4504.

5. Barber G.A., Hassid W.2. The formation of guanosine diphosphate D-glucose by enzymes of higher plants. Biochem., Biophys. Acta, 1964, vol. 86, N°1, pp. 397-399.

6. Barber G.A., Elbe1n A.D., Hassid W.2. The Synthesis of Cellulose by Enzyme Systems from Higher Plants. J. Biol. Chem., 1964, v. 239, N° 12, pp.4056- 4061.

7. Elbein A.D., Barber G.A., Hassid W.2. The syshthesis of cellulose by an Enzyme System from a Higher Plant. J. of Amer. Chem. Soc., 1964, vol. 86, N°2, pp. 309-310.

8. Barber G.A., Hassid W.2., Synthesis of Cellulose by Enzyme Preparations from the Developing Cotton Boll. Nature, 1965,vol. 207, No. 4994, pp. 295-296.

9. Wedehra I.Z., Manley John St.R. Irreversible Exchange of Hydrogen in the Drying of Cellulose at High Temperature. J. Appl. Pol. Sci, 1965, vol. 9, No. 10, pp. 3499-3354.

10. Dennis D.T., Preston R.D. Constitution of Cellulose microfibrils. Nature, 1961, vol. 191, 1° 4789, pp. 667-668.

11. Freg-Wyssling A., Majhlethaler K. Dle Elementarfibrillen der Cellulose. Makromol. Chem., 1963, v. 68, s. 227-231.

12. Preston R.D. in: The interpretation of ultrastructure.Vol. 1, Academic Press, New-York, 1962, p. 325 ff.

13. Usmanov Kh. U., Sushkevitch T.1. Study of polydisperst1dy of cotton Cellulose According to Molecular Weigh. J. Pol. Scl., 1962, vol. 58 (196), p. 1325-1331.

14. Marx-F1gini M. Kinetische Untersuchungen zur Biosynthese der Cellulose in der Baumwolle. Chem., 1963, Bd. 68, s. 227-231.

15. Grolen N., Svedberg T. Molecular We1ght of Nature Cellulose. Nature, vol. 152, № 3865, p. 625.

16. Tsvetkov V.N., Eskin V.E., Frenkel S.A. The Structure of Macromolecules in Solutions. Moscow: Nauka, 1964, p. 105.

17. Kargin V.A., Leipunskaya d.i. Diffraction o² fast electrons of hydrate cellulose. Zhokh, 1940, vol. 14, issue 3, pp. 32-39.

18. 18. Manley R. Crystallizat1on of Cellulose Triacetate from solution. J. Pol. Scl., 1960, vol. 47 (149), pp. 509-512.

19. Usmanov Kh.U. International symposium on macromolecular chemistry "Macro-78". Chem. Fibers, 1979, No. 2, pp. 56-60.

20. Kargin V.A. Structure of cellulose and its place among other polymers. Vysokomol. soed. 1960, vol. 2, no. 3, pp. 465-467.

21. Mikhailov N.V., Kargin V.A., Bukhman V.M. Radiographic studies of artificial fiber orientation. KFH, 1940, vol. 14, issue 2, pp. 205-207.

22. Manley R. Crystallization of Cellulose Triacetate from sclution. J. Pol. Scl., 1960, vol. 47 (149), pp. 509-512.

23. Hess K. Chemistry of cellulose and its satellites (translated from German), GCTI, 1939, - 490 p.

24. Ranby B., Noe R. Crystallization of Cellulose and Cellulose Derivatives from Dilute solut1on. 1. Growth of Single Crystals. J. Pol. Scl., 1961, vol. 51, w° 155, pp. 337-347.

25. Nelson M. O'Connor R, Relation of certain infrared Beuds to Cellulose Crystallin1ty and Crystal Latt1ce Type. Part 1. Spectra of Lattice Types, 1, 11, 111 and of Amorphory Cellulose. J. Appl. Pol. Scl., 1964, vol. 8, 1° 3, pp.1311-1324.

26. Alince B., Kuniak L., Beurteilung der übermolekularen Struktur der Cellolose mittels Dichteänderungen bel varschiedenem Fenchtegchalf. Faserforch. und Textiltechnik, 1967.18, s. 17-21.

27. Zaides A.L., Sinitskaya I.G. Electronographic study of natural rami cellulose. DAN USSR, 1957, v. 80, 82, p. 23-24.

28. Nikitin N.I. Chemistry of wood and cellulose. M.-L.: Izd. AS USSR, -711c.

29. Smith I.K., Kitchen W.I., Mutton D.B., Structural Study of Cellulosic Fibers. J. Pol. Sci., 1963, pt., C., No. 2, pp. 499-513.

30. M1kha1lov N.V. On the Phase structure of Cellulose. J. Pol. Sc1., 1958, vol. 30 (121), pp. 259-269,

31. Warwicker 1.0. Structures of Native and Mercerized Celluloses. Nature, 1954, vol. 174, № 4420 (17), p. 135.

32. Stepanov B.I., Zhbankov R.G., Rosenberg A.Y. Infrared spectra of cellulose during viscose production. PhC, 1959, v. 33, 29, p. 1907 -193.

33. Hurtubise F.G., Krässig H. Classification of Fine Structural characteristics in Cellulose by infrared spectroscopy. Analytical Chemistry, 1960, vol. 32, N. 2, pp. 177 - 181.

34. Mann J., Marrinan H.I. Polarized ifrared spektre of Cellulose. 1 J. Pol. Sci., 1958, vol. 27, no. 115, pp. 595-596.

35. Marrunan H.J., Mann J. A Study by infrared spectroscopy of hydrogene bonding in cellulose. J. Appi. Chem., 1954, 1°4, pp. 204-211.

36. O'Connor R.T., Du Pre E.T., Micall E.R. Applications of infrared Absorption Spectroscopy to investigations of Cotton and Modified Cottons. Text. Res. J., 1958, vol. 28, no. 7, pp. 542-554.

37. Zueva R.V., Zhbankov R.G., Ivanova N.V., Kozlov P.V., Podgorodetsky E.K. Cellulose acetates obtained under heterogeneous conditions of acetylation and partial saponification. High-molecular solids, Cellulose and its derivatives, 1963, pp. 124-130.

38. Kozlov P.V., Zhbankov R.G., Zueva R.V., Ivanova L.V., Podgorodetsky E.K. Nature of intermolecular bonds in cellulose acetates obtained by homogeneous and heterogeneous methods. High-molecular solids, Cellulose and its derivatives, 1963, pp. 1Z1-138.

39. Nikitin V.N. Study of cellulose and its derivatives by means of IR absorption spectra. ZhPH, 1949, vol. 23, 27, pp. 775-785.

40. William H. Alken. Effect of Acetylation on Water-Binding Properties of Cellulose. J. Ind. Eng. Chem., 1943, vol. 35, no. 11, pp. 1206-1211.

41. Tsvetaeva I.P., Petropavlovskii G.A., Iopanova T.N. Influence of the supramolecular structure of cellulose and its solubility during Emulsion xantogenization, ZhT I X, 1966, vol. 39, № 7, c. 1549-1554.

42. Petropavlovsky G.A., Tsvetaeva I.B., Ionanova T.N. The effect of the supramolecular structure of cellulose on its solubility during carboxymethylation. ZHLCH, T. 40, N3, p. 648-652.

43. Petropavlovskii G.A., Stolyarova L.I. On the difference of properties of cellular and summer parts of Siberian Larch pulpwood Cell. Chem. Technol., 1968, 4, pp. 423-436.

44.Bytensky V.Y., Kuznetsova E.A., Production of cellulose ethers. L.: Chemistry, 1974. - 208 c.

CHAPTER II. WATER-SOLUBLE CELLULOSE ETHERS

2.1 Carboxymethylcellulose.

Among cellulose ethers containing carboxyl groups, carboxymethyl cellulose (CMC) has the greatest practical application due to the availability of reagents used for the synthesis.

Many reviews have been published in journals and monographs on the production and application of carboxymethylcellulose (CMC).

Here we will focus mainly on the change in solubility during carboxymethylation and on those main specific features (structure, properties of aqueous solutions, viscosity, etc.) of CMC that distinguish it from other water-soluble esters.

The formula for CMC is as follows:

$$[C_6 H_7 O_2 \text{-}(OH)_{3\text{-}x} \text{-}(OSN_2 SoNa)]_x$$

In commercial samples of carboxymethyl cellulose, the degree of substitution ranges from 0.5 to 1.2 carboxymethyl groups per anhydroglucose unit. The average molecular weight ranges from 50,000 to 500,000.

The distribution of substituents (carboxymethyl groups) in different samples of CMC was studied by Timell and Spurlin [1]. These authors showed that the chain molecule of CMC contains both unsubstituted and monoand dysubstituted anhydroglucose units. It turned out that in all the studied products only one of the two neighboring secondary hydroxyl groups of cellulose was replaced with carboxymethyl radicals. The authors explain this fact by the influence of repulsive forces between negatively charged chloroacetate ions and carboxymethyl radicals located in position 2 or 3. Similarly, based on steric electrostatic considerations, one cannot expect the simultaneous substitution of hydroxyl groups at positions 2, 3, and 6.

The relative reactivity of hydroxyl groups in the reaction of alkaline cellulose with monochloroacetic acid was also studied in /2/. The author obtained 10 samples of CMC having a degree of substitution from 0.2 to 0.36, whose solubility in water increased from 0.08 to 29.65%. In these samples, the ratio of the number of substituted primary hydroxyl groups to the number of secondary groups was found to be 0.91-1.28. This ratio for the water-soluble fraction of CMC ranged from 0.92-1.35. Thus, within the studied degrees of substitution, the ratio of the number of substituted primary hydroxyl groups to the number of secondary ones was approximately equal to one in all cases.

On this basis, we can assume that in the carboxymethylation reaction of cellulose, its primary hydroxyl group has about 2 times greater reactivity than the secondary one.

The solubility of various CMCs is mainly determined by their degree of esterification (Y) degree of molymerization (SP) [2].

However, often carboxymethylcellulose with the same values of Y and SP can differ significantly in solubility [3], which is explained by the chemical heterogeneity of these products.

The effect of chemical heterogeneity of the product on solubility is observed not only for CMC, but also for other cellulose ethers. Therefore, in the following chapters we will focus separately on the studies related to the study of the heterogeneity of the esterification products cellulose.

In cold water, CMC swells first to form a gel, with water molecules penetrating the interior of the fibers, causing micellar swelling inside before the outer layers of the fiber can dissolve. A highly concentrated solute layer is formed around the swollen material, from which the dissolved molecules diffuse very slowly, so intensive agitation greatly accelerates the dissolution process. Under a microscope, typical spherical forms of swelling can be observed [4]. Often chains of link spheres are formed, gradually breaking up into separate sections and then dissolving completely.

As mentioned above, the solubility of CMC depends on the degree of substitution. Products with a very low degree of substitution of ~0.1 per anhydroglucose unit are soluble in 6% NaOH only at low temperatures. Products with a higher degree of substitution (about 0.3) are also soluble in 6% NaOH, but at room temperature. Products with a degree of substitution of 0.4 or more dissolve partially or completely in cold and also in hot water (unlike methylcellulose).

In general, depending on the degree of substitution, preparations of 0-carboxymethylcellulose, as well as other O-alkyl derivatives, can be divided into low-substituted and highly substituted ones. Obtaining CMC preparations with a degree of substitution greater than 1.0 (Y =100) is, however, very difficult because of the electrostatic repulsion effects of the charged homonymous groups (chloroacetate ion and carboxymethyl group). Therefore, almost "highly substituted" preparations of CMC are products with a degree of substitution of 0.5-1.0 (Y =50÷100).

In general, low-substituted cellulose esters and ethers have some new unique physicochemical properties compared with the original cellulose.

Thus, cellulose esters and ethers are soluble in alkali solutions at low temperature and more accessible to the influence of reagents. These properties are explained by destruction of supramolecular structure: its amortization and weakening of hydrogen bonds.

Changes in the properties of low-substituted cellulose ethers and ethers are explained by the specific role played by hydroxyl groups in the structure of cellulose. These groups, as we have already noted, form intra- and intermolecular bonds.

Replacement of part of hydroxyl groups by other radicals leads to quantitative and qualitative changes in the system of hydrogen bonds. It is noteworthy that changes in the structure of cellulose and in the system of hydrogen bonds occur during alkylation of a small part of hydroxyl groups.

Figure 2 shows X-ray radiographs of CMC taken from [5]. For comparison, an X-ray radiograph of lint mercerized with 18% NaOH solution is given.

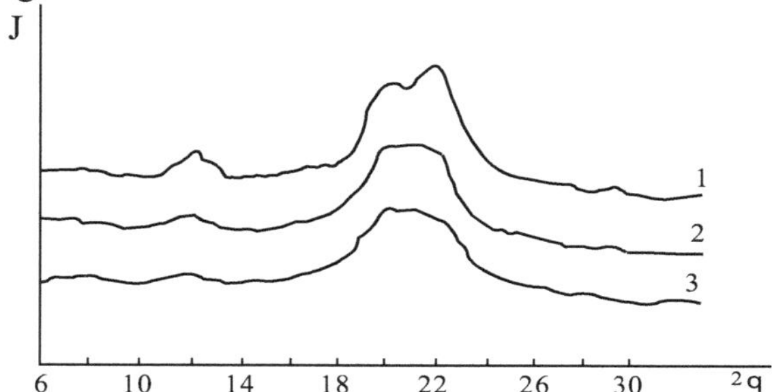

Fig.2. Effect of partial carboxymethylation and mercerization on the cellulose structure.
1. lint mercerized with a solution of 18% (NaOH).
2. CMC from lint with Υ =5.
3. CMC from lint with Υ=10.8.

The carboxymethylation reaction affects the structure of cellulose much more than methylation under heterogeneous conditions [5].

This is due to the fact that the reagent (monochloroacetic acid) is soluble in 18% alkali, and the alkali itself gives considerable swelling of the pulp and leads to a change in its structure, thus forcing the reagent to reach the deep regions of the fiber. This leads to greater depreciation of the starting product compared to the methylation process. The addition of a small amount of $-OSH_2 COONa$ groups (Υ=5) leads to a significant

22

change in the diffractogram of mercerized cellulose. The crystallinity index K_{OTH} also changes dramatically depending on the degree of substitution: i.e., at $Y=5$ it is 0.49, and at $Y=10.8$ $K_{OTH} =0.28$.

The pH value of pure aqueous CMC solution is 7.0-7.5, i.e., it is practically neutral. The aqueous solutions of commercially available CMC samples can vary in viscosity (at the same concentration) by tens of times. Consequently, viscosity is one of the important properties of CMC, which plays a major role in the formation of many of its physico-chemical properties. Much work in this direction was carried out by G.A. Petropavlovskii [6].

The main factors affecting the viscosity of such solutions are: concentration, degree of polymerization, temperature, nature of the solvent and pH value of the medium.

Figure 3 shows the effect of pH value on the viscosity of aqueous CMC solutions [6].

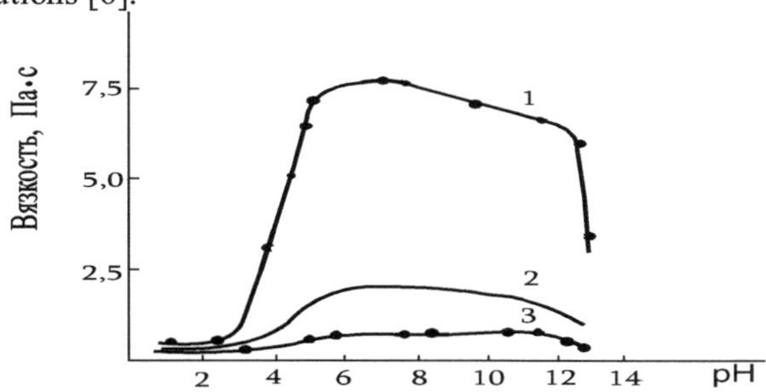

Fig.3. Viscosity changes of 1% Na-CMC solution depending on pH value. 1-High viscosity, 2-Middle viscosity, 3-Low viscosity.

Figure 3 shows that the viscosity of the Na-CMC solution has a maximum in the pH range of 6 to 9. Below pH 6, it drops rapidly due to the gradual precipitation of free CMC, which ends completely at pH = 2.5. The viscosity of the solution also begins to decrease above pH 9, slowly at first, and then more rapidly when it reaches 11.5. The very significant decrease in viscosity in the strongly alkaline region is not, however, the consequence of the precipitation of the CMC, since even at pH = 13 it remains completely dissolved.

It turns out that these viscosity changes are reversible and that the initial viscosity of the solution can be restored to its maximum value again by adding acid and setting the pH value between 6 and 9. Thus, there is a certain specific effect of the alkali on the viscosity here.

At present, it is believed [6] that this effect is apparently due to a change in the configuration of the CMC molecules in solution under the influence of the electrolyte. In the absence of an electrolyte, the ionized groups of the polymer chain repel each other, causing the macromolecule to become more straightened. When the electrolyte is added, the repulsion between the neighboring ionized groups of the polymer chain becomes less due to the screening effect of the electrolyte and the macromolecule becomes more coiled. In addition to the pH of the medium, the viscosity of solutions is also strongly influenced by the temperature of the medium. If solutions are heated to different temperatures and then cooled again, an interesting course of viscosity curves during heating and cooling is observed. This dependence is shown in Fig. 4.

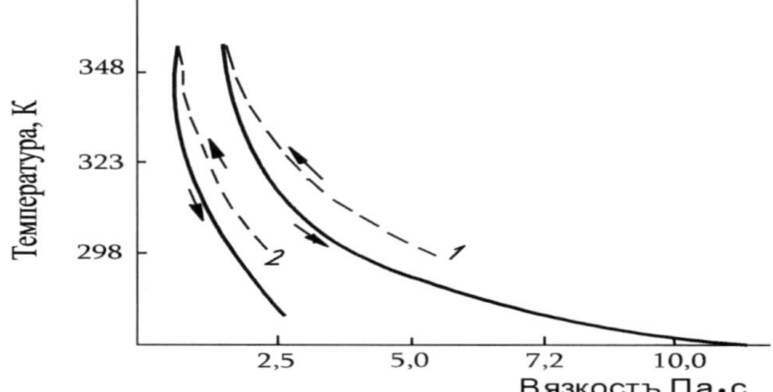

Fig.4. Viscosity changes of 1% Na-CMC solution upon heating and cooling 1.-high-viscosity; 2-medium-viscosity

Figure 4 shows that the final viscosity of the Na-CMC solution determined at room temperature after the heating-cooling cycle is less than half of its initial value.

The consistent decrease in viscosity with increasing temperature and time is also shown in Fig. 5.

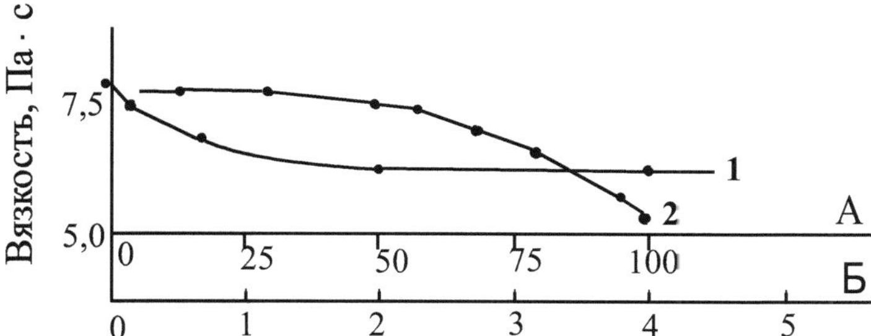

Fig.5. Variation of viscosity of 1% Na-CMC solution in depending on the temperature (1) and heating time (2).

A-temperature at which the solution was heated for $12\text{-}10^2$ sec.

The viscosity of solutions increases rapidly with increasing concentration, and the viscosity of samples with high molecular weight increases most intensively. This, in addition to the natural influence of the size of the molecules, is due to the emergence of the so-called secondary structural formations in the solution "grid" of macromolecules, the "cells" of which contain bound molecules of the solvent.

The viscosity of CMC solutions does not obey Newton's law [7], i.e. in such solutions the shear rate is not in direct proportion to the shear stress, and for the reason mentioned above these solutions are also thixotropic.

If we graphically depict the character of such flow on the "shear rate-shear stress" rheogram, then the resulting curve will be shifted in the direction of the shear rate axis Fig. 6.a.

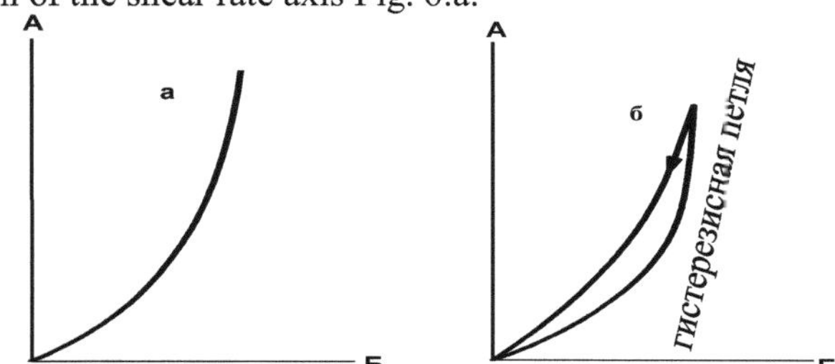

Fig.6. Deograms of netixotropic pseudoplastic (a) and tictotropic (b) CMC solutions

-A-shear rate, B-shear stress.

The apparent viscosity (ratio of shear stress to shear rate) decreases when the shear rate increases. This effect for pseudoplastic systems is

25

independent of time. In contrast, in thixotropic systems, time is important. The apparent viscosity decreases with time if the shear rate is constant. Structural failure of a thixotropic material occurs under these conditions in a measurable time. When the shear stress is removed, the structure of the thixotropic material reappears. Depending on the system available, this reversal time may vary from a few seconds to several days. A rheogram was obtained for thixotropic CMC solutions using a rotary viscometer by De-Butts, Hudy and Elliotton (Fig. 6.b.) [8].

It should also be pointed out that carboxymethylcellulose dissolves in water almost never completely and in the obtained solutions there are residues of the so-called "primary structure", i.e. fragments of fibers or particles of these fibers poorly esterified and having a high degree of crystallinity [4,9].

These crystals act as gel centers, capturing relatively large amounts of molecularly soluble CMC and forming a three-dimensional lattice through electrostatic or Van der Waals forces. Durig and Banderet [10] isolated such gel particles from Na-CMC solutions, which showed a clearly pronounced crystalline structure under fluoroscopic examination.

From the above it can be assumed that the tikstropy of CMC solutions is caused by the influence of the gel centers. When this system is subjected to shear, some of the gel centers and their aggregates by individual macromolecules are subjected to disintegration. This dispersion takes place at the measured time, and therefore the apparent viscosity of the system will decrease with time. As the solution stands, the gel centers may aggregate again. The apparent viscosity of the system will increase. This circumstance is one of the reasons for the formation of gels in the CMC solution. The formation of gels in these solutions may be a consequence not only of a low degree of esterification in some crystalline areas of the cellulose fiber, but also a consequence of the presence of multivalent metals in the original cellulose or in the solvent. These cations can form a number of cross-bridges between macromolecules of CMC carrying univalent carboxyl anions.

In CMC solution, secondary aggregates, as well as free macromolecules, are in equilibrium, while primary aggregates hardly disintegrate to any appreciable extent and rather serve as nodes of the macromolecule grid mentioned above. Thus, three related structural elements can combine in the gel phase, namely, crystallites, labile aggregates, and single molecules. Adsorption equilibria between these elements pollutes the volume and apparent viscosity of the gel phase.

Therefore, when such gels are diluted, the equilibrium shifts toward an increase in the number of single non-aggregated molecules and the overall viscosity of the gel phase decreases. The properties of diluted solutions (i.e., solutions in which there is no macromolecule interaction) have been studied by a number of authors.

In dilute solutions of Na-CMC in water, in the presence of an extraneous electrolyte, the value of the characteristic viscosity is related to the molecular weight by the Kuhn-Mark-Hauvinck relation $/\eta/=K \cdot M^{\alpha}$

The constants K and α included in this equation were found by Schurz, Strozit, and Wurz [11] by comparing viscometric data with osmotic molecular weight determinations. For CMC solutions in 2% NaCl $/\eta/- 0.225 \cdot P^{-1.128}$ and for solutions in 6% NaOH$/\eta/= 1.05 \cdot P^{-0.93}$ [12].

In all cases, the value of $/\eta/$ found for the salt solutions was 1.3-1.9 times higher than for the alkaline solutions. This can be explained by the fact that in salt solutions the CMC molecules are more elongated, while in alkaline solutions they are curled due to the different nature of electrostatic interactions.

Brown, in his original review article [13] on the determination of the conformation of cellulose molecules and their derivatives in solutions, makes interesting conclusions. These conclusions are as follows:

1) by comparing the characteristic viscosity of water-soluble cellulose derivatives (Na-CMC) and cellulose itself in the common solvent cadoxene comes to the conclusion that cellulose and Na-CMC have almost the same conformation in cadoxene. While cellulose obviously forms a complex in this solvent, Na-CMC almost does not;

(2) A study of the effect of the degree of substitution from 0 to 1 Na-CMC showed that for small degrees of substitution no significant change in the increase in the size of the tangle was detected. This indicates that the substitution group has little electrostatic effect on the size of the molecules and affects the chain properties through polymer-solvent interactions. As a consequence, in good solvents, which are usually used to obtain true cellulose solutions, the molecules are strongly expanded, which accounts for high viscosity.

Rudy and his collaborators [14] studied the viscosity and electrical conductivity of solutions of various CMC salts. The results obtained confirm that all the salts belong to polyelectrolytes. Moreover, their electrical conductivity is greater the greater the mobility of the salt cation.

When Na-CMC is attacked by a dilute mineral acid (HCl or $H_2 SO_4$), free CMC (H-CMC) is formed, which is insoluble in water. Free CMC is a weak acid and is close to acetic acid in this respect, having a dissociation constant of $5 \cdot 10^{-5}$. Other CMC salts can be obtained by treating Na-CMC with a solution of the desired metal salt. When dried for a long time at high temperature, CMC is much more difficult to dissolve in alkali.

It is also interesting to note that when Na-CMC aqueous solution is acidified with any strong mineral acid, the precipitation of free CMC occurs within pH values starting from pH=6 and ending completely only at pH=2.5.

Application of water-soluble carboxymethylcellulose.

There are a number of review articles indicating extensive applications of CMC.

CMC solutions, which have good adhesive and film-forming properties and long-term storage stability, are successfully used for sizing and finishing of fabrics and as dye thickeners for printing dyes.

CMC is also successfully used in the paper industry. It is used in the form of aqueous solutions to treat the surface of the finished paper or is added during milling.

CMC additives increase the resistance of paper to the penetration of hydrophobic substances-oil, grease, wax-and increase the degree of its adhesion. The strength of the paper increases significantly. Application of CMC when drilling is of great importance. It is an effective stabilizer of flushing fluids when drilling oil and gas wells, as well as a good cementing agent for walls of these wells.

CMC solutions have good adhesive properties and can also be successfully used in industries where this property is required.

In the ceramic industry, CMC is used as a binder, thickener and plasticizer in glazes, refractories, etc.

The use of CMC in food, perfume and pharmaceutical industries is of great importance. High purity preparations of CMC can be successfully used for preparation of blood substitutes.

2.2.Methylcellulose.

Of the cellulose ethers and aliphatic monatomic alcohols, methylcellulose is the most used. General methods of 0-alkylation of cellulose in the presence of alkali or alkaline cellulose with alkylsulfates (dimethylsulfate) or alkyl halides (methyl chloride or iodide) can be used to obtain methylcellulose.

The formula for methylcellulose is as follows:

$$[FROM_6 \, N_7 \, O_2 \, (HE)_{3-x} - (BAS)]_{3x}$$

The degree of substitution in industrial samples of methylcellulose ranges from 1.6 to 20.

In the process of obtaining methylcellulose with increasing the degree of esterification to $Y=50$ the hygroscopicity of the obtained ester increases. This is explained by the fact that in cellulose macromolecules there is mutual saturation of most hydroxyl groups with formation of hydrogen bonds. When a bulk radical is introduced, although it is more hydrophobic than hydroxyl groups, the distance between macromolecules increases, hydrogen bonds are broken, and the number of hydroxyl groups capable of hydration increases, which leads to an increase in the hygroscopicity of cellulose ether. This effect in the initial stages of uniform etherification is the more noticeable, the larger the size of the introduced radical /14 p.327,62/.

Figure 7. shows diffractograms of methylcellulose with a low degree of substitution containing 11.7% OSH_3 ($\gamma=64$) and linter treated with 12% alkaline solution. A comparison of the two patterns shows that methylcellulose with a low degree of substitution corresponds to an intermediate structure in the transition from cellulose I to cellulose II. However, the positions of the diffraction maxima at $2\theta=15°$, $16°$, and $22°$ corresponding to the cellulose structure remain unchanged, although their intensity changes. If cellulose I is converted to cellulose by carboximethylation. II, the lattice of cellulose I remains initially (i.e., at $low\bar{u}$ γ) unchanged during methylation. In order to dissolve methylcellulose in water, a much higher degree of substitution is necessary.

The solubility of methylcellulose primarily depends on the degree of substitution (Y), the degree of polymerization (SP), and the homogeneity of the product by degree of substitution. Methylcellulose is soluble in water at $Y=130-200$ [15], with products obtained in a homogeneous environment dissolving at lower values. To increase the solubility of low-substituted methylcellulose ($\gamma=10-20$) in alkali G.A. Petropavlovsky and N.I. Nikitin [16] used their proposed method of dissolution in 4-6% alkali solutions at solution freezing temperature (261^0 K) followed by thawing.

Later, this dissolution method proved to be applicable to other cellulose ethers as well [17]. Unlike other water-soluble

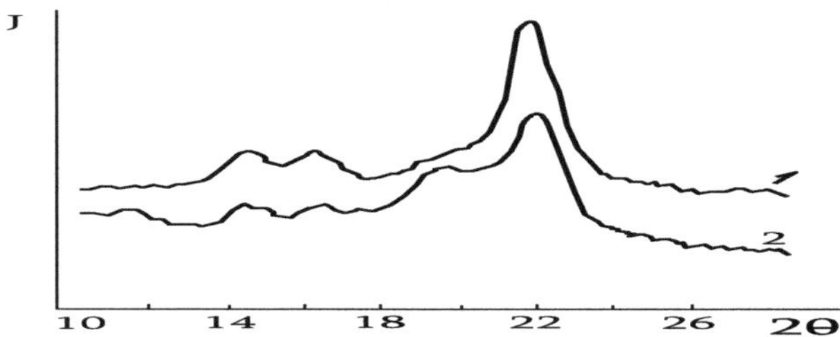

Fig.7. Changes in the structure of cotton cellulose (linter) treated with 12% NaOH solution (1) after partial methylation (γ=64) (2) /14/.

esters, aqueous solutions of methylcellulose gelatinize when heated to a temperature of 323-328°K [18, p.326].

When the temperature drops, the original solution is easily reconstituted. The reasons for the precipitation from solution and reverse dissolution of methylcellulose can be interpreted differently. A number of researchers [19] explain this phenomenon by the formation of oxonium compounds during dissolution according to the following scheme:

$$\overset{|}{\underset{|}{C}} - O - CH_3 \xrightarrow{K} \overset{|}{\underset{|}{C}} - OCH_3 + H_2O$$

These oxonium compounds are relatively unstable and decompose when heated, releasing water, which reduces the hydration of macromolecules.

According to another assumption [20], methylcellulose molecules are strongly solvated at low temperature, and the solvates break up when the temperature rises. Thus deprived of the hydrate layer, methylcellulose molecules combine with each other, and the formation of aggregates in methylcellulose with a low degree of polymerization, according to the authors, occurs at a lower temperature. According to other authors [21, p. 403], a decrease in the molecular weight of methylcellulose increases its resistance to gelatinization. The presence of salts in aqueous solutions of methylcellulose leads to a decrease in gelatinization temperature, since the hydration of macromolecules decreases.

The gelatinization temperature is usually between 323°K and 363°K. It is the lower the greater the degree of polymerization and concentration of the solution. This coagulation is reversible and the original solution is easily recovered when the temperature is increased. This is one of the interesting features of methylcellulose. The temperature at which reverse dissolution of the precipitate occurs (upon cooling) is called the "hydration temperature.

Methylcellulose with low content of methoxyl groups is soluble in dilute alkali solutions. When a higher degree of substitution in the region of 26.5% to 32.5% of the content of methoxyl groups is soluble in water. When the methoxyl groups content is further increased to 38% or higher, it loses its solubility in water (at room temperature and above).

Highly methylated products are also soluble in organic solvents. The given data on solubility of methyl cellulose refer to products obtained by reaction of alkaline cellulose with dimethyl sulfate and methyl chloride. If we compare the value of the degree of substitution stated above at which methyl cellulose is water soluble (30% - OSH_3 ; $Y=180$) with the value of the degree of substitution of Na-CMC soluble in water ($Y =50-80$), we see that in the latter case this value is much lower. This circumstance is explained by the fact that the carboxymethylation reaction occurs with reagents which are mutually soluble (aqueous alkali solutions of monochloroacetate-Na). In this case, the cellulose swells in the reagent solution. In the case of the methylation reaction, the reagents aqueous solutions of NaOH and alkyl sulfates or halides are mutually insoluble. Although cellulose swells in alkali solution, the reaction is carried out, as already mentioned, by slow diffusion of the alkylating reagent in the pulp fiber mass. In this case, one should distinguish between the reaction in one macromolecule and the possibility of each macromolecule to carry out the reaction. In order for a macromolecule to pass into solution, some minimal degree of its chemical transformation - esterification of hydroxyl groups - is required. In order for cellulose to be completely transferred into solution, all of its macromolecules must receive this minimal degree of esterification.

Thus, if the reaction is heterogeneous and occurs due to diffusion of the reagent into the cellulose fiber, then a higher degree of sorification is required for complete dissolution of the latter than in a homogeneous reaction or close to it. The fact that the solubility of obtained esterification products is significantly determined by these circumstances is well confirmed by the well-known fact of obtaining methylcellulose well soluble in water with 15% -COH_3 , by carrying out the reaction in solutions of quaternary ammonium bases.

In dilute solutions, methylcellulose has the properties of non-ionic high molecular weight substances. Its viscosity in these solutions can characterize its molecular weight and is expressed by the following equation:

$$/\eta/ = KP_w^{\alpha}$$

Wink [22] carried out destruction of methylcellulose by hydrolysis to determine the change in the characteristic viscosity, depending on the molecular weight and to determine the constants of this equation. Methylcellulose was first purified by precipitation from a water-ethanol solution with ether. The degree of substitution of the original cellulose was C3=1.74 and the degree of polymerization SP=200. Based on the measurements of absolute values of molecular weight using osmometry and determination of end groups, the dependence of the characteristic viscosity of solutions of the obtained fractions of methylcellulose on the molecular weight was determined. It turned out to be as follows:

$$/\eta/ = 0.016 * P_V^{0.86}$$

Wink also found that the characteristic viscosity of methylcellulose is the same as the "c" constant in the Schulz-Blaschke equation:

$$\eta = \frac{\dfrac{\eta_{y6\text{д}}}{c}}{1 + B\eta_{y\text{д}}}$$

do not depend on the presence of a foreign electrolyte-acid in the solution. Thus, for example, the values of these parameters were:

in $H_2 O$ $/\eta/=11,0$ в=0,36

in 2MNSl $/\eta/=10.9$ in=0.36, etc.

It should be noted that other authors (who determined the absolute molecular weights using sedimentation on an ultracentrifuge and light scattering) obtained for methyl cellulose somewhat different values of the degree index "α" in the Kuhn-Mark equation. Thus, in [23] $\alpha = 0.63$, and in [24] $\alpha = 0.55$. The authors themselves explain these discrepancies by the great ability of methylcellulose to aggregate in aqueous solutions.

In [25] we studied the viscosity of methylcellulose solutions of various concentrations with a capillary viscometer at $298\pm0,2°K$. It was found that at concentrations of $1\div1.8$ g/l methylcellulose, Newton's law is observed, in the area of higher concentration there is a deviation from this law, indicating the non-ionic character of the polymer chain of methylcellulose.

Interesting work was done by Foverake and Geningo [26] to study the electrical double ray refraction of aqueous solutions of methylcellulose. They found that the dependence of the length of the equivalent rod Дn , determined from the values of relaxation times π-derived in turn from the rates of incidence Δn, on molecular weight is a saturation curve that coincides with the linear dependence of elongated chain length Z on molecular weight only at very small values of molecular weight $(<10\ )^4$

Hence, according to the authors, methylcellulose macromolecules are not rigid rods in the interval of molecular weights under study. The data obtained from r also do not correspond to the models related to persistence length. It was proposed to use the parameters $B= Dr/LE^2$, where E is the field strength, to characterize the flexibility of methylcellulose macromolecules, since the dependence of Dn on B, is linear in a sufficiently wide range regardless of the molecular weight of methylcellulose.

Temperature-viscosity relationships have received much attention because these studies for aqueous solutions of cellulose ethers are of great practical importance, as their use in many cases depends on it.

Application of water-soluble methylcellulose.

In most cases, methylcellulose is used to thicken aqueous media. The effectiveness of thickening depends on viscosity (i.e. the degree of polymerization). Methylcellulose allows water-soluble substances to be converted into a stable fine-dispersed state in aqueous medium, as it forms hydrophilic monomolecular protective layers around the individual particles.

Valuable properties of methylcellulose are its high binder effect for pigments, its high adhesion in the dry state and its ability to form films. These interesting properties are used in the preparation of aqueous molar paints and adhesives.

Methylcellulose with a low viscosity value is particularly suitable for this purpose, as it can be applied to a wide variety of substrates.

In the textile industry, methylcellulose is used as a sizing agent for wool base and for soft textile finishing, in order to obtain elegant griffin and gloss. Methylcellulose is successfully used in the soap industry. In pharmaceutical practice, methylcellulose is used as a non-greasy base for so-called mucilage and oil/water emulsion ointments, which serve to protect the skin or to protect against light burns. In addition, methylcellulose serves as an independent medication.

In cosmetics, water-soluble cellulose ethers are used to make toothpastes and elixirs, protective emulsions and non-greasy skin creams.

In all possible emulsions, methylcellulose is used as emulsifiers and stabilizers for vegetable oils.

It is also very widely used in the food industry. For example, in the production of ice cream its use provides the necessary opulence, stability and taste. Methylcellulose is used in aromatic emulsions, gravies for fruit juices, preserves, etc.

A curious application in the food industry is the ability of methylcellulose solutions to gel when heated. For example, adding methylcellulose to fruit pie fillings or sweet jam fillings prevents these components from leaking out during baking, which greatly improves the appearance and preserves the taste of the products.

In pencil factories, methylcellulose is used instead of tragacanth for colored copying rods for pastel rods, school crayons and paints, etc.

2.3. Oxyethyl ethers of cellulose.

Oxyethyl ethers can be obtained by the action of ethylene chlorohydrin on alkaline cellulose according to the scheme:

$$[C_6H_7O_2(OH)_3]_n + CH_2ClCH_2OH \xrightarrow{NaOH} [C_6H_7O_2(OH)_2OCH_2CH_2OH]_n + nNaCl + nH_2O$$

However, most of these esters are obtained by a different method, significantly different from the usual methods of obtaining cellulose ethers.

When oxyethylene interacts with the hydroxyl groups of cellulose, ethylene oxide is attached to the hydroxyl groups of cellulose to form cellulose ethers and glycol (cellulose oxyethyl ethers) according to the scheme:

$$[C_6H_7O_2(OH)_3]_n + nCH_2\text{-}CH_2 \longrightarrow [C_6H_7O_2(OH)_2 OCH_2CH_2OH]_n$$

Oxyethyl ethers can be obtained in the presence of both acids and alkalis, which are catalysts for this reaction.

The reaction of obtaining cellulose oxyalkalic ethers is usually carried out in the presence of alkalis or by the action of olefin oxides on alkaline cellulose, since the use of cellulose strongly swollen in alkalis as a starting material facilitates diffusion of the olefin oxide into the fiber and thus obtaining a uniformly esterified product.

When ethylene oxide (or other oxides) interact with cellulose, a side process of polyglycol formation occurs along with the main esterification process:

$$CH_2\text{-}CH_2 + H_2O \longrightarrow CH_2OHCH_2OH$$
$$\diagdown O \diagup$$

$$CH_2OHCH_2OH + CH_2\text{-}CH_2 + H_2O \longrightarrow CH_2OCH_2OCH_2CH_2OH$$
$$\diagdown O \diagup$$

The question about the ratio of the rates of the reaction of cellulose esterification with ethylene oxide and the side process of polyglycol formation, as well as the possibility of changing the ratio of these rates, has not been studied so far.

When the esterification process was carried out for 10 hours at 100^0 C in the presence of catalysts (acids or diluted alkali solutions), low esterified products were obtained, not dissolving but only swelling in a 10-15% solution of caustic soda. In the presence of more concentrated caustic soda solutions, the rate of the etherification process increases significantly.

When the reaction was carried out at 50^0 for 2 hours in the presence of 6% or 7% caustic soda solution, cellulose oxyethyl ethers with $\gamma=225$-250 were obtained.

Oxyethyl ethers of highly esterified cellulose may have different structures. For example, a preparation with $\gamma=200$ may have the structure of dioxyethyl cellulose ether:

$$[C\,H\,O_{672}\,(OH)(OCH_2\,CH_2\,OH)\,]_{2n}\quad (I)$$

However, the second ethylene oxide molecule can esterify not the cellulose hydroxyl but the glycol residue hydroxyl, forming a polyglucol grouping. In this case the ester will have the following structure:

$$[C\,H\,O_{672}\,(OH)\,OCH_{22}\,CH\,OCH_{22}\,CH_2\,OH]_n\quad (II)$$

Both products (I and II) have the same elementary composition and contain the same number of free hydroxyl groups. Therefore, determination of the structure of highly esterified oxyethyl ethers of cellulose is very difficult. Apparently, the esterification process produces a mixture of both products.

The introduction of small amounts of oxyethyl groups into the cellulose macromolecule causes a decrease in the number of strong bonds between hydroxyl groups in the cellulose macromolecule and thus significantly changes the solubility and reactivity of cellulose.

Already in the presence of one oxyethyl group per 4-5 elementary links of the macromolecule, preparations of oxyethyl cellulose are soluble in diluted alkali, the preparation with $y=50\text{-}100$ is soluble in water. This property of oxyethyl ethers has been used (pure or mixed with cellulose xanthogenate) to produce artificial fiber.

Artificial fiber obtained from cellulose oxyethyl ethers and from mixed xanthogen oxyethyl cellulose ethers has lower strength (especially in the wet state), which is explained by reduced bond strength between macromolecules as a result of the introduction of oxyethyl groups.

As noted above, oxyethyl ethers of cellulose of high ethification can have different secondary hydroxyl groups of cellulose, the total number of primary hydroxyl groups in the cellulose chain increases and, consequently, the reactivity of the cellulose macromolecule increases.

If only the most reactive primary hydroxyl groups are attached, the total number of primary hydroxyl groups may not change.

The increase in the reactivity of the cellulose macromolecule in this case, however, also occurs due to the loosening of the cellulose structure by the ethylene oxide groups 27].

Analysis of all these products is difficult because some amount of ethylene oxide esterifies not only the hydroxyls of cellulose, but also the hydroxyl group of glycol residue, forming a polyglycol grouping. This can be seen from the following diagrams:

$$\text{I. } C_6H_7O_2(OH)_3 + 2CH_2\text{-}CH_2 \underset{O}{\diagdown\diagup} \longrightarrow C_6H_7O_2(OH)\cdot(OCH_2CH_2OH)_2$$

$$\text{II. } C_6H_7O_2(OH)_2\cdot(OCH_2CH_2OH) + 2CH_2\text{-}CH_2 \underset{O}{\diagdown\diagup} \longrightarrow C_6H_7O_2(OH)_2\cdot(OCH_2CH_2OCH_2CH_2OH)$$

The products formed by these two schemes have the same elementary composition and contain the same amount of free hydroxyl groups.

Thus, characterization of such products by degree of substitution is not always correct, since for cellulose oxyethyl ethers the degree of substitution (i.e. the number of hydroxyl groups of the glucose link substituted for oxyethyl groups) may not coincide with the amount of chemically bound oxyethylene, as a result of the reasons described above, i.e. esterification of hydroxyls of previously introduced oxyethyl groups. Therefore, to determine the exact number of introduced oxyethyl groups, a number of researchers [28,29] suggested that the number of moles of chemically bound oxyethylene per one glucose residue, rather than the degree of substitution (C3), should be determined and designated as the substitution modulus (M3).

To determine the chemically bound oxyethylene content in oxyethyl cellulose, this ether is heated with hydrogen iodide acid at its boiling point.

Ethylene iodide is formed as an intermediate product according to the scheme:

$$ROCH_2CH_2OH+3HJ=RJ+JCH_2CH_2J+2H_2O$$

The formed ethylene iodide can further decompose to form ethylene and iodine, or react with hydrogen iodide according to the schemes:

$$JCH_2CH_2J \longrightarrow CH_2=CH_2+J_2$$
$$JCH_2CH_2J+HJ \longrightarrow CH_3CH_2J+J_2$$
$$CH_2=CH_2+HJ \longrightarrow CH_3CH_2J$$

The amount of ethylene and ethyl iodide released is determined. The total amount of released substances is equivalent to the amount of ether glycol groups in oxyethylcellulose.

In addition to determining the modulus of substitution (M3), the degree of substitution (C3) can also be determined under special conditions [30]. The degree of substitution is determined by the degree of phthalation of oxyethyl cellulose, in which only the hydroxyl groups of the oxyethyl groups are subjected to phthalation and the hydroxyl groups of the cellulose are not. Always C3 is less than the modulus of substitution.

One of the characteristic features of oxyethyl cellulose is that it can be used as a semi-product for obtaining mixed cellulose ethers. This makes use of its ability to strongly swell or even dissolve in dilute solutions of alkali or in water, with a small content of oxyethyl groups. This property allows the secondary esterification reaction to be carried out under milder conditions and more homogeneously, which is very important in obtaining products of good quality. This principle was successfully applied in the preparation of the secondary esterification reaction by N.I. Nikitin and co-workers [31-34]. This principle was successfully applied in obtaining cellulose acetates.

The increase in reactivity is characteristic not only for oxyethyl cellulose, but is also observed in other low-substituted cellulose ethers - this is their common property, which was noted in [35]. In a number of cases the increase in reactivity as a result of oxyethylation is a consequence not only of the effect on the structure of cellulose, but also of the introduction, as already mentioned, of reactive primary hydroxyl groups.

The increased reactivity of oxyethyl cellulose can be illustrated by the data taken from the work of Mirkamilov et al. [36] (Table 2).

Table 2.

**Acetylation rate constants of the original cellulose and oxyethylated
samples.**

Degree of substitution (ɣ) by oxyethyl groups	$K \cdot 10^{-2}$ min^{-1}
0 (original pulp)	4,50
8,7	5,55
10,5	6,03
12,4	7,32
13,6	8,74
15,2	9,08

The data in the table show that partial oxyethylation significantly increases the rate of acetylation of cellulose.

In [37], the authors, using 20% sulfur-carbon from the weight of cellulose, obtained mixed xanthogen-oxyethyl cellulose ethers. On the basis of this ester silk was obtained, which did not differ in its physical and mechanical properties from the usual viscose silk. Although this silk did not have high strength (due to very strong lyophilicity), but it was better dyed than ordinary viscose silk and was quite stable. One of the advantages of this method is that it allows a much lower consumption of carbon disulfide, as well as eliminating harmful production.

A water-soluble ethyloxyethyl cellulose, which has interesting properties, was also obtained on an industrial scale on the basis of oxyethyl cellulose [38,39]. The ethylation reaction in this case also proceeds more easily.

Solubility and other properties of oxyethyl cellulose, as well as other cellulose ethers, are mainly determined by the degree of esterification (ɣ), the degree of polymerization (SP) and the uniformity of substituents distribution. These characteristics are important in shaping the basic properties of this product.

Oxyethyl cellulose with a degree of substitution from 0.05 to 0.15 dissolves in dilute alkali solutions when frozen, with a degree of substitution from 0.2 to 0.9 in dilute alkali solutions at room temperature. Water-soluble products have a degree of substitution of 0.4 or higher. Products having a degree of substitution of 2.0 to 3.0 are soluble in organic solvents.

During oxyethylation of cellulose the substitution of hydroxyl groups by oxyethyl groups in the anhydroglucose link is uneven. Thus, in [40] it was shown by hydrolysis and paper chromatography that oxyethyl cellulose with a substitution modulus of 1.33 was an inhomogeneous product in terms of substitution in each glucose residue. It was shown that the distribution of substituents in the glucose links found experimentally was in good agreement with the distribution calculated from the assumption of equal availability of all glucose residues and the ratio of rates of esterification of hydroxyls at C_2 , C_3 and C_6 and the primary hydroxyl in the oxyethyl group - 3:1:10:20.

Oxyethyl cellulose is also a heterogeneous product in terms of molecular weight. By fractional precipitation of it from an aqueous solution with acetone different fractions were obtained. In this way 11 fractions with values $[\eta]_{25}$ =8.4 (in water) were obtained from an initial product having $[\eta]_{25}$ from 11 to 1.5. Several of these fractions were used to determine their absolute molecular weights (but by sedimentation, diffusion, light scattering, and osmosis).

As a result of viscosity and molecular weight measurements, the relationship of these quantities in the Kuhn-Mark formula was established, which turned out to be as follows:

$$[\eta]_{25} =1.03*10^{-3}\dot{M}\frac{0,70}{\vartheta}$$

As for other properties of CEC in dilute aqueous solutions, we should point out the detailed work by Brown (40), which allows us to estimate the configuration of oxyethylcellulose molecules in aqueous solutions. Thus, for example, from the above equation, as well as the relations between the molecular weight and the diffusion and sedimentation coefficients:

$$S_{25} =1.0\cdot10^{-15}\ \dot{M}\frac{0,46}{\omega\cdot w}$$
$$S_{25} =8.2\cdot10^{-5}\ \dot{M}\frac{0,54}{\omega\cdot w}$$

It follows that the excipient α in them lies in the range 0.4-0.7. In accordance with Flory's theory, these values characterize chaotically coiled chains.

Application of oxyethyl cellulose.

Water-soluble oxyethyl cellulose is mainly used in the same applications as water-soluble methyl cellulose and carboxymethyl cellulose.

Aqueous solutions of oxyethylcellulose are used as a thickener and stabilizer of latexes, paints, adhesives, are part of paper sheets, etc. In the textile industry oxyethylcellulose is used for sizing yarn, for preparation of stable emulsions, as a binder, etc. We believe that the use of low-substituted oxyethyl cellulose as an intermediate product in the production of other ethers, such as acetyl cellulose and cellulose xanthogenate, may have special prospects. Weak oxyethyl cellulose, as mentioned above, significantly increases its reactivity in the above reactions and improves the properties of the final product solutions.

The use of water-soluble cellulose oxyethyl esters may be of known interest for the textile industry, where they can be used to replace starch in sizing and aprathering.

2.4. Glycerol esters.

These esters can be obtained in two ways:

1.　　　By the action on alkaline cellulose of α-monochlorohydrin glycerol:

$$[C_6H_7O_2(OH)_3]_n + nClCH_2CHOHCH_2OH \xrightarrow{NaOH} [C_6H_7O_2(OH)_2OCH_2CHOHCH_2OH]_n + nNaCl + nH_2O$$

2. By the action on the alkaline cellulose of the glycid alcohol:

$$[C_6H_7O_2(OH)_3]_n + nCH_2\text{-}CHCH_2OH \xrightarrow{NaOH} [C_6H_7O_2(OH)_2OCH_2CHOHCH_2OH]_n$$
$$\diagdown O \diagup$$

Introduction of a glycerol residue into the cellulose molecule increases the solubility of the obtained cellulose ethers in water and in alkali to an even greater extent than the introduction of oxyethyl groups. For example, monoglycerin cellulose ethers are soluble in water.

The esterification of glycerol ethers of cellulose can produce corresponding derivatives of these ethers. Acetates and nitrates of glycerol cellulose ethers were obtained. Acetylation was performed by treating the glycerin esters with acetic anhydride in the presence of sulfuric acid, but the acetates of these esters can probably be obtained by acetylation with acetic acid alone.

Acetylation of cellulose monoglycerin ester yields tetraacetate of this ester:

$$[C_6H_7O_2(OH)_2OCH_2CHOHCH_2OH]_n + 4n(CH_3CO)_2 \longrightarrow$$

$$\longrightarrow [C_6H_7O_2(OCOCH_3)_2OCH_2CHOCOCH_3CH_2OCOCH_3]_n + 4nCH_3COOH$$

If glycerin cellulose ethers obtained by the action of pure glycerol monochlorohydrin are completely soluble in water and in dilute alkalis, the addition to monochlorohydrin of 1% dichlorglycerol yields glycerin cellulose ethers of the same degree of esterification, but completely insoluble in various solvents.

This can be explained by the fact that glycerol dichlorohydrin, which is a polyfunctional compound, reacts simultaneously with two elementary links of neighboring cellulose macromolecules, which leads to the formation of chemical bonds between macromolecules according to the scheme:

$$\left.\begin{array}{ll} -OH- & OH- \\ -OH- \\ -OH- & OH- \\ -OH- & OH- \end{array}\right| + ClCH_2CHOHCH_2Cl + OH- \longrightarrow \left|\begin{array}{ll} -OH & OH- \\ - O \quad CH_2CHOHCH_2\text{-}OH- \\ -OH & OH- \end{array}\right| + 2HCl$$

The formation of chemical bonds between macromolecules results in insoluble products.

No insoluble compounds are formed during esterification with glycerin alcohol.

LITERATURE

1. Timell T.E., Spurlin H.M. Svensk Papperstidning. 1953,56.p.483-490. Citation: Nikitin N.I. Chemistry of wood and cellulose. M.-L.: Izd. AS USSR, 1962, pp. 348-349.

2. Spurlin H.M. B KH.: E. Ott, Cellulose and cellulose derivatives, 11, N.Z., 1954.

3. Dürlg G., Banderet A. Surlastructure des solutionsagnerses de carboxymetyllal. Helvetica Chimica Acta. 1950, vol. 33, fasc. 4, N 143, pp. 1106-1118.

4. Stawitz J. Die wasserlösiichen Methyl und Carboxymethylzellulosen. Prakt.chem. , 1953, 1g. 4, Heft 7, pp. 181-184.

5. Nikitin N.I., Petropavlovsky G.A. Preparation and properties of low-substituted methyl- and carboxymethylcellulose. Ix, 1956, Vol. 29, № 10, c. 1540-1549.

6. Petropavlovsky G.A. Carboxymethylcellulose and its chemical and physical-chemical properties. J1X, 1959, v. 32, 42. p. 241-253.

7. Zhigach K.F., Finkelstein M.Z., Timokhin I.M. Structural Viscosity of aqueous solutions of carboxymethyl cellulose. DAN of the USSR, 1959, Vol. 126, No. 5, pp. 1025-1028.

8. Butts E.H., Hudy J.A., Elliott J.H.. Rheology of Sodium Carboxymethyicellulose Solutions. industrial and Eng. Chem., 1957, vol. 49, no. 2, pp. 94-98.

9. Schulz J., Kienzl E. Untersuchungen an Carboxymethylcellulose. 1V. Optische Untersuchungen Monat shefte für chemie 1957, Bd. 88/6, pp. 1017-1023.

10. Dirig G., Banderet A. Surla structure des solutions of uenses de carboxymethyicellulose. Helvetica Chimica Acta, 1950, vol. 33, No. 4, pp. 1106-1118.

11. Schulz Jo, Stretzig H., Wurz E. Untersuchungen an Carboxymethyicellulose. 1. Viskosität - Molekulargewichtsbezichung in verschiedenen Lösungsmitteln. Monatshefte für chemie. 1950, Bd. 87/4, pp. 520-525.

12. Kosyreva I.K., Glickman S.A. On nature of solutions and gels of carboxymethylcellulose. Vysokomol. soed. 1961, vol. 3, pp. 1564-1580.

13. Brown W.I., The Configuration of Cellulose and Derivatives in Solution Tappi, 1966, v. 49, 98, pp. 367-373.

14. Rudy V.P., Albeta N.K., Safronova N.I. Viscosity and electrical conductivity of solutions of carboxymethylcellulose salts. Ukr. chim. zh., 160, v. 26, p. 7L6 - 718.

15. Nikitin N.I., Klenkova N.I. Physico-chemical properties of mercerized and weakly alkylated cellulose. In: Research in the field of high molecular weight compounds. M.-L.: Publishing House of Academy of Sciences of USSR, 1949, pp. 138-145.

16. Petropavlovsky G.A., Nikitin N.I. Properties of low-substituted carboxymethylcellulose from Dahurian larch wood. Proceedings of the Forest Institute. L., 1958, v. 35, p. 93-102.

17. Petropavlovsky G.A., Nikitin N.I. Preparation and properties of low-substituted. Cellulose esters. In book: Chemistry and technology of cellulose derivatives. Vladimir, 1964, pp. 208-231.

18. Nikitin N.I. Chemistry of wood and cellulose. M.-L.: Izd.AS USSR, -711 p.

19. Rogovin Z.A., Shorygina N.N. Chemistry of cellulose and its companions. M.-L.; Goskimizdat, 1953, - 678 p.

20. Kuhn W., Moser P., Majer H. Aggregation von Methyicellulo- se in Lösung. Helvetica Chemica Acta, 1961, vol. 44, fase 3, N° 89, p. 770-792.

21. Heuser E. The Chemistry of Cellulose, New-York, 1947, - 660 p.

22. Vink H. Degradatlon of Cellulose and Cellulose Derivatives by Acid Hydrolysis. Makromol. Chem., 1966, Bd. 94, pp. 1-14.

23. Uda K., Meyerhoff G. Hydrodynamische Eigen Schaften Methyicellulosen in Lösung. Makromol. Chem., 1961, Bd.47, von pp. 168-184.

24. Neeby B.W.. Solution Properties of Polysaccharides. IY. Molecular Weight and Aggregate Formation in Methyicellulose Solution. J. Pol. Sci., 1963, pt A, N° 1, pp. 311-320.

25. Kumagen Yo, Nagasowa Shin. "Nihon shokuhin kocho gakkai shi. Viscosity of diluted methylcellulose solution. J. Food Sc1. and Technol., 1965, v.12, N° 12, N°8,p.408-412. RGM., 1967, 2c373.

26. Foweraker A.R., Jennings B.R., Electr1c birefringence of agneonssolutions of methyicellulose. Polymer science and technology of polymers and biopolymers. 1976, vol. 17, no. 6, pp. 508-510.

27. Ushakov S.N., Klimova O.M., E.I. Feingor. JPX 27, 1.71 (1954).

28. Morgan R.W., Jud Eng. Chem. ? Anal. Ed., 18, 500 (1946).

29. Quinchon J., Hebd C.R., Seances Acad Sci., 148, 125 (1959).

30. Practical works on the chemistry of wood and cellulose. M., 395 (1966).

31. Quinchon J., Hebd C.R., Seances Acad Sci., 250, 1258 (1960).

32. Akim E.L., Nikitin N.I. ZhPH,36,1075 (1963).

33.Akim E.L., Nikitin N.I. Tr. Lenin Gr. of the Technological Institute of the Pulp and Paper Industry, 12, 199 (1964).

34. Klenkova N.I., Kulakova O.M., Volkov L.A., ZhPH,37,9,2023 (1964).

35. Tsvetaeva I.P., Petropavlovsky G.A., Ionanova T.N. ZhPH,39,7,1549 (1966).

36. Mirkomilov G.M., Autoreferat,p.20, Tashkent (1966).

37. Kargin V.A., Kurilchikov E.A., "Tr. Research Institute of Artificial Fibers" 44 (1955).

38. Gullender J.Jud. Eng Chem, 49,3,364 (1957).

39. Harry D. "Soap and Chem. Specialtice" 40,11. 165 (1964).

40. Brownell H.H., Purvis C.B. Canad. S.Chem, 35, 677 (1957).

CHAPTER III. WATER-SOLUBLE CELLULOSE ESTERS

3.1. Cellulose sulfates.

Cellulose, the most abundant natural polymer on earth, is a very important object because its application is expanding every year, and the available reserves, if rationally used, can be inexhaustible.

Cellulose ethers-nitrates, acetates, phosphates, sulfates, carboxymethyl ethers and numerous others are of great interest for the national economy, cellulose processing is largely due to obtaining its simple and complex ethers. There are numerous studies in the chemistry and technology of cellulose and its processing. Hydrophilic cellulose derivatives, in particular carboxymethylcellulose (CMC) and methylcellulose, are produced by industry and have wide application in various branches of the economy.

Among hydrophilic cellulose derivatives, cellulose sulfates (SC) have limited application due to their inaccessibility. The latter is due to the complexity of production of SC, which requires special reaction conditions, in particular, the conduct of the reaction at low temperature and the difficulty of regulating the side reaction - destruction of the cellulose macromolecule, leading to charring of the reaction product.

SO has unique physical, chemical and biomedical properties. There is evidence that SC was once used as an ingredient in drilling fluids. In recent years, it has been found that SC, which has a degree of substitution (SZ) of 3.0 and a molecular weight (MM) of $2.3\text{-}1.9\text{-}10^6$ with a polydispersity of 4.5, inhibits HIV infection and exhibits antibacterial activity against trichomatosis, gonorrhea, etc.

There are no systematic studies of cellulose sulfation in the scientific literature, there are no methods of obtaining SC with given molecular parameters, its medical and biological activity and the influence of molecular parameters on medical and biological activity have not been studied. The solution of these questions contributes to the elucidation of the mechanism of action of SC on the living organism and to the obtaining of new original antiviral and antibacterial drugs.

In this regard, it is relevant to conduct systematic studies of the reaction of cellulose sulfation and obtaining SC with predetermined chemical, physical and chemical, and medical and biological properties.

Peculiarities of chemical activation and its effect on chemical and physicochemical characteristics of cellulose.

Cellulose reactivity is a complex concept reflecting various aspects of its chemical processing process: kinetics of heterogeneous reaction, uniformity of esterification, dissolution rate, quality of solutions and end products. The purpose of cellulose activation is to increase and average the reactivity of cellulose fibers. The problem of its activation, is directly related to the elucidation of the mechanism of penetration of reagents into cellulose fibers.

The capillary transport of liquids through its pores is the decisive factor in determining the rate of penetration of reagents into cellulose materials. Fluid transport into the pulp preparation thickness is governed by the laws of fluid flow through the pores. The porous structure of cellulose obtained in the processing of natural fibers and materials is due to the peculiarities of the processing process and the content of amorphous, crystalline structures, as well as the content of defective areas appearing in the process of processing [1].

The authors [2] note that hydroxyl groups in the cellulose macromolecule, exhibit low reactivity in reactions characteristic of alcoholic hydroxyls. The latter is primarily due to the inaccessibility of hydroxyl groups for the reaction due to the heterogeneity of the process and its macromolecular nature. To increase the reactivity of hydroxyl groups in esterification reactions, their pre-activation is required.

Activation consists of treating the cellulose with liquids that cause the fibers to swell. This causes loosening of the supramolecular structure, especially in the disordered areas of the fiber. Macromolecules move away from each other, hydrogen bonds between hydroxyl groups are destroyed, as a result, porosity and total internal surface of fiber capillaries increase and, consequently, access of reacting mixture to hydroxyl groups of cellulose is facilitated [3].

The weakening and destruction of intermolecular hydrogen bonds increases the reactivity of hydroxyl groups and provides macromolecules with a certain degree of freedom and, apparently, ensures the transition of cellulose from the glassy to the highly elastic state [4].

A number of chemical compounds are known to activate cellulose in esterification reactions. They include water, formic and acetic acids, alcohols, glycols, surfactants, various amines and many others. In production conditions activation is usually carried out with acetic acid with or without various additives [5,6].

The efficiency of cellulose activation depends on many factors, in particular on the composition of the activating mixture, temperature, process duration and reaction module. Processes of cellulose activation to increase its reactivity and the resulting structural changes have been studied in many papers [2,6,8]. Many methods of cellulose activation for chemical processing have been developed [2,7,8].

To find out the mechanism of cellulose activation and deactivation, we studied the change in its structural and physical state under the action of various NMVs: ammonia, monoethanolamine, ethylenediamine, water, ethylene glycol and glycerol [9]. Wood, cotton and regenerated cellulose were used as an object of study.

So far organic solvents in the vast majority of cases have been used only for pretreatment of cellulose ("soaking") or for dilution of the reaction medium. The main reason for increasing the reactivity of cellulose was seen either in increasing the degree of swelling of cellulose and, accordingly, increasing its availability and transition to the highly elastic state, or in a change in the chemical activity of its reaction centers during their interaction with the solvent [2, 10].

According to the authors [11], the activating role of the organic solvent consists mainly not in the fact that the degree of swelling of cellulose increases, but in the fact that water is removed from its micro voids and amorphous regions, interfering with the course of the main reaction.

Therefore, the use of organic liquids to increase the availability of cellulose in specific technological processes can be effective only when water chemically interacts with the agents of the reaction mixture and thus reduces their concentration or shifts the equilibrium of the main reaction in the opposite direction.

Cellulose can be made highly reactive by treating it with water and then displacing the water (if it is harmful to the reaction). Even very small amounts of water absorbed by cellulose fibers loosen its structure [12].

To remove thin films that partially overlap the pores in the cellulose, in principle, media that cause dissolution of low molecular weight fractions of the cellulose can be used, which can be achieved by hot water treatment [12].

As it was shown, under the influence of water the transition of amorphous regions of cellulose into highly elastic state occurs, and during drying - their partial vitrification. At alternation of moistening-drying cycles the duration of cellulose in highly elastic state increases, which contributes to fuller course of relaxation processes leading to compaction of cellulose [13].

Mershant M. V. [14] showed that the values of internal surface determined by nitrogen vapor sorption data differ significantly depending on which solvents are used to displace water from the swollen fiber. He carried out a study of the inner surface values and total pore volume in bleached sulfite cellulose fibers by gas absorption. Cellulose after its extraction from wood, without drying, was treated first with methyl alcohol and then with non-polar liquid hydrocarbon and dried in a vacuum. When water is displaced by different hydrocarbons, the internal pore surface increases in the following order:

Benzene..............................43,0 m^2/g.
Toluene.............................48,1 m/g^2
Cyclohexane......................88,5 m/g^2
n-Hexane...........................108,0 m/g^2

The drying temperature after treatment of pulp, hydrocarbons also affects the value of the inner surface of the pulp [14].

For full activation of cellulose it is necessary to weaken the intermolecular interaction of macromolecules, at which a developed network of the finest submicroscopic capillaries is formed. The latter is achieved by treating cellulose with such substances, which, with a sufficiently small molecular volume, tend to have strong donor-acceptor interactions with hydroxyl groups of cellulose [3].

It is reported [15,16] that treatment of cellulose fibers with glacial and especially 80% acetic acid leads to increase of internal surface of fibers and increase of volume of micropores in their structure, mainly due to small radius pores. The radius of the predominant pores after swelling in acetic acid and its displacement by changing solvents is in the range of 12-15 Å. Liquids that can cause strong swelling of cellulose and break hydrogen bonds between cellulose molecules themselves include formic acid and many amines [2].

Molecules such as pyridine or acetic acid, when introduced into the cellulose system before the introduction of the entire reaction mixture, interact with the OH groups and thus contribute to the activation of the hydroxyls themselves and their subsequent more active reaction with other molecules [17]. If pyridine or acetic acid are already in the reaction medium before reacting with cellulose, their role as activators of the OH-groups of cellulose will be increased because their molecules are in less interaction with other components of the reaction mixture.

It is assumed that in acylation reactions where organic bases (e.g., pyridine) are used, hydroxyls acquire more pronounced alcoholic properties under the influence of these activators, which facilitates their interaction with the esterifying agent by weakening the intermolecular hydrogen bonds.

Comparative evaluation of structural changes in cellulose under the influence of pyridine, DMSO, DMF and methylene chloride, conducted by Matveeva N.A., Kutsenko L.I., Klenkova N.I. [18], showed that a large value of the inner surface of cellulose appears in the treatment of cellulose with pyridine.

Palit S. and Amis E. [19,20] formulated the basic principle of the effect of solvent activation on the reaction rate of cellulose. If the active center involved in the reaction is blocked by hydrogen bonds with the solvent, the reaction rate decreases. On the contrary, if the solvent is involved in the release of hydroxyl groups of hydrogen bonds, the reaction rate increases.

Inorganic and organic bases, in particular tertiary amines are more active activators of cellulose, especially its low molecular weight fractions. When treated with water at high temperature and under the action of bases simultaneously with the dissolution of low molecular weight fractions the restoration of capillaries previously closed due to collapse occurs [21].

A large number of studies are devoted to the effect of ethylamine on cellulose [2,22], and here more attention is paid to the state of cellulose structure, which is achieved after removal of the amine, i.e. after destruction of its complex with cellulose. Amines were displaced by changing solvents - methanol, chloroform, n-hexane [23]. This achieved complete displacement of the amine and preservation of the altered structure, which was achieved after activation. It was shown that aniline and diphenylamine have low activation ability, and diethylamine is more active. Ethylenediamine has the greatest effect. However, it should be noted that the SP of cellulose decreases sharply when it is activated by primary amines [23].

In [24] it was shown that organic solvents (ethylene chloride, chloroform, carbon tetrachloride, n-hexane, cyclohexane, n-heptane) have a significant effect on the activity of primary and secondary hydroxyl groups of cellulose. Nitric acid cellulose ethers with small NW obtained in the medium of ethylene chloride, chloroform, carbon tetrachloride, n-hexane, cyclohexane and n-heptane have different properties. This is due to the different arrangement of nitrate groups in the polymer elementary chain. For example, in normal hydrocarbons, the rate of esterification of primary hydroxyls exceeds that of secondary hydroxyls. This difference is leveled in halogenated hydrocarbons, especially in methylene chloride.

Treatment of cellulose with monoethylamine [25] leads to a change in the structure of native cellulose fibers-almost to its complete amorphization (according to X-ray data) and to the formation of a developed inner surface. As a result, the reactivity of cellulose increases sharply. However, monoethanolamine has only a very weak activating effect on cellulose. This is apparently explained by the fact that during the self-association of monoethanolamine molecules, the intermolecular interaction due to the hydroxyl and amino groups is very strong. Probably, the formation of intramolecular hydrogen bonding of monoethanolamine leads to loosening and removal of intermolecular hydrogen bonds in the cellulose macromolecule.

Activation of cellulose with acetic acid alone requires a significantly longer reaction time to obtain fully soluble cellulose triacetates, even at the same temperature conditions and the same composition of the reaction mixture. This is because the process of cellulose activation by acetic acid cannot deeply affect the fiber structure and most of the cellulose system remains unavailable for reaction at the beginning of the process. Pre-swelling of hydrate cellulose fibers in water enhances the subsequent swelling in glacial acetic acid and dramatically increases their reactivity during acetylation.

In the American patent [26] for the production of SC, first the cellulose was activated with water, then dehydrated with an organic solvent, such as isopropyl alcohol or another type of alcohol.

The authors of the patent [27] for the synthesis of SC cellulose (purified wood pulp or cotton lint) were activated by treating with water and then dehydrated with alcohols. In the American patent [28] chemical pulp was used for sulfation, which was accordingly activated for esterification with sulfuric acid, activating the pulp with water by any suitable means and then replacing the water in the pulp with anhydrous acetic acid.

The patent [29] used a combined activation method. The cellulose was treated in a mixture of acetic anhydride and glacial acetic acid. The water in the cellulose was converted into acetic acid. The patents /30-34/ show methods of cellulose activation before sulfation by boiling in water, treatment with dimethylformamide, acetone, nitrogen-containing bases. In the latter case there is a high NW and SP of the final product. This method has great advantages, but there are great difficulties in purification of the product from a nitrogen-containing base.

In [35] cellulose was activated with acetic acid before sulfation. For this purpose, wood pulp, dried to 105 °C, was soaked in glacial acetic acid. The resulting suspension was vacuumed and filtered at atmospheric pressure. The filtered pulp containing twice the amount of acetic acid was placed in a polyethylene bag in a freezer and held at 10°C., for 2 hours. The reactivity of cellulose in the sulfation reaction was quite high.

It is reported [36,37] that high-SC can be obtained from water-activated cellulose by direct action of aqueous sulfuric acid solution or sulfuric acid diluted in an inert organic solvent such as toluene, hydrocarbon tetrachloride or lower alcohols. Activation of cellulose for its sulfation by mercerization [38, 39] have a number of disadvantages because of the difficulty of complete removal of alkali from the reaction medium. Activation of cellulose by mineral acids leads to destruction of the original cellulose.

SC obtained by activation of cellulose with tertiary amines or a mixture of tertiary amines and amine hydrochlorides [40] has poor solubility.

Thus, based on the analysis of the scientific literature available in the results, we can conclude that the problem of cellulose activation in order to increase its reactivity in various chemical reactions comes down to creation:

-thin supramolecular structure, which would contribute to the maximum use of the mechanism of capillary transport of reagents inside the cellulose fiber;

-preventing the effects of overlapping pores with monolithic thin films, through which the reagent very slowly penetrates into the inner regions of the material by the mechanism of molecular diffusion.

-Necessity to take into account the activation factors: processing conditions - water content, hydromodulus of activators, processing temperature, molecular characteristics of the original pulp sample, drying temperature, etc.

Peculiarities of sulfation of cellulose, its physical and chemical properties.

Cellulose sulfate (CS) is one of the most widely used cellulose ethers. It is readily available and can be obtained from various types of cellulose. There is a prospect of wide introduction of SC instead of other polysaccharides used today. Due to its greater availability and processability in the medical and food industry, it appears that SC could partially replace natural sulfated polysaccharides such as heparin and carrageenans. In medicine, SC can be used to make special coatings and obtain blood anticoagulants [41], antiviral [42], and anti-inflammatory substances [43].

The main way to obtain SC is the treatment of cellulose with sulfuric acid in the presence of inert diluents or aliphatic alcohols [44].

SCs are cellulose esters, and their composition can be expressed by the following general formula:

$$[C\ H\ O_{672}\ (OH)_{3-x}\ \text{-}(OSO_3\ H)\ ]_{xn}$$

The esterification of cellulose with sulfating agents can theoretically produce acidic (I) and intermediate (II) cellulose ethers:

$$Целл - O - SO_2 - OH \quad (I) \qquad\qquad Целл - O - SO_2 - O - Целл \quad (II)$$

However, all methods of producing SO always produce only acidic esters (I).

Sulfuric acid cellulose ethers can be obtained from cellulose by the action:

-concentrated and dilute sulfuric acid;

-concentrated sulfuric acid in the presence of inert diluents

$$[C_6H_7O_2(OH)_3]_n + nxH_2SO_4 \rightarrow [C_6H_7O_2(OH)_{3-x}(OSO_3Na)_x]_n$$

-sulfur dioxide (gaseous; dissolved in carbon tetrachloride, carbon disulfide, liquid sulfur dioxide; as complexes with dimethylformamide, dimethyl sulfoxide, tertiary alkylamines, pyridine, poly-2-vinylpyridine);

$$[C_6H_7O_2(OH)_3]_n + nxSO_3 \rightarrow [C_6H_7O_2(OH)_{3-x}(OSO_3Na)_x]_n$$

- HSC or sodium chlorosulfonate in the presence of pyridine or acetic acid and acetaldehyde;

$$[C_6H_7O_2(OH)_3]_n + nxClSO_3Na \rightarrow [C_6H_7O_2(OH)_{3-x}(OSO_3Na)_x]_n + nxHCl$$

$$nxHCl + 3nxC_6H_5N \rightarrow nx3C_6H_5N \cdot HCl$$

sodium or ammonium fluorosulfonate in the presence of caustic soda;

$$[C_6H_7O_2(OH)_3]_n + nxNaSO_3F + nxNaOH \rightarrow$$

$$\rightarrow [C_6H_7O_2(OH)_{3-x}(OSO_3H)_x]_n + nxNaF + nxH_2O + 24.3 ккал/моль$$

SO with different NWs - up to three.

The sulfation reaction is an equilibrium reaction. The degree of esterification of the resulting product depends on the ratio of the rates of forward and reverse reactions. From this point of view it is advisable to esterify cellulose with concentrated sulfuric acid, but such acid causes charring of cellulose. The lower the concentration of sulfuric acid, the lower the possibility of charring, but the more intensive is cellulose degradation and saponification of the resulting ether and the lower is the degree of esterification [27].

Usually cellulose sulfating is carried out with 60-70% sulfuric acid at lower temperatures (0-15 °C) to reduce degradation. In a longer action of sulfuric acid the resulting sulfated cellulose dissolves in 60-70% sulfuric acid; consequently, the esterification reaction proceeds in a homogeneous environment.

Sulfate cellulose obtained by the action of 70% sulfuric acid for 11 hours at 15 ° C has CP 0.20 - 0.30. When increasing the time of esterification, products of higher Cp are obtained, but at the same time there is destruction of cellulose. The maximum SP, which is achieved in the etherification of cellulose with sulfuric acid, corresponds to 2.0 /48/.

In [45] esterification of regenerated or swollen cellulose was carried out with an aqueous sulfuric acid solution in the presence of Na_2 SO_4 , $(NH)_{42}$ SO_4 and urea as a catalyst in the esterification reaction [46,47]. The SZ of the final product is 0.54 and it is water soluble. A disadvantage of these methods of obtaining SO is a side reaction of degradation by $(1 \rightarrow 4)\beta$ - glucopyrnose bonds due to a large excess of sulfuric acid, as a result of which SP of the product decreases. According to this method, sulfuric acid esterifies cellulose nonuniformly, and most of the cellulose remains unreacted. Another disadvantage of the method is that the product is obtained with low SP and SP, and therefore in the future this method has not found wide application.

The authors proposed [48-51] to esterify cellulose with a mixture of H_2 SO_4 with aliphatic alcohols. The obtained product was washed with alcohol, ether or alcoholic alkali. The reaction yielded a product with a Cp of 0.41. A 2.0% aqueous solution of this product had a viscosity of 47.9 cP at 25°C (for comparison, the viscosity of 100% glycerol at 25°C is 945 cP [52].

In other patents [53,54] cellulose was treated with a mixture of H_2 SO_4 with aliphatic alcohols in the presence of liquid SO_2 .

In [55-58], wood pulp was treated with a mixture of aliphatic alcohol and H_2 SO_4 in the presence of $(NH)_{42}$ SO_4 at -15— -4°C for 45-90 min. and 10% aqueous solutions of this product had a viscosity of 2936-4408 cP.

G.A. Petropavlovsky and his collaborators [59-64] studied the reaction of cellulose with mixtures of sulfuric acid and aliphatic alcohols. A mixture of concentrated sulfuric acid (97 %) and methane-type aliphatic alcohols (from methyl to octyl) was used for cellulose (cotton lint) esterification. The effect on the degree of esterification of such factors as the alcohol chain length, its structure, the molar ratio in the alcohol/acid mixture, reaction time and temperature was studied. In separate experiments the effect of pure alkylseric acids on cellulose was studied [59].

The yield of SO by the method of obtaining by a mixture of sulfuric acid with aliphatic alcohols is small and amounts to 40-45% per cellulose, but it can be increased up to 70-85% by introduction into the reaction mixture of surfactant (OP-10). The lowest degradation of cellulose occurs when sulfuric acid ethers are produced by the action on cellulose with sulfuric acid in the presence of aliphatic alcohols, inert diluents and surfactants [43,65-78].

Cellulose can be esterified with a mixture of H_2SO_4 and acetic acid [79,80]. For this purpose, the cellulose is pre-activated by treatment with water, which is then displaced by acetic acid.

In [81] it was proposed to sulfate cellulose with a mixture of H_2SO_4 with chlorinated alcohols. The S of the product was 0.4-0.5. The reaction was carried out at -4 -25°C. In works [82,83] mixed H_2SO_4 with chlorinated hydrocarbons without alcohol groups.

These methods make it possible to obtain SO with CH no more than 1.5, which is related to the equilibrium of the reaction. Another group of methods is associated with obtaining more highly substituted products. The esterification of cellulose with both sulfuric anhydride SO_3 , and sulfuric acid in a medium of liquid SO_2 as a solvent was studied [84]. Sulfuric acid, unlike other reagents, is poorly soluble in SO_2 . The esterification of cellulose with a mixture of H_2SO_4 and liquid SO_2 proceeds under heterogeneous conditions, and the fibrous structure of the cellulose is not retained. A gel is formed and the yield is 60-75% of the theoretical yield, as the SO swells and partially dissolves in the etherifying mixture. The authors assume that this is due to the presence of water in H_2SO_4 . In 1.5 h at 8-10°C, SO is formed with a Cp of up to 0.9. The apparatus design of the reaction is rather complicated.

It was proposed to esterify cellulose with sulfuric anhydride [85]. SP of the obtained product was 375-600. The yield is up to 65 %. In [86] cellulose esterification with a stable solution of SO_3 in $\tilde{N}\tilde{N}_2$ with SO concentration$_3$ 5-6 % was carried out. The same authors suggested mitigation of SO_3 by esterification in the presence of pyridine taken in excess with respect to SO_3. In this method SO is obtained in the form of its pyridine salt - [C H O_{672} $(OSO_2 O\text{-}PYR)_3$]n.

The esterification of cellulose with sulfuric anhydride dissolved in liquid SO_2 has been studied [84]. The esterification proceeds much better than with $H_2 SO_4$ under the same conditions. The fibrous form of the product is retained and the yield is in the range of 60-90%. The SP does not exceed 2.0 due to the presence of moisture not completely removed by drying the pulp before the reaction (the cotton contained about 5% moisture). Increasing the concentration of SO_3 does not lead to an increase in Cp, and the Cp thus falls by a factor of about 5. The apparatus design of the reaction, as in the case of esterification of $H_2 SO_4$ in liquid SO_2, is rather complicated.

In order to avoid cellulose degradation when obtaining CC it is advisable to carry out the reaction in the presence of pyridine [87]. For cellulose esterification, treatment with 5-6% sulfuric anhydride solution in carbon disulfide for 20-30 min with continuous shaking of the mixture can also be used [88]. However, the reaction products were not investigated in detail in this work.

In works [89-95] production of SC by etherification of cellulose by a complex of sulfuric anhydride with dimethylformamide was considered. At that, SC with S 2.2-2.6 was obtained. The samples were high molecular weight: 1% aqueous solution had a viscosity of 150-550 cP. Significantly less degradation of cellulose occurs when sulfuric acid esters are obtained by the action on cellulose of CSC in the presence of pyridine [96-102]. The reaction proceeds according to the following scheme:

$$[C_6H_7O_2(OH)_3]_n + 3nClSO_2OH \xrightarrow{PYR} [C_6H_7O_2(OSO_2O-PYR)_3]_n + 3nPYR \cdot HCl$$

The released hydrogen chloride binds with pyridine. Since no water is released in this method of esterification, the reaction proceeds to the end and cellulose trisulfate is formed. When precipitated with alcohol or an alcoholic alkaline solution, cellulose trisulfate is released as a pyridine salt, which is a solid white hygroscopic substance. This product is soluble in hot water, forming a colorless viscous solution, and insoluble in organic solvents. The properties of this ester, particularly SP, have not been investigated so far. When neutralized to pH=8.0-9.0 this salt transforms into a sodium salt of SC:

$$[C_6H_7O_2(OSO_2O-PYR)_3]_n + 3nPYR \cdot HCl + 6nNaOH \xrightarrow[-3nH_2O,\, 3nNaCl]{}$$

$$\longrightarrow [C_6H_7O_2(OSO_2O-Na)_3]_n + 3nPYR$$

The formation of a viscous solution is indirect evidence of relatively minor degradation of cellulose - the sample with Cp 1.8 had a viscosity of 1% aqueous solution of 25 cP. When treating HSC in the absence of pyridine a product of the same degree of esterification is obtained, but giving very low viscosity solutions (viscosity = 8.4 cP), which is explained by strong degradation of cellulose in interaction with the released hydrochloric acid. In [102] it was also proposed to esterify HSC cellulose in an excess of a weak base - pyridine, which was to mitigate the destructive effect of acid. This mixture, as well as the mixture SO_3 + pyridine, leads to the formation of N-pyridinosulfonic acid, which sulfates cellulose. In this case, the maximum CP reaches 2.7-2.8. For a fully substituted product, it was necessary to perform repeated sulfation.

The firm "Pharbenindustry" [86] proposed a method of cellulose esterification with a mixture of HSC and pyridine under milder conditions: with a large excess of pyridine at a temperature not exceeding 180°C. In this case a homogeneous mass was obtained, as in [102], and the product, to a certain extent preserving the fibrous structure, had a SP of 2.5. The authors [86] increased the reaction time to several weeks in order to obtain cellulose trisulfate. However, they failed to obtain cellulose trisulfate with a CH greater than 2.9.

We studied the sulfation of HSC cellulose in liquid SO_2 [84]. The maximum SZ in this case was 1.8. The final product had a fibrous structure and its yield was 60-95%.

In [103] a method of obtaining sulfuric acid cellulose ethers was developed, which differs significantly from previously known methods and is based on the treatment of alkaline cellulose with sodium or ammonium fluorosulfonate by the reaction:

$$C_6H_7O_2(OH)_3 + nFSO_2ONa \xrightarrow{nNaOH} C_6H_7O_2(OH)_{3-n}(SO_3Na)_n + nNaF + nH_2O$$

As can be seen from the reaction equation, this method is similar to the method of obtaining the well-known CMC [104] by the suspension method. Its advantage is that the reaction is carried out in aqueous-alkaline medium at normal (or slightly higher) temperature and with a fairly simple apparatus design of the process.

Sodium fluorosulfonate sulfonation of cellulose is more effective than ammonium fluorosulfonate: sodium fluorosulfonate utilization factor is 48% (СЗ СЦ is 0.96) and ammonium fluorosulfonate is 33% (СЗ СЦ 0.64-0.67). This is due to the fact that sodium fluorphonate is more resistant to the hydrolytic action of caustic soda solution than ammonium fluorosulfonate. SC was regulated by changing the ratio of cellulose to sodium fluorosulfonate.

In [105,106] a method of obtaining cellulose sulfoacetate was proposed, which consists in the treatment of cellulose with acetic anhydride and Na bisulfate in acetic acid. The reaction proceeds according to the following scheme:

$$\text{Целл (OH)}_3 + (CH_3CO)_2 + NaHSO_4 \longrightarrow \text{Целл} \begin{cases} (OCOCH_3)_2 \\ OSO_3Na \end{cases}$$

Thus, the existing methods of cellulose sulfation are mainly reduced to the production of SC with low SP and SP. In the literature there are no systematic studies of methods of synthesis of CP, the effect of cellulose activation methods on the molecular parameters of the obtained CP, there are no specific studies on obtaining CP with CP over 2.5. The physicochemical properties of cellulose ethers depend primarily on the nature of the substituent, SP, and MM. The information on the properties of SC salts is often contradictory. This is explained by the fact that each of the researchers describes the product obtained, as a rule, only by one method, while these methods give products of different degrees of esterification, and different degrees of destruction, which should affect their properties.

SC can be obtained both as an acid (H-CP) and as various salts. The most important is the sodium salt of SC (Na-SC), which is a white solid substance. Dry SO with a moisture content of 3 - 5 % is a powdery or fibrous mass, nonflammable [107], pale yellow in color, odorless, with bulk density 480-720 kg/m^3 and true density 1752 kg/m^3 , it is not hygroscopic. SC is insoluble in common organic solvents, but soluble in hot and cold water [102].

The viscosity of the SC solution increases rapidly with increasing concentration, eventually the viscous solution becomes a gel. Viscosity decreases with increasing temperature [102].

H-CA is a strong acid with a dissociation constant of $1.23 \cdot 10^{-2}$ [108]. It and its salts are insoluble in low molecular weight alcohols and ketones. Salts of alkali metals and ammonium with SP up to 0.10 dissolve in 6% NaOH or NH_4 OH solution when frozen, salts with SP up to 0.14 at room temperature; salts with SP above 0.16 are soluble in water. SC salts at SP of 0.55-0.95 are soluble in water.

Na-CC does not precipitate from aqueous solutions when heated. Its aqueous solutions are characterized by high viscosity values that do not obey Newton's law [109]; viscosity increases sharply with increasing concentration and SP and decreases with increasing temperature. In aqueous solutions, salts of CC exhibit the properties of polyelectrolytes [110,111]. Na-CC has film-forming properties. Film tensile strength is 7.5-12.5 kgf/mm^2 , relative elongation at break is 10-22%. SO films are transparent, flexible and insoluble in organic solvents and nonflammable [112]. The sulfate group in SO is very resistant to hydrolysis by acids and bases at low temperatures. By boiling SO with hydrochloric acid solution, sulfuric acid can be quantitatively reduced [110]. The polyelectrolyte nature of Na-SC causes considerable difficulties in its fractionation. In [61] 10 fractions of low-substituted samples of SC (CP up to 0.6) were obtained, their characteristic viscosity and sulfur content were determined.

A large number of studies are devoted to the solubility of SO salts [70,86,102,113]. A number of authors conclude that the salt is insoluble based on the fact that it precipitates when the corresponding cation is introduced into a solution of H-CS or a water-soluble salt of SC. In most cases, however, this involves concentrated solutions of SC salts. Other researchers [85,114,115] conclude that SC salts are insoluble when dealing with preparations that have undergone the drying process. In this case, the solubility pattern can be changed by the process of "keratinization" of the CC salt.

The solubility of various SC salts has been studied in more detail by S. Inokava [115]. It was shown that the solubility increases in the following order Fe-CC < Al-CC < Pl-CC < Ba-CC < Ca-CC < Cu-CC < Mg-CC < Hg-CC < Na-CC. The author also concluded that the solubility of SC is primarily dependent on NW rather than SP of these salts.

NW SO is calculated in the literature by various methods. One of them is to determine the content of sulfate groups [116].

Methods for the study of cellulose sulfates.

The objects of the study were cellulose of different origin (cotton, wood, cotton fiber, lint, MCC), sulfating agents differing in their chemical nature (sulfuric, ethylsulfonic, chlorosulfonic acid).

Synthesis of SC using the pyridine-HSC complex.

a) Direct sulfation without pre-activation of cellulose (lint, fiber, cotton pulp) in pyridine -HSO complex$_3$ Cl.

In a three-neck flask equipped with a mechanical stirrer, a reflux condenser and a dropping funnel, anhydrous pyridine is poured and CSK is added by addition at a certain ratio of pyridine : HSO$_2$ Cl, under constant stirring, at -20°C. Then the sulfating mixture is heated until the complex dissolves, then pre-dried inactivated cellulose is added to the reaction mixture. The synthesis is carried out at +75°C for 3-6 hours. At the end of the reaction the excessive pyridine is filtered out of the reaction mixture by vacuuming out in a Schott filter. The filtrate is washed first with ethanol, then with acetone several times. The resulting filament is neutralized by the addition of saturated aqueous sodium bicarbonate solution and the pH of the medium is adjusted to 9.0. The system stratifies and the upper layer consisting of pyridine is decanted. The mass is then dissolved in an aqueous-alkaline solution and left at room temperature for 2 hours at a pH of about 9.0. The solution was dialyzed through a cellophane membrane until it was negative for -SO ions$_4^{2-}$ in the dialysate. The dialyzed clear Na-CC solution was dried on a rotary evaporator. The dried product is analyzed.

Sulfation of pyridine-activated cellulose (lint, fibers, cotton cellulose) by pyridine - HSO complex$_3$ Cl was carried out similarly according to the above method, taking into account the residual pyridine content in activated cellulose. The reaction product was isolated from the reaction mixture by filtration followed by treatment of the filtered product with saturated solution of sodium bicarbonate in ethyl alcohol. The final product is a Na salt of SC (hereinafter referred to as SC)

Determination of the structure of modified polysaccharides by methylation method. To determine the structure, aqueous solutions of sodium salts of cellulose sulfate (~2.0%) were passed through columns with Amberlite IR-120 ion-exchange resin in triethylammonium form to obtain the target and soluble salt in DMSO and lyophilized:

where, R= -SO$_3$ Na or -PO$_3$ Na$_2$ or -H.

The obtained samples in triethylammonium salt form were dissolved in DMSO and permethylated with 1.6 M lithium methylthionylmethanide solution, with the addition of methyl iodide (30 eq/AGE):

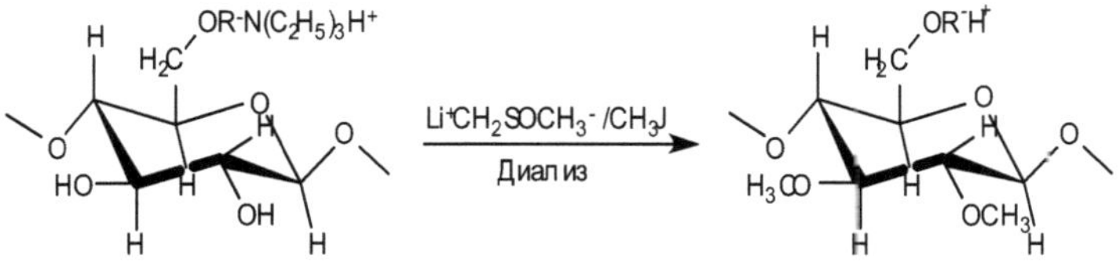

OR-N(C2H5)3H+
H2C
HO
OH
Li+CH2SOCH3- /CH3J
Диализ
OR-H+
H3CO
OCH3

The methylated samples were hydrolyzed with 2 M trifluoroacetic acid (TFCA) solution at 120°C for 2-3 h:

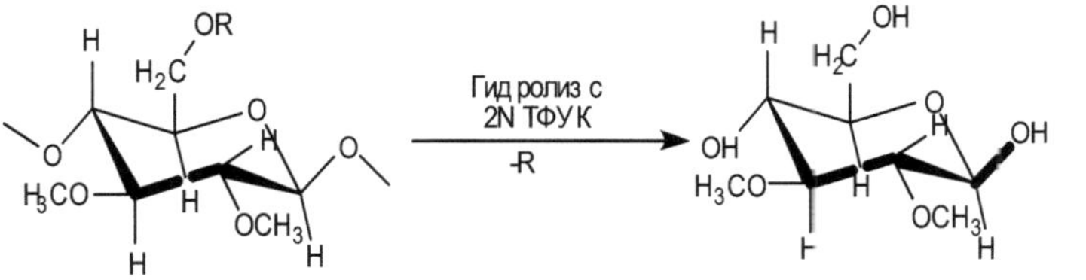

OR
H2C
H3CO
OCH3
Гидролиз с
2N ТФУК
-R
OH
H2C
OH
H3CO
OCH3

The hydrolysis product was reduced with a 0.5 M $NaBH_4$ solution in ammonia.

Acetylation of free hydroxyl groups was performed with pyridine (50 μl) and acetic anhydride (200 μl) at 90°C for 3 h:

OH
H2C
OH
H3CO
OCH3
Восстановление
гидридами
NaBH4
Ацетилирование
(CH3CO)2O/Pyr
CH2OAc
OCH3
OAc
OAc
CH2OAc

Partially methylated and substituted glucoacetates were extracted with dichloromethane, then washed with aqueous saturated NaHCO solution3 and demineralized water. The obtained samples were analyzed by gas chromatography.

IR spectra of the samples under study were taken on a Perkin Elmer FTIR spectrometer system 2000 in the frequency range of 400-4000 cm^{-1} in a tablet with KBr.

Viscosity was determined on an Ubellode viscometer (d=0.6 mm), at +25°C (±0.1°C) and the flow time of the solution was determined.

The molecular weight distribution of the samples was determined by high-performance gel permeation chromatography in a Knauer liquid chromatograph with a refractometric detector.

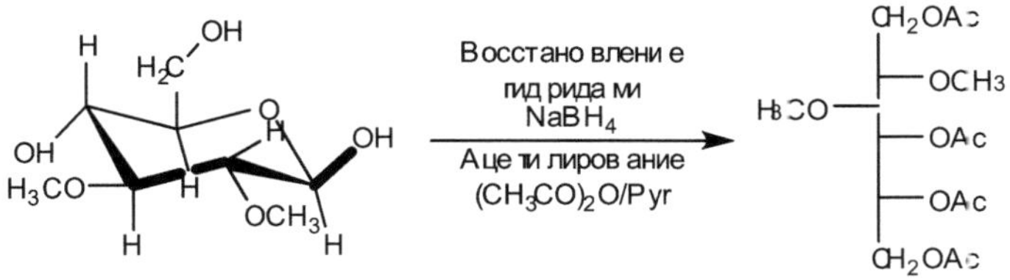

Studies of structural changes in the samples were carried out using a complex of physical and physicochemical methods: radiographic, microscopic and sorption methods.

X-ray radiographic studies were performed on a Dron-3M X-ray diffractometer with monochromatized CuKα radiation at 22 kV and 12 mA using the powder method. Imaging was performed on reflection in the interval $2\theta=10\text{-}40°$.

The degree of crystallinity (CK) was estimated using Segal's formula, using the ratio of reflex crystallinity (Ik) and amorphous scattering (Ia) intensities:

$$CK(\%) = \frac{Ik - Ia}{Ik} \times 100\%$$

Microscopic studies were performed on an optical microscope MBI-6 in transmitted polarized light. Pharmacological studies of SC samples were performed in the Laboratory of Pharmacology of the Institute of Bioorganic Chemistry in accordance with the existing requirements.

The data obtained were subjected to statistical processing.

Directed synthesis of modified polysaccharides of polyanionic nature in a heterogeneous medium.

Direct sulfation of cellulose.

The reaction of cellulose sulfation was carried out in the presence of primary alcohols, and the ratio alcohol: sulfuric acid is 1:3. It was found that this, i.e. the mixing of primary alcohols with sulfuric acid results in the formation of ethylsulfuric acid:

$$
\begin{array}{ccc}
\mathrm{C\,H\,O\text{-}_{25}\boxed{\text{-}H + H\text{-}O}} \quad O & & \mathrm{C\,H\,O_{25}} \quad O \\
\backslash\!/\!/ & & \backslash\!/\!/ \\
S \quad \rightarrow & & S\ (\text{ethylsulfuric acid}) + H_2 \\
/\,\backslash\backslash & & O\,/\,\backslash\backslash\backslash \\
\text{H-O} \quad O & & \text{H-O} \quad O
\end{array}
$$

However, ethylsulfuric acid is not involved in cellulose sulfation reactions, but it plays the role of binding water released into the reaction medium.

Another sulfating agent used in this work is chlorosulfonic acid (CSA), which is in equilibrium according to the following scheme:

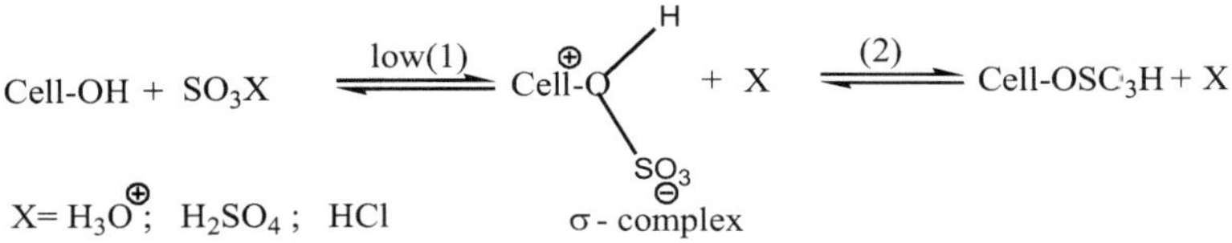

Cell-OH + SO$_3$X $\xrightleftharpoons{\text{low(1)}}$ Cell-O$^{\oplus}\Big\langle^{\displaystyle H}_{\displaystyle SO_3^{\ominus}}$ + X $\xrightleftharpoons{\text{(2)}}$ Cell-OSC$_3$H + X

X= H$_3$O$^{\oplus}$; H$_2$SO$_4$; HCl   σ - complex

The reaction of cellulose sulfation proceeds by the mechanism of electrophilic attack of the sulfating agent, with the formation of a sulfate complex, the cleavage of which leads to the formation of cellulose sulfate.

In all known methods, along with the sulfation reaction, a process of hydrolytic cleavage of the polysaccharide macromolecule takes place, which reduces the MM of the reaction mixture. Regulation of this reaction makes it possible to obtain high-molecular-weight products. Regulation of the chemical nature of the sulfating agent and carrying out the reaction under mild conditions makes it possible to suppress the macromolecule degradation reaction to a certain extent. When obtaining esterified cotton cellulose derivatives, the interaction of the reagents with the cellulose mass occurs under heterogeneous conditions. Therefore, the overall rate of the esterification reaction depends on the rate of penetration of the esterifying reagent from the reaction medium into individual sections of the cellulose.

As can be seen from the sulfation reaction diagram, the limiting stage of the reaction rate is the concentration of hydroxyl groups and the sulfating agent. Since the sulfation reaction is an equilibrium reaction, the reaction rate is also significantly affected by the concentration of the resulting products, namely, depending on the nature of the sulfation agent, water content, hydrogen chloride, which must be bound to reduce the rate of the reverse reaction.

The molecular structure and supramolecular structure of cellulose is a complex system, and it has a significant influence on its reactivity [103,117], particularly in sulfation reactions.

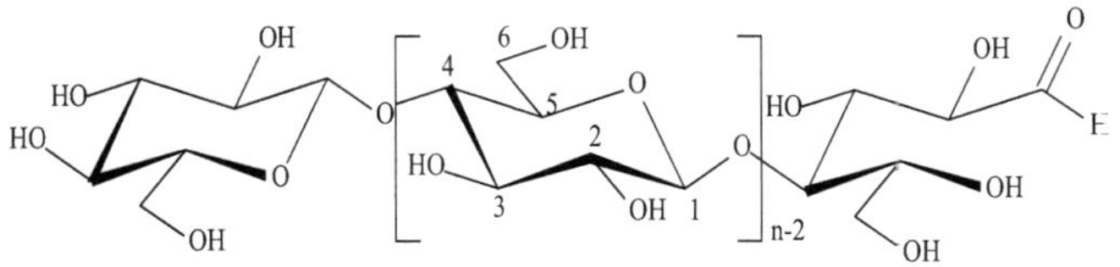

Fig. 8. Molecular structure of cellulose.

Cellulose consists of D-anhydroglucopyranose units bound by β-(1,4)-glucoside bonds, as shown in Figure 9.

The hydroxyl groups are located at the C-2, C-3, and C-6 positions, so chemically they exhibit the properties of primary and secondary alcohols [104].

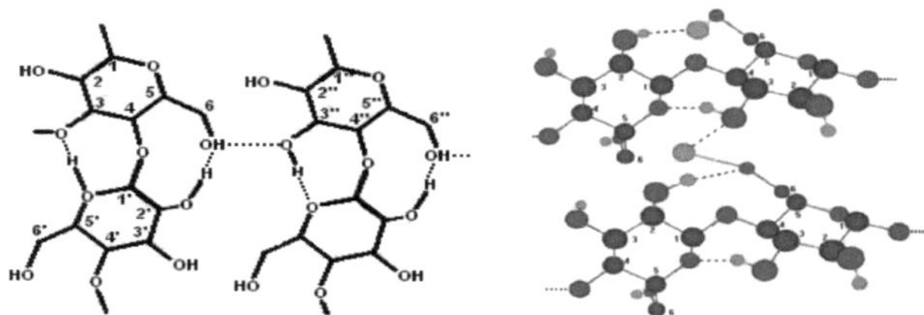

Fig.9.Schematic representation of the probable intra- and intermolecular hydrogen bonds of cellulose.

Hydroxyl groups in the cellulose macromolecule form a network of intra- and intermolecular hydrogen bonds (H-bonds), which determine the rigidity and stability of cellulose chains [118].

The supramolecular structure of cellulose, which, as proved by numerous studies [105,106], consists of low-ordered amorphous and highly ordered crystalline parts, has a significant role in determining its reactivity during chemical transformations.

In reactions of SC synthesis, along with the sulfation reaction, a process of hydrolytic cleavage of the cellulose macromolecule takes place, which leads to the formation of a low-molecular-weight product. Regulation of the chemical nature of the sulfating agent and carrying out the reaction under mild conditions allows the macromolecule degradation reaction to be suppressed to a certain extent.

The reactivity of cellulose in the sulfation reaction, i.e. the reach of hydroxyl groups by the sulfating agent, is significantly influenced by the supramolecular structure of the cellulose as well as the presence of fibrillar structure, cracks, porosity, capillarity and other structural parameters. Cellulose of different origin differs from each other in supramolecular structure and molecular characteristics.

We studied the reaction of cellulose sulfation in a mixture of sulfuric acid with ethyl alcohol (binary system) at a ratio of 2:1-3:1 and in a medium of marginal hydrocarbons (suspension method) (reaction module

4-5 ml/g) without activation by two-step activation - first in water and then in ethyl alcohol.

Study of cellulose sulfation with sulfuric acid in the medium of saturated hydrocarbons.

The reaction of cellulose sulfation with sulfuric acid belongs to electrophilic substitution reactions [107] and proceeds according to the following scheme:

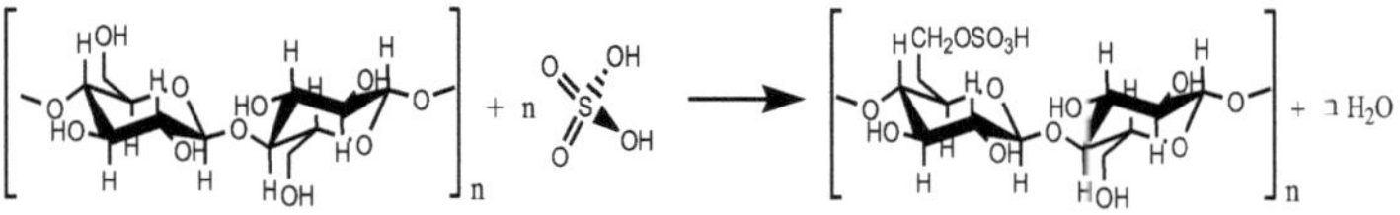

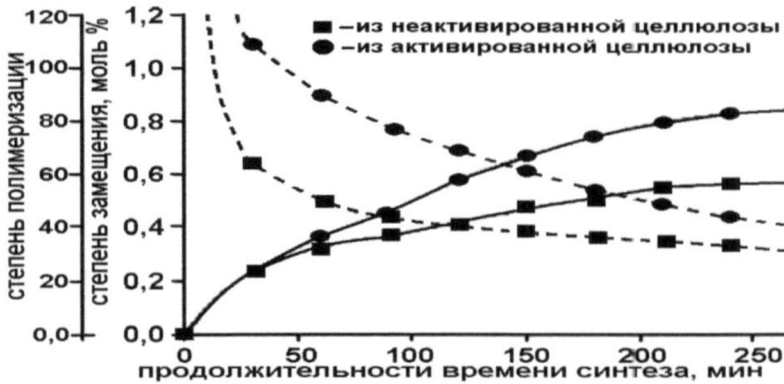

Figure 10. Reaction of cellulose sulfation with sulfuric acid.

The synthesis was carried out at +2 - +4° C, with constant stirring for 0.15-3.0 hours at various cellulose: sulfating agent ratios. At the end of the synthesis the acidic form of SC was neutralized by a mixture of alkaline solution with ethyl alcohol.

Kinetic curves of the cotton cellulose sulfation reaction indicate that preliminary activation of cellulose leads to an increase in the reaction rate and to the synthesis of SC with higher SP (Fig. 6).

Fig. 11. Variation of SP and NW of SC as a function of the synthesis time.

The results of the study also show that the reaction rate depends on the acid concentration and the activation of cellulose in all cases leads to a decrease in the degradation of the main chain.

It was found that during the reaction in a mixture of sulfuric acid: ethyl alcohol, the SP of SC (SP equals 0.23) decreases from a value of 1400 (for the original cotton pulp) to a value of 110 in the reaction with pre-activation of the pulp (Fig. 12).

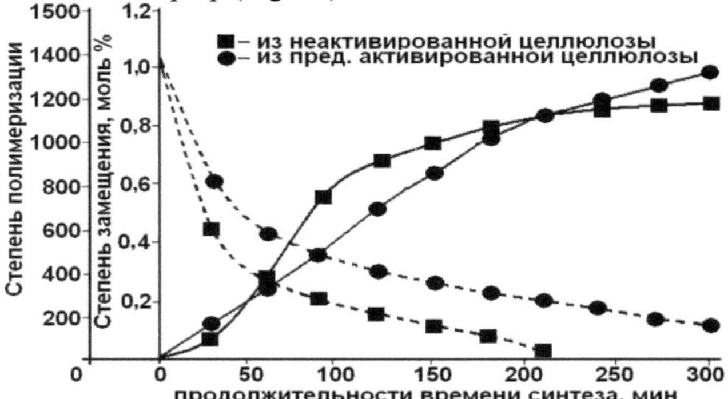

Fig.12. Variation of SP and NW in Na-CC suspension synthesis as a function of the synthesis time.

Whereas, carrying out the reaction in the medium of saturated hydrocarbons leads to the value of SP SC (SP equals 0.24) 440. The latter is due to the fact that the limit hydrocarbons - an inert diluent provides the most uniform penetration of sulfating agent deep into the thickness of the pulp material (which is proved by structural studies) and sulfating agent is consumed for direct reaction.

Figure 13. shows the dependence of change of SP of SC on NW in the reaction of sulfation of cellulose of different origin, by preliminary two-step activation, in a medium of marginal hydrocarbons.

As can be seen from Figure 13, the highest SP is observed in the case of sulfation of cotton lint. The lowest SP, as would be expected, is observed in the case of MCC sulfation. The decrease of SP with NW in the case of cotton lint is explained by the high content of amorphous areas in it.

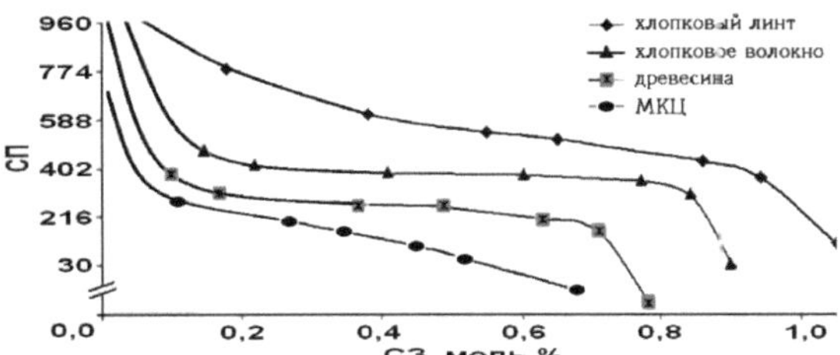

**Fig. 13. The difference of changes in the ratio of NW and SP SC at
the time of sulfation of cellulose masses of different origin.**

Whereas in the case of sulfation reaction of cotton and wood cellulose up to C3 0,7-0,9, SP changes insignificantly, i.e. in this limit there is mainly sulfation reaction. Such a difference in sulfation reactions of cellulose of different origin is caused by their structural features.

Thus, it was found that the sulfation of cellulose with sulfuric acid, even with pre-activation of cellulose, leads to the formation of SC with SP 25-600 and C3 0,10-1,10.

**Study of cellulose sulfation with chlorosulfonic acid in a medium of
tertiary amines.**

The disadvantage of cellulose sulfation with concentrated sulfuric acid is a side reaction of degradation by $(1 \rightarrow 4)\beta$ - glucopyranose bonds because of the large excess of sulfuric acid, resulting in a significant decrease of SP of the product.

It is known that during heterogeneous etherification cellulose is not dissolved immediately in sulfuric acid and, thus, OH-groups of cellulose are inhomogeneously replaced by sulfate groups, and most of the cellulose remains unreacted. As shown above, the synthesis of CC with sulfuric acid produces products with low SP and Cp(<1,2), even if the original cellulose is pre-activated.

Another sulfating agent used in this work is chlorosulfonic acid (CSA), which is in equilibrium according to the following scheme:

$$\text{Cell-OH} + \text{SO}_3\text{X} \xrightleftharpoons{\text{low(1)}} \text{Cell-O}^{\oplus} \overset{H}{\underset{\text{SO}_3^{\ominus}}{\diagdown}} + X \xrightleftharpoons{(2)} \text{Cell-OSO}_3\text{H} + X$$

$$\text{X} = \text{H}_3\text{O}^{\oplus}; \ \text{H}_2\text{SO}_4; \ \text{HCl} \qquad \sigma\text{ - complex}$$

The sulfation reaction of polysaccharides, particularly cellulose, proceeds by the mechanism of electrophilic attack of the sulfating agent, with the formation of a sulfate complex, the cleavage of which leads to cellulose sulfate.

For deep sulfation, it is more convenient and suitable to sulfate cellulose with a complex of CSC and pyridine, which proceeds with little destruction of the macromolecule.

The use of CSC in aprotonic solvents leads to different results than sulfation with sulfuric acid or oleum. The disadvantage of HSC sulfation is that hydrogen chloride is released during the reaction. The sulfation mechanism can be represented as follows:

$$\text{Cell-OH} + \text{ClSO}_3\text{H} \underset{\text{low(1)}}{\rightleftharpoons} \text{Cell-}\overset{\oplus}{\underset{\underset{\sigma\text{-complex}}{\text{SO}_3^{\ominus}}}{\text{O}}}\diagup^{\text{H}} + \text{HCl} \underset{(2)}{\rightleftharpoons} \text{Cell-OSO}_3\text{H} + \text{HCl}$$

The cellulose sulfation reaction proceeds by the mechanism of electrophilic attack of the sulfating agent, with the formation of a sulfate complex, the cleavage of which leads to cellulose sulfate

The sulfation of cotton cellulose with the complex of chlorosulfonic acid HSC:pyridine was investigated in order to obtain SC with higher values of SP and SP.

Pyridine has four functions in the sulfation reaction:

- solvent, which provides swelling of the pulp;
- to bind highly reactive CSCs as a complex, releasing the sulfating agent during synthesis;
- neutralizing agent of the formed acidic form of SC (resulting in the pyridine salt of SC ;
- trapping hydrochloric acid released during the interaction of HSC and cellulose

The pyridine hydrochloride formation reaction is as follows, pyridine binds free HC1 as soon as it is formed in the reaction medium and effectively competes with the cellulose degradation by free hydrochloric acid.

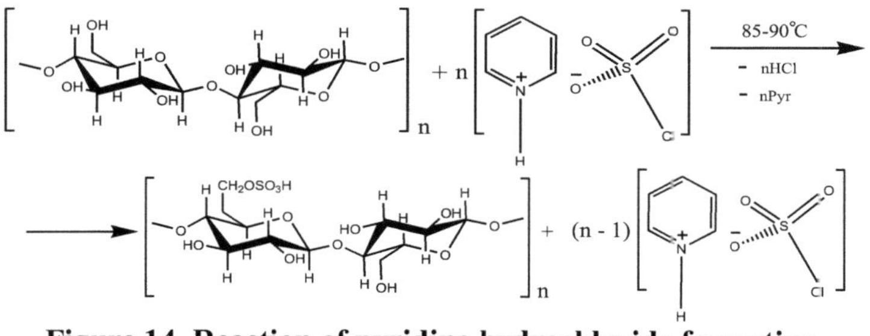

Figure 14. Reaction of pyridine hydrochloride formation.

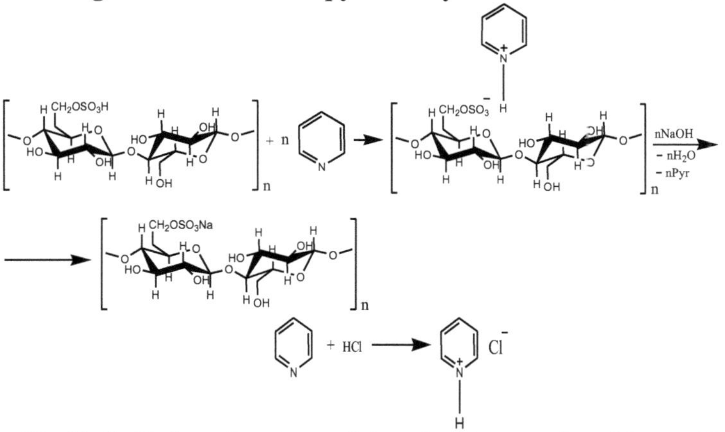

Fig.15. Sulfation of cotton cellulose by the complex chlorosulfonic acid HSC:pyridine.

Thus, pyridine provides cellulose activation, binds CSCs as a complex, releasing the sulfating agent during synthesis, and neutralizes the acidic form of SC by binding the released hydrochloric acid.

The effect of temperature and duration of cotton pulp sulfation reaction without activation and by pre-activation with pyridine on NW and SP of SC was studied (Table 3).

As can be seen from Table 1, as the reaction duration increases, both an increase in NW SC and a decrease in SP are observed. The same picture is observed in the case of temperature dependence. In any case, the sulfation rate of pre-activated pulp is much higher than that of non-activated pulp, and the SP drops much less.

Conducting the sulfation reaction by the complex CSC:pyridine allows to obtain SC with SP 150-800 and C3 0.1-2.8. Achievement of such effect is caused by influence of nature of sulfating agent. During sulfation of cellulose with complex CSC-pyridine is replaced by SO_3 Cl and the released hydrogen chloride instantly binds with pyridine and the course of the side reaction of degradation is much delayed.

Thus, maximum sulfation was achieved by sulfation of cellulose with HSC-pyridine complex in pyridine medium at high temperatures. Pre-activated (in water) and swollen (in pyridine) cellulose is in the most accessible and reactive form.

Table 3.

Dependence of NW and SP of SC during sulfation of cotton cellulose (reaction conditions: molar ratio CSC : cellulose = 9.21:1.0; pyridine:CSC = 3.51:1.0; reaction module = 19.8 ml per 1 g of cellulose).

№	Duration of synthesis, min	synthesis temperature, °C	sedimentation, hour.	without activation		Pre-activated	
				SZ, mole %	SP	SZ, mole %	SP
1	10	85	12	0,07	664	0,11	778
2	90	85	12	0,54	622	0,66	680
3	140	85	12	0,96	586	1,17	610
4	170	85	12	1,22	562	1,52	541
5	230	85	12	1,81	514	2,35	460
6	252	85	12	2,36	396	2,58	424
7	275	85	12	2,65	152	2,67	388
8	240	15	12	0,84	754	1,58	394
9	240	35	12	1,12	652	1,86	484

10	240	55	12	2,14	571	2,27	480
11	240	85	12	2,27	496	2,53	430
12	240	125	12	2,35	124	2,64	292
13	240	85	0	2,04	682	2,25	550
14	240	85	2	2,12	650	2,38	520
15	240	85	6	2,16	592	2,46	490
16	240	85	10	2,20	538	2,50	466
17	240	85	12	2,28	482	2,53	442
18	240	85	16	2,32	400	2,57	325
19	240	85	0	2,21	550	2,25	550
20	270	85	0	2,24	340	2,44	430
21	300	85	0	2,27	220	2,56	260
22	330	85	0	2,32	40	2,61	100

Selective sulfation of polysaccharides.

Molecular characteristics of cellulose sulfate play an important role in the manifestation of antimicrobial activity: the length of the macromolecule, the content of sulfate groups, their location within the link, leading to changes in the macromolecule conformation and to features of complementary interaction with microbial cell receptors.

However, there are no systematic studies of cellulose sulfate determining the regularities of the influence of the chemical nature of the macromolecular basis, molecular weight, content and arrangement of sulfate groups in the anhydroglucopyranose link on the antibacterial activity. These studies allow us to establish the molecular basis of the antibacterial activity of cellulose sulfate.

For the first time, based on the results obtained on the synthesis and analysis of the structural characteristics of cellulose sulfates on the test of their antimicrobial activity, the correlation of antibacterial activity depending on:

- the chemical nature of the macromolecular basis;
- the degree of substitution of sulfate groups;
- molecular weight;
-positions of sulfate groups in the 2, 3, and 6 carbon atoms of the anhydroglucopyranose link.

Conventional cellulose sulfation produces samples with an uneven distribution of sulfate groups, which does not help to establish the mechanism of their biological activity. Therefore, it is necessary to find a suitable method of sulfation with selective arrangement of sulfate groups.

The following selective sulfation methods were used to synthesize polysaccharide sulfates with the required molecular parameters:

-sample with CP 1.0 with the arrangement of sulfate groups at the 6 carbon atom was synthesized by mixed sulfoacetate cellulose ether with a mixture of acetic and chlorosulfonic acids. The duration of the synthesis was 8 hours, the mixed cellulose sulfoacetate ester was neutralized by an alkaline solution in ethanol. The final product was analyzed for NW and distribution of substituted groups. The results indicated that the samples had a SP of 0.91 and contained mostly monosubstituted samples in which the sulfate groups were located mainly in the 6th position of the hydroxyl groups.

When comparing the molecular parameters with other samples synthesized by direct sulfation with pre-activated and non-activated cellulose samples, it was determined that the sample had a substitution in the 6th AGE position. In addition, it contained a small amount of di- and trisubstituted groups.

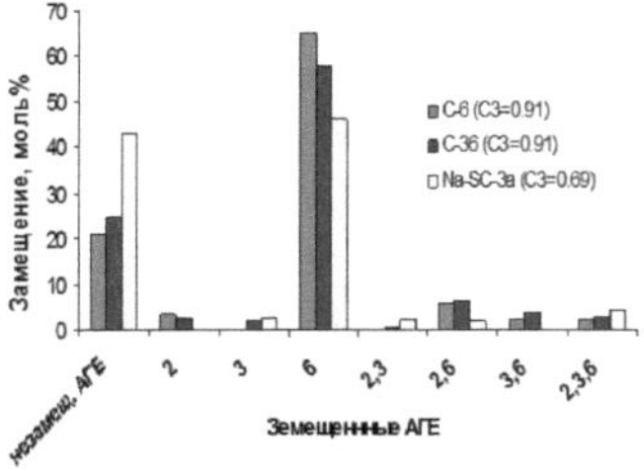

Rhee.16. Analysis by NW and distribution of substituted groups.

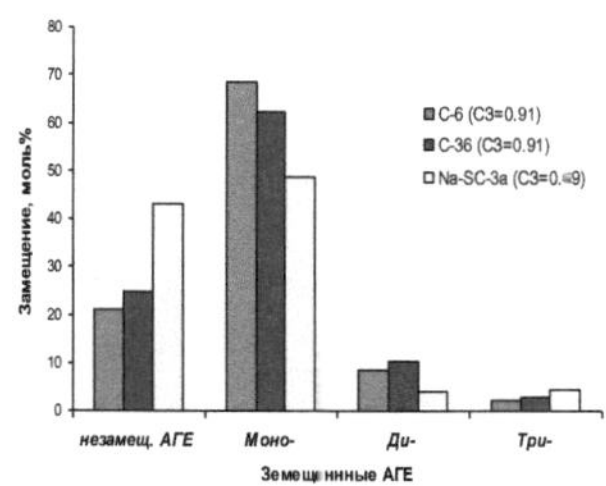

Figure 17. Analysis by the distribution of substituted units.

-fluoroacetate to produce samples with CP 1.0 and with the arrangement of sulfate groups in the 3 carbon atom (by obtaining difluoroacetate cellulose with CP 2.0 in fluoroacetate groups, their sulfation by appropriate sulfation methods, and complete dephluoroacetylation at the end of synthesis by neutralizing the final product), and to obtain samples with CP 2.0 with the arrangement of sulfate groups in 2 and 3 carbon atoms (by obtaining monofluoroacetate cellulose with CP 1.0 by fluoroacetate groups, their sulfation by appropriate methods of sulfation and complete desulfurization at the end of synthesis by neutralization of the final product), O-6 desulfation of cellulose trisulfate by N,O-bis-(trimethylsilyl) acetamide (BTSA). The study found that with esterification there is rapid hydrolysis due to destruction of the main chain of polysaccharides, as a result, it is impossible to obtain samples of sulfated polysaccharides with a molecular weight greater than 150000-180000 with CP 2.0 and with the arrangement of sulfate groups at 2 and 3 carbon atoms (ie, fluoroacetate cellulose and sulfoacetate cellulose).

Study of the structure and physicochemical properties of cellulose sulfates.

Structural studies were performed on samples of CC obtained by different methods, which allowed us to determine the effect of changes in the supramolecular structure in the process of pre-activation of cellulose and in the reaction of sulfation (Table 4).

The samples were studied using a complex of physical and physical-chemical methods: X-ray diffraction, microscopic and sorption methods. X-ray diffraction studies showed that while the X-ray diffractogram of cotton cellulose shows maximums at 2θ=22.6°, 14.7°, 16.8° and 34.4° characteristic of cellulose-I in crystalline form, all the samples of CC have a slight rise only in the area 2θ=20°, which indicates their amorphous nature.

Table 4.

Conditions for obtaining SO and their molecular parameters.

№	Specimen code	Method of obtaining the sample	SP	NW, %	CK, %
1	Cotton pulp	-	-	0	76
2	SC-Na-3	Alcohol + $H_2 SO_4$	280	45,4	30
3	SC-Na-3a	Pre-Active (alcohol), alcohol + $H_2 SO_4$	220	78,2	0
4	SO-K11	Hydrocarbon + alcohol + $H_2 SO_4$	400	58,1	18
5	CC-K11a	Pre-Active (alcohol), hydrocarbon + alcohol + $H_2 SO_4$	340	82,7	0
6	TCC-17	HSC + pyridine	388	197,2	3
7	TCC-16	Pre-Active (pyridine), HSC + pyridine	376	266,5	0

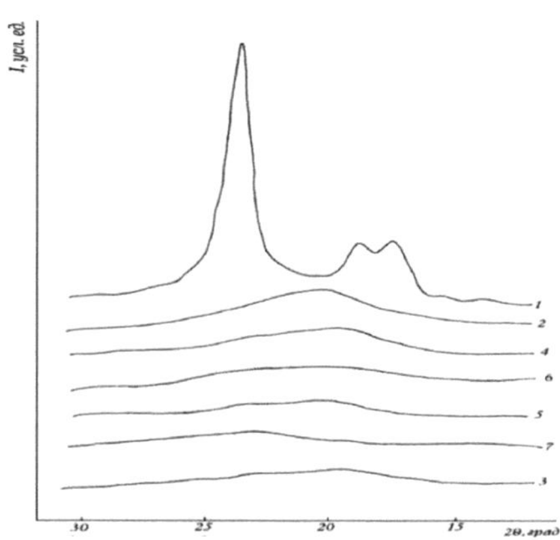

Fig.18. Diffractograms of samples of cotton pulp and CC: 1) cotton pulp; 2)CC-Na-3; 3)CC-Na-3a; 4)CC-K-11; 5)CC-K-11a; 6)CC-17; 7)CC-16.

X-ray diffraction study (Fig. 13, Table 2) showed that while the X-ray diffractogram of cotton cellulose shows maximums at $2\theta=22.6°$, $14.7°$, $16.8°$ and $34.4°$ characteristic of cellulose-I in crystalline form, all the samples of CC have a slight rise only in the area $2\theta=20°$, which indicates their amorphous nature.

With the help of electron microscopic studies, it was obtained that the surface of the original cotton cellulose is folded. The fibrillar structure of the primary fiber wall is not formed.

Table 5.

Description of optical images and sample size interval the original HC and SO.

№	Specimen code	Description of the optical image	Interval dimensions (μm)	
			width	length
0	cotton pulp	long winding fibers of different thicknesses and lengths, with a bright glow in polarized light	7-25	long
1	SC-Na-3	Short fibers of varying thicknesses and lengths, with a slight glow in polarized light, with partially preserved tortuosity	7-25	60-160
2	SC-Na-3a	even shorter fibers and pieces, no tortuosity, the fibers are round, straight, generally thicker, glow weakly in polarized light	8-28	25-80
3	SO-K11	tortuosity is rare, the fibers are shorter, there are long ones with obvious fibrillation, slightly glowing	10-28	100-250
4	SO-K11a	tortuosity disappears, fibers are shorter, thicker, many small particles, weakly glowing	4-30	20-80

| 5 | TCC-17 | slightly tortuosity preserved, thick fibers of different lengths, slightly glowing | 16-28 | 180-450 |
| 6 | TCC-16 | there is no tortuosity, the fibers are defective, shorter and of different thicknesses, they do not glow | 12-20 | 40-280 |

Characteristic for cotton cellulose wriggles of different thicknesses and lengths and defects with a bright luminescence in polarized light, strongly change during esterification (in Table 5, Figure 19). The fibers become shorter, thicker, straighter, with sizes and degrees of tortuosity depending on the conditions of production, which is accompanied by a decrease in the ability to glow in polarized light, that is, a decrease in the degree of anisotropy or crystallinity of the sample. This correlates with the X-ray data.

It should be noted that all the changes observed for cotton fibers are intensified for samples of SC with pre-activation for all conditions of production: the fibers are even shorter, defective and the tortuosity disappears, which probably provides a deeper esterification process with the achievement of a higher value of C3.

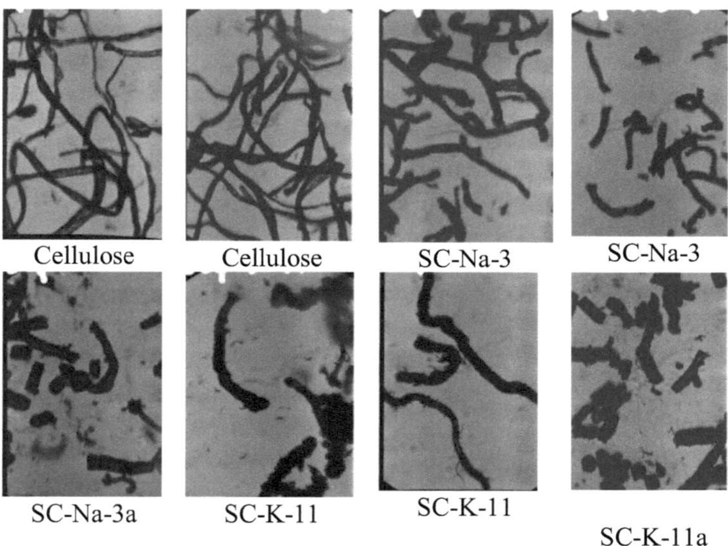

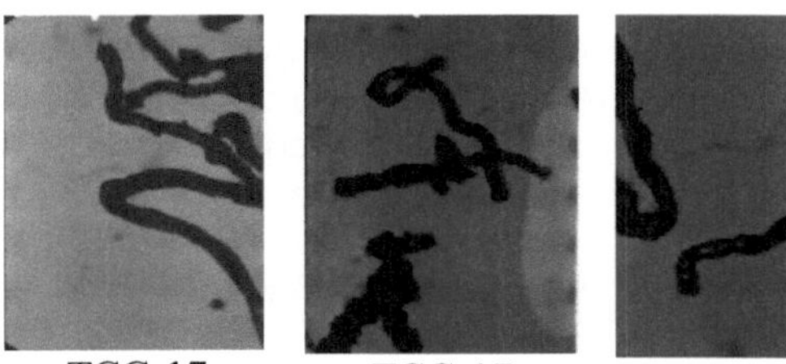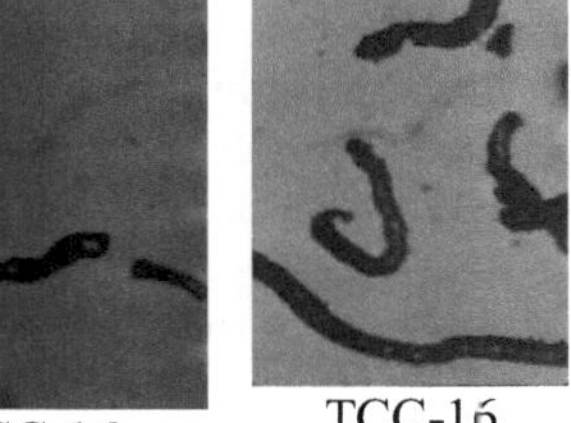

TCC-17   TCC-17   TCC-16   TCC-16

Fig.19. Changes in fiber thickness and length before and after the sulfation reaction.

This is evidenced by the results of fiber size estimation, the width of which increases up to 20%, apparently due to swelling during activation, and the length is shortened by 3-4 times, indicating that the static process of fiber splitting during esterification of cotton cellulose is intensified.

Using electron microscopic studies, it was found that the structural changes observed for cotton fiber are amplified for samples of SC with pre-activation under all production conditions: fibers are even shorter, defective, tortuosity disappears, which probably provides a deeper process esterification with achievement of higher value of C3. In SC, structural changes are characterized by an increase in the specific surface area, specific volume and pore radius of cellulose, and the values are 45.0-109.1 m^2 /g, 0.164-0.410 cm^3 /g; 36.2-133.3 A°, respectively, which accounts for their increased hydrophilicity.

Next, the content of sulfate groups in SO with different NW and MM was determined by three independent methods: elemental analysis, infrared spectroscopy, and conductometric titration.

We found a calibration curve of the dependence of the optical density of the absorption band at 800 and 1240 cm^{-1} SC on the change in its NW, which obeys the Lambert-Bera law.

The conductometric titration method is based on the difference in the electrical conductivity of the system before and after the reaction of CC with $BaCl_2$: the electrical conductivity of the system before the reaction of CC with $BaCl_2$ changes insignificantly, while a jump increase occurs with an excess of the latter. On the basis of these data the dependence of conductivity on the added amount of $BaCl_2$ is compiled. The point of intersection of sections of the curve is equivalent.

Table 4 shows the results of determining the PP of SO samples by different methods of analysis. Reliability of determination of HC by three

75

independent methods is in the following order: method of elemental analysis, conductometric titration, infrared spectroscopic determination.

The molecular-mass and conformational characteristics of SC were studied by viscometry and gel permeation chromatography.

Table 6.

The results of determining the NW of SO by different methods.

SP	Degree of substitution		
	By elemental analysis	By conductometry	By infrared spectroscopy (at D800)
388	2.665±0.012	2.625±0.046	2,551±0.045
350	1.197±0.011	1.961±0.023	1,253±0.046
820	1.856±0.015	1.836±0.042	1,932±0.039
465	1.806±0.012	1.782±0.045	1,724±0.045
436	1.530±0.014	1.545±0.018	1,582±0.037
394	1.166±0.013	1.160±0.043	1,211±0.039
694	0.934±0.015	0.911±0.027	1,010±0.058
616	0.766±0.014	0.779±0.031	0,750±0.018
208	0.377±0.009	0.374±0.016	0,040±0.074

The MM and molecular weight distribution (MMD) of SO were determined by high-performance gel permeation chromatography (HE-GPC). The results of these studies are shown in Table 7 and Figure 20.

Table 7.

The results of the MMR determination of some samples of SO.

Samples	Terms and conditions for receiving SO	MMR
TMSC-1	MCC, no activation	1,6
TSC-4	Chl. fiber, activation	1,9
TSC-1	Chl. fiber, activation	2,3
TSC-8(2)	Chl. fiber, activation	2,3
TSC-5	Chl. fiber, no activation	2,5
DSC-2	Chl. fiber, no activation	2,6

It can be seen that the MMR of the SC obtained from the synthesis is related to the origin of the initial cellulose samples as well as to their pre-activation. The MMR of the MCS samples obtained from MCC is in a narrower range than the MMR of samples obtained from cotton cellulose.

76

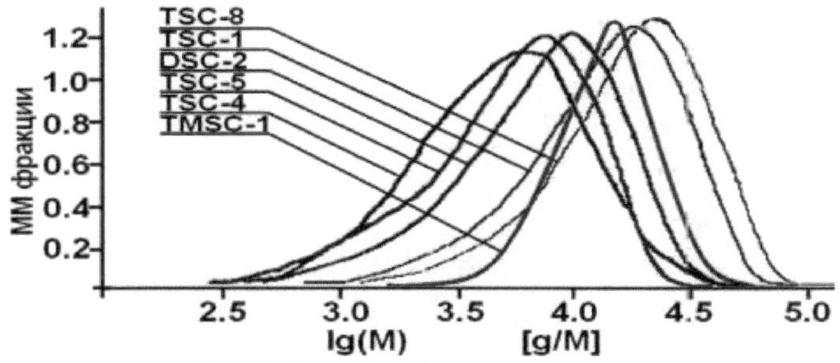

Fig.20. MMR values of SO samples.

Using the example of SO obtained from pre-activated cellulose samples, one can see that their MMR is in a narrow range than that of SO synthesized from non-activated cellulose. This is explained by the difference in the ratios of crystalline to amorphous sections of cellulose.

Studies of initial cellulose and samples of CC at high magnifications by scanning electron microscopy revealed more profound changes at structural level of CC samples. This is reflected not only in shortening and thickening due to swelling of fibers, reduction and even absence of tortuosity, but in disappearance of their fibrillarity, manifestation of more obvious signs of defects in the form of shapeless particles of different sizes, often destroyed or flattened, grained structure of fibers, obviously, due to small fracture particles, etc.

The structure of the obtained samples was studied by IR and NMR spectroscopy, elementary analysis.

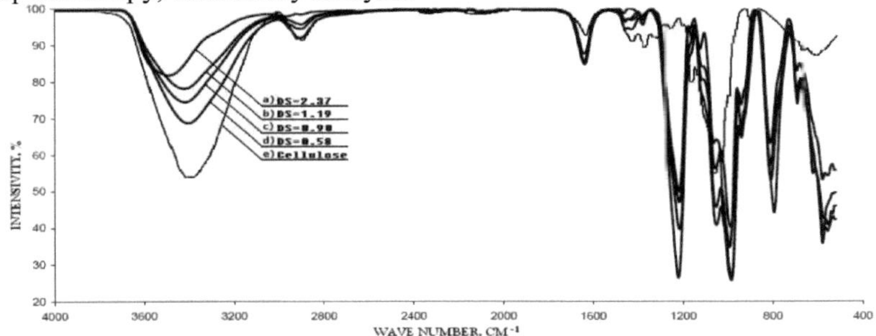

Fig. 21. IR spectra of the obtained samples.

IR spectroscopy is a sensitive method for various functional groups in polymer molecules and is widely used to determine cellulose and its modified derivatives. The presence of sulfate groups in modified cellulose has been established by IR spectroscopy (Fig.). During sulfation reaction some specific absorptions of cellulose macromolecules are sharply reduced and others are increased, also new bands of strong intensity are observed at 590-600, 800, 1220-1240 cm^{-1} , which refer to sulfated esters. The bands at 590-600 cm^{-1} belong to ethane absorptions indicating different vibrations of sulfated ester groups, and their width increases with increasing total NW of the products. The absorbance at 800-805 cm^{-1} indicates a stretching of the C-O-S bond characteristic of sulfate groups located predominantly in the axial-equatorial position (C-2, and C-3) in glucose residues [108]. A sharp band at 900 cm^{-1} corresponds to the β-glycosidic bond between the glucoside units [110]. The intensely increasing series of bands in the 1000-1100 cm range^{-1} indicates asymmetric intraplanar vibration of the glucoside ring. This increase is explained by the disruption of intramolecular hydrogen bonds during substitution. The spectra of the samples showed the presence of changes at 1230-1250 cm^{-1} , which is associated with the valence vibrations of the sulfate groups S = O. Consecutive bands at 1380-1420 cm^{-1} are due to strain vibrations of CH groups [111]. The peaks at 1640 cm^{-1} correspond to adsorbed water molecules in the samples [109]. The intensity of the bands at 2900 cm^{-1} is attributed to the stretching and/or deformation vibration of asymmetric CH participants and the alkyl chain bond [112], which is reduced in the spectrum of fully sulfated cellulose samples. The OH valence vibrations in cellulose are observed at 3400 cm^{-1} (Fig., d). In four samples (Fig.a, b, c, and d), the spectra of SC become wider and shift to a higher number of waves in the spectra; presumably, this is associated with the disturbance of intermolecular hydrogen bonds in cellulose during the sulfation process [113].

The synthesized polysaccharide samples were examined by[13] C-NMR spectroscopy in $D_2 O$. The peak values were determined according to Nehls et al. [114].

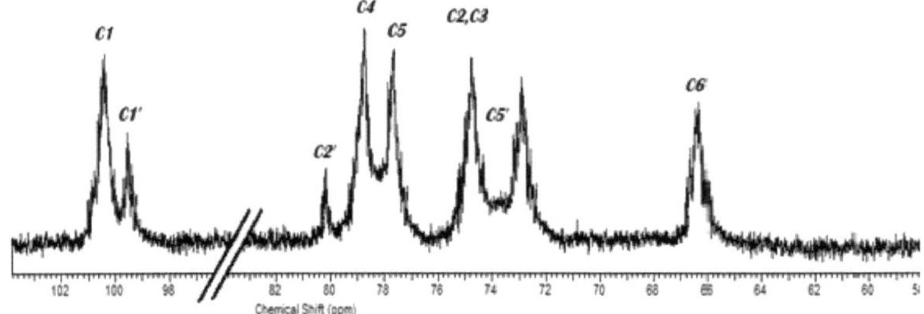

Fig. 22. NMR spectroscopy of the obtained samples of SO with NW 2.6

Figure 22. shows the C1-C6 chemical shifts at 102, 73.2, 74.6, 79.2, 77.3, and 61.0 m.d., respectively. The C6 absorption of Na-CC appears with a relative shift from 61.0 to 67.0 indicating a substitution of the OH group at the sixth carbon atom (C6'). It can be seen that in the highly substituted Na-CC samples all OH groups at the sixth carbon atom C6 undergo esterification (C6 '). The chemical shifts at 81.5 and 77.1 m.p.a. show that the substitution with sulfate groups occurs at the C2' and C3' positions.

Study of sulfation of cellulose samples in a homogeneous medium, their structure and physical and chemical properties.

During homogeneous sulfation of cellulose samples (SP=900-6000) by SO_3 /Py complex in DMA/LiCl medium mono-, di- and trisulfate derivatives with different molecular parameters (C3=0.23-2.56, SP=584-5540) were obtained. From the results of the study it is clear that the main factors controlling the value of SP are the amount of sulfating reagent and temperature. During the first 1-4 hours, a high sulfation rate is observed. Due to the weakly alkaline reaction of the medium, little depolymerization was observed in the polysaccharide chain (at 4 hours, an average of 23.7%). Homogeneous sulfation of cellulose can produce mono-, di- and trisulfate derivatives in high yields (70-80%) (Fig. 23.).

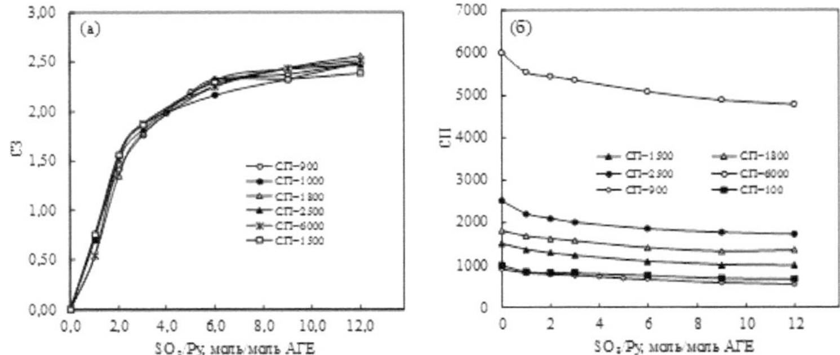

Fig.23. Variation of Cp (a) and CP (b) with increasing amount of sulfating reagent during homogeneous sulfation of cellulose samples by SO₃ /Py complex in DMA/LiCl medium (temperature 80° C, duration 4 hours).

During homogeneous sulfation the sulfation ability of carbon atoms in AGE of cellulose also decreases in the series C-6 > C-2 > C-3, by this method it is possible to obtain C-6, C-2, C-3 substituted sulfate derivatives. Due to the flow of reaction in cellulose solution, sulfating ability of cellulose samples with different SP (500-6000) is the same.

The results of comparative analysis of heterogeneous and homogeneous synthesis of cellulose sulfate derivatives showed that homogeneous sulfation reaction in DMA/LiCl medium is superior to heterogeneous sulfation reaction in all respects.

In the studied heterogeneous and homogeneous sulfation reactions a high sulfation rate is achieved during the first 1-6 hours. The reactivity of hydroxyl groups of carbon atoms decreases in the series C-6 > C-2 > C-3. By IR and[13] C NMR spectroscopy the structure of sulfate derivatives of cellulose was established. The results showed that no by-products were formed during the reaction. The degree of sulfation of C-2, C-3, C-6 carbon atoms of sulfate derivatives of cellulose was determined by[13] C NMR spectroscopy. Absorption bands corresponding to the valence vibrations of O-H (3400 cm⁻¹), H-C-H (2900 cm⁻¹), C-O-C (1060 cm⁻¹), C-O-C β-glucoside bond (894 cm⁻¹), C-O-S (818 cm⁻¹), O=S=O (1257 cm⁻¹) were determined in the infrared spectra of sulfate derivatives of cellulose. In[13] C NMR spectra of sulfate derivatives of cellulose, signals corresponding to carbon atoms C-1 (102.2 mu.), C-1' (100.2 m.s.), C-2s, C-3s (79-82 m.s.), C-4 (78.3 m.s.), C-4' (77.8 m.s.), C-2, 3, 5, 5' (71-76 m.s.), C-6s (66.7 m.s.), C-6 (60.3 m.s.).

As a result of studies, viscometric constants for Na-CC sample solutions (C3=0.3-3.0) (0.1 M $NaNO_3$, 25° C) were determined and the following MHKS equations were made on their basis:

$$
\begin{array}{lll}
\text{C3 (0,3-0,5)} & [\eta]=5{,}85\times10^{-6}\cdot M^{1{,}0} & \text{дл/г} \\
\text{C3 (0,5-1,0)} & [\eta]=6{,}22\times10^{-6}\cdot M^{0{,}99} & \text{дл/г} \\
\text{C3 (1,0-1,5)} & [\eta]=9{,}10\times10^{-6}\cdot M^{0{,}95} & \text{дл/г} \\
\text{C3 (1,5-2,0)} & [\eta]=1{,}14\times10^{-5}\cdot M^{0{,}92} & \text{дл/г} \\
\text{C3 (2,0-2,5)} & [\eta]=1{,}85\times10^{-5}\cdot M^{0{,}88} & \text{дл/г} \\
\text{C3 (2,0-3,0)} & [\eta]=2{,}25\times10^{-5}\cdot M^{0{,}86} & \text{дл/г}
\end{array}
$$

Using the data of MHKS equation, the method of capillary viscometry can express determination of MM of Na-CC samples. It was found that the values of the exponent α of Na-CC samples are in the limit of 1.0-0.86, and in aqueous solutions they are represented by a semiflexible conformation. With the increase of NW values, the flexibility of the chain increases.

Using the fractionation of sulfated polysaccharides by gel-filter chromatography, it is possible to obtain samples with lower MM with preservation of sulfate groups. However, due to the large number of fractions, they were obtained in low yields. In addition, in order to obtain low-molecular-weight fractions, the SP value of fractionated sulfate derivatives should be close to the SP value of the obtained oligosaccharides.

By radical depolymerization of sulfate derivatives of cellulose sulfated oligosaccharides with MM from 32200 to 6500 Da were obtained. Depolymerization under acidic conditions (pH 2.5-4.0) compared to alkaline conditions (pH 7.5-8.0) is almost 2 times faster. Also under acidic conditions (pH 2.5-4.0) oligosaccharides with low SP and short molecular weight distribution were obtained (Fig. 24).

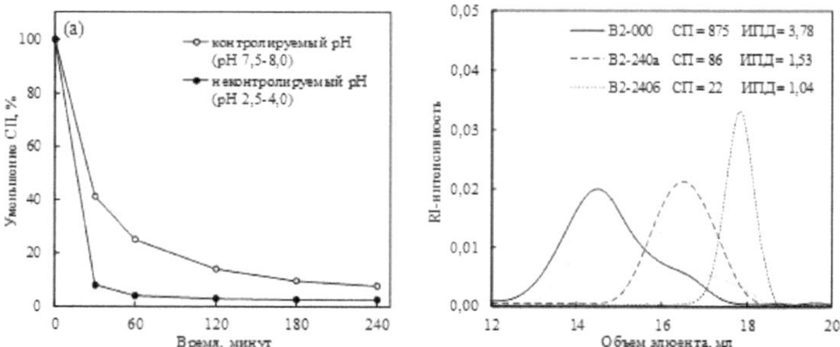

Fig.24. Reduction of SP during depolymerization of sulfate derivatives of cellulose (a), gel-chromatograms of the original sulfate cellulose and products of its depolymerization (eluent, water/0.1 M NaNO).₃

In the process of depolymerization a slight desulfation is observed. The ratio of depolymerization rates (according to the average SP value) and desulphation under alkaline (pH 7.5-8.0) and acidic conditions was 3650:1 and 3840:1, respectively. During the depolymerization process, a change in the SP$_{час}$ was also observed depending on the pH of the medium. The sulfate groups of carbon atoms at C-6 show high stability. It was found that the desulfation of carbon atoms in alkaline conditions (pH 7.5-8.0) occurs in the range C-6<C-3<C-2, and in acidic conditions (pH 2.5-4.0) - in the range C-6<C-2<C-3. The results on the study of the structure of the obtained samples showed that under the conditions studied no oxidized derivatives and side reactions of oxidation in the polysaccharide chain are observed.

The structure of sulfate cellulose oligosaccharides was established by IR and[13] C NMR spectroscopy. The degree of sulfation of carbon atoms C-2, C-3, C-6 was determined by[13] C NMR spectroscopy. Absorptions corresponding to the valence vibrations of O-H (3400 cm^{-1}), H-C-H (2900 cm^{-1}), C-O-C (1060 cm^{-1}), C-O-C β-glucoside bond (894 cm^{-1}), C-O-S (818 cm^{-1}), O=S=O (1257 cm^{-1}) were determined in IR spectra. In[13] C NMR spectra, signals corresponding to carbon atoms C-1 (102.2 m.u.), C-1'(100.2 m.u.), C-2s, C-3s (79-82 m.u.), C-4 (78.3 m.u.), C-4' (77.8 m.u.), C-2, 3, 5, 5' (71-76 m.u.), C-6s (66.7 m.u.), C-6 (60.3 m.s.), C-1 (β- NVPE) (94.9 m.s.), C-1' (β- NVPE) (92.9 m.s.), C-2, 3, 5, C-5' (α- NVPE) (68.5-71 m.s.), C-1 (α- NVPE) (90.5 m.w.), C-1' (α- NVPE) (87.5 m.w.), C-4 (α- NVPE, β- NVPE) (77-79.5 m.w.), C-4 (NVPE) (67.3 m.w.).

Thus, a systematic study of cellulose sulfation reactions of different origin in heterogeneous and homogeneous media, by sulfating agents of different chemical nature, has been conducted. It was found that the chemical nature of the sulfating agent determines the direction of the main and side (destruction) reactions. The highest NW and SP of SC is observed in the series: chlorosulfonic acid > sulfuric acid > ethylsulfuric acid. It was found that the structural parameters of cellulose (tortuosity, fiber length, defectiveness) undergo significant changes in the process of activation and sulfation reaction. For the first time the possibility of sulfation of all three hydroxyl groups in the anhydroglucopyranose chain of cellulose and the production of SC with CP from 0.50 to 2.95 and SP from 275 to 480 has been established.

LITERATURE

1. B.A. Kargin.// J.Polymer Sci., 1958, vol. 30, p. 247.

2. Klenkova N.I. Structure and reactivity of cellulose. L., Nauka, 1976.

3. Klenkova N.I. Autoref. doctoral dissertation. IVS AS USSR, 1967.

4. Akim E.L., Perepechkin L.P. Cellulose for acetylation and cellulose acetates. Moscow, Lesnaya Promyshlennaya Promyshlennost', 1971. 232 c.

5. U.S. Patent 2835665, 1958.

6. Kuznetsova E.P., Klenkova N.I. ZhPH, 1964, T 37, p. 399-408.

7. Boguslavskaya L.V. et al. - Textile Industry, 1967, No. 6, pp. 65-67.

8. Roussel M.-A., Nielson M.-L.-Textile Reserch J.,1976, vol 46, № 9, p. 648-653.

9. I.F. Kaimin, M.Y. Ioelovich, L.I. Slysh "Influerce of activators on structural and physical state of cellulose" Proceedings of the Conference on "Physical and Physicochemical Aspects of Cellulose Activation. Riga, 1981. p. 76.

10. Kuznetsova N.Y., Zhavoronkov V.E., Ivanov V.I. Some aspects of catalysis by reaction of o-polyheterocyclic compounds. - Izv. of AS Latvian SSR. Chem., 1971, No 4, pp. 406-410.

11. D.D. Grinshpan, F.N. Kaputsky "The role of organic solvent in changing the reactivity of cellulose" Proceedings of the conference on "Physical and physical-chemical aspects of cellulose activation. Riga - 1981. p. 114.

12. Klenkova N.I. // ZhPH, 1963, vol. 36, p. 836.

13. Klenkova N.I., Kulakova O.M., Matveeva N.A., Volkova L.A. //ZhPH, 1965, vol. 38, p. 919.

14. Merchant M. V. - TAPPI, 1957, v. 40, p. 926.

15. Klenkova N.I.//ZhPH, 1956, v.28, p. 393.

16. Klenkova N.I., Kulakova O.M., Tsimara N.D., Khlebosolova E.N. // ZhPH, 1962, vol. 35, p. 2770.

17. Reutov O.A. Theoretical Bases of Organic Chemistry. M., 1964, Ch. 1, VIII.

18. Matveeva N.A., Kutsenko L.I., Klenkova N.I. // ZhPH, 1973, vol. 46 p. 1813.

19. Palit S. // J.Organ. Chem., 1957, v. 12, p. 752.

20. Amis E. Effect of solvents on the rate and mechanism of chemical reactions. "Mir, Moscow, 1968.

21. Segal L., Loeb L., Creely J.J. //J.Polymer Sci., 1954, v. 13, p. 193.

22. Spedding H. Polymer, 1962, v. 3, p.195.

23. Z.Sh. Sharshenalieva, E.N. Tarasova "Influence of organic solvents on the reactivity of cellulose during etherification" Proceedings of the Conference on "Physical and physical-chemical aspects of cellulose activation. Riga - 1981. p. 136.

24. Thinius K., Thümmler W. // Makromol. Chem., 1966, v. 99, p. 117-125. 25. Sidikov A., Kadyrova S., Tyagai E. et al. "Study of structural changes in cellulose when treated with monoethanolamine"// High-Molecular Compounds, 1976, vol. 18B, no. 9, pp. 658-661.

26. U.S. Patent 2,539,451, 1951.

27. U.S. Patent 2,675,377, 1953.

28. U.S. Patent 2,862,922, 1958.

29. U.S. Patent 2,582,009, 1952.

30. U.S. Patent 3,528,963, 1972.

31. Canadian Patent 2,582,009, 1952.

32. Japanese Patent Application No. 75154, 1972.

33. U.S. Patent 3,720,659, 1973.

34. Japanese Patent Application No. 89787, 1974.

35.U.S. Patent 6,500,947, 2003.

36.WO #96/15137, 1986.

37. "Cellulose Chemistry and Its Applications," Ed. T. P. Nevell and S. H. Zeronian, Halstead Press, John Wiley and Sons, 1985, page 350.

38. U.S. Patent 2,042,484, 1940.

39. U.S. Patent 2,969,356, 1961.

40. U.S. Patent 4,064,342, 1975.

41. Rogovin Z.A., Myrlas D.I. Chem. nauka i prom., 1958, vol. 3, no. 6, pp. 831-832.

42. Petropavlovsky G.A., Krunchak M.M., Vasilyeva G.G. ZhPH, 1967, vol. 40, c. 2209.

43. Timokhin I.M., Lopatin V.A., Marakhina M.S., Ismailov K.K. In: Chemistry and technology of cellulose derivatives. Vladimir, 1971, p. 218.

44. Finkelstein M.Z., Borisov I.L. Izvestiya vuzov. Oil and Gas, 1959, No. 11, p. 49.

45. Ward F. //C. A. 4042c. 1951.

46. Nenitsescu C. D. Organic Chemistry. M., 1963. 863 c.

47. Ward F. // S. A. 8371c. 1952.

48. Frank G. // C. A. 7117g. 1947.

49. Frank G. // C. A. 8770h. 1951.

50. Lambiotte //C. A. 3597f. 1951.

51. Klug E.// S.A. 15083b. 1956.

52. Short Reference Book of the Chemist. Moscow: Goskhimizdat, 1952. T. 3.

53. Malm C, Crane C. // C. A. 9621i. 1954.

54. A. Rogovin, N. N. Shorygina, Chemistry of cellulose and its satellites, Goskhimizdat, 1953, p. 353.

55. Malm C, Crane C. // S. A. 4453e. 1959.

56. Jullander E. // Chem. Ztbl. 1956. S. 6550.

57. Rogovin 3. A., Myrlas D.Y. // Chem. nauka i prom. 1958. T. 3. C. 832.

58. Finkelstein M. 3. // Izv. vuzov. Oil and gas. 1959. № 11.

59. Petropavlovskii G.A., Krunchak M.M. // ZhPH. 1963. T. 36, N 11. C. 2506.

60. Petropavlovsky G.A., Krunchak M.M. // ZhPH. 1966. T. 39, N 1. C. 170.

61. Petropavlovsky G.A., Krunchak M.M. // ZhPH. 1966. T. 39, № 10. C. 2347.

62. Petropavlovskii G.A., Krunchak M.M. // ZhPH. 1966. T. 39, № 12. C. 2779.

63. Petropavlovsky G.A., Vasilyeva G.G., Vasilyeva O. A. //GPH. 1966.T. 39, № 9. C. 2053.

64. Petropavlovskii G.A., Krunchak M.M., Vasilyeva G.G. // ZhPH. 1967.T. 40, № 10. 2209.

65. Timokhin I.M., Lopatin V.A., Marakhina M.S., Ismailov K.K. In: Chemistry and technology of cellulose derivatives. Vladimir, 1971, p. 218.

66. U.S. Patent 6,500,947, 2002.

67. Dautzenberg, H., Lukanofff, B., Holzapfel, G., Tiersch, B., Joger, W., Schellenberger, A., Forster, M., DE 4 021 050(1992).

68. Shanjing M.S., Cho Man-gi, Buchholz R. An environmentally friendly method for the production of cellulose sulfate for immobilization of microorganisms and enzymes in biotechnology// Chem.-Ing.-Techn. - 1993. - 65, N 9. - C. 1124-1125.

69. Finkelstein M.Z., Borisov I.L. Izvestiya vuzov. Oil and Gas, 1959, No. 11, p. 49.

70. Rogovin Z.A., Myrlas D.I. Chem. nauka i prom., 1958, vol. 3, no. 6, pp. 831-832.

71. U.S. Patent 2,539,451, 1951.

72. U.S. Patent 2,969,356, 1961.

73. U.S. Patent 3,075,963, 1963.

74. U.S. Patent 3,528,963, 1970.

75. U.S. Patent 2,560,611, 1951.

76. U.S. Patent 2,675,377, 1954.

77. U.S. Patent 2,753,337, 1956.

78. Diecman S.F., Jarrell J.C., Voris R.S. // Ind. Eng. Chem. 1953. Vol. 45. p. 2287.

79. Klug E., Spurlin H. // Chem. Ztbl. 1956. S. 8240.

80. U.S. Patent 2,714,591, 1955.

81. Belgian Patent 553,396, 1957.

82. Blaser B., Rusenstein M.//C. A. 1608g. 1958.

83. Klug E., Spurlin H. //C. A. 16432. 1955.

84. Asami R., Tokura N. // J. Chem. Soc. Jap. Ind. Chem. Sect. 1959. Vol. 62, N 10. P. 1593.

85. Traube W., Blaser B. // Ber. 1928. Bd 61. S. 754.

86. Traube W., Blaser B. // Ber. 1932. Bd 65. S. 603.

87. U.S. Patent 2,675,377, 1954.

88. Canadian Patent 921,903, 1974.

89. Schweiger R. // Chem. Industry. 1966. N 22. P. 900.

90. Roy L. Whistlaer, Alan H. King// Biochemistry and Biophysics, Volume 121, No.2, August. 1967

91. Carbohid. Polymers 14 (1991) 53-63.

92. U.S. Patent 3075963, 1963.

93. U.S. Patent 4064342, 1977.

94. U.S. Patent 5679375, 1997.

95. I. Yamamoto, K. Takayama, K. Honma, T. Gonda, K. Matsuzaki Synthesis, Structure and Antiviral Activity of Sulfates of Cellulose and its Branched Derivatives Carbohydrate Polymers 14, 1991, p. 53-63

96. S. A. Glickman, N. G. Shubtsov, Colloid, J., 19, no. 2,172; no. 3, 1957, p. 281

97. S. E. Bresler, Advances in Chemistry and Technology of Polymers, Goskhimizdat, 1957, p. 110.

98. U.S. Patent 5,378,828, 1995.

99. C. A. 15030g, 1956

100. Helv. Chem. Acta. 1943, v. 26, p.1296

101. J. Amer. Chemistry. 1958, v. 80, 14, p. 3700-3702

102. George P. Tozeu //Modern Plastics, 29 no. 3, 1951, p.109, 110, 112, 114, 183

103. Tolkunova V. V. Research in the synthesis of sulfuric acid and carboxymethyl sulfuric acid cellulose esters: Author's degree of Candidate of Chemical Sciences.M., 1969.

104. Petropavlovskii G.A., Nikitin N.I. // Proceedings of the Institute of Forestry of the USSR Academy of Sciences. Vol. 45, 1958, pp. 93-94

105. Touey G. // RGC 6P760. 1962.

106. Touey G., Gearhart W. // J. Chem. Eng. Date. Vol. 6. 1964, p. 566.

107. Dusen S. // Chem. Ztbl. 1957. S. 4853.

108. Terayame H. // J. Polymer Sci. 1955. Vol. 15. P. 575.

109. Tager A. A. "Physico-chemistry of polymers" M.:Chemistry, 1968

110. Gagnon P. //Canad. J. Technol. 1957. N 6. P 477.

111. Gagnon P. //Canad. J. Chem. 1958. Vol. 36. P. 1039.

112. Rogovin Z.A., Myrlas D.I. Chem. nauka i prom., 1958, vol. 3, no. 6, pp. 831-832.

113. The Chem. Age. 1951. Vol. 64, N 1645. P. 137.

114. Weizmann A. // Artificial fiber. 1932. No. 1. p. 6.

115. Inokawa S. // Chem. high polymers. 1955. Vol. 12. P. 353.

116. Steyermark, A., "Quantitative Organic Microanaiysis," p. 276. Academic Press, New York (1961).

117. Sparrow D. // C.A. 5672h. 1959.

118. Thomas J. // C. A. 8657c. 1950.

3.2. Acetates of cellulose

Water-soluble acetyl cellulose (WPC) can be obtained only by hydrolysis of the original highly substituted acetyl cellulose in a homogeneous medium to a well-defined degree of substitution.

Direct acetylation of cellulose does not produce water-soluble products at any degree of substitution.

As you know, when producing acetyl cellulose by homogeneous or heterogeneous methods, the final product is always cellulose triacetate ("primary" acetate). This so-called "primary" cellulose acetate is soluble in some organic solvents (chloroform, methylene chloride, glacial acetic acid, aniline, pyridine, etc.). Most of these solvents, however, are not technically applicable, since they belong to the number of expensive, poisonous or not widely used substances. The most technologically and economically acceptable solvent for cellulose triacetate is methylene chloride, which has recently been widely used. To improve solubility, "primary" acetates are first converted to "secondary" acetates by partial hydrolysis, which gives products that are soluble in acetone and other more accessible organic solvents. Partial hydrolysis of triacetyl cellulose to obtain secondary acetates can be performed in either acidic or alkaline media. Due to the great technical importance of this process, it has been studied in many works.

The regularities of acetyl group detachment in both acidic and alkaline environments are important for understanding the properties of partially deacetylated products, so these regularities will be described below.

A detailed study of acetylation and hydrolysis of secondary cellulose acetates was carried out by Hiller [1]. Cellulose acetates with different contents of bound acetic acid-61.5%; 54.4%; 44.2% were chosen for the study. The product with the content of bound acetic acid 54.4% (industrial sample) was the starting material. Cellulose acetate with bound acetic acid content of 61.5% was obtained by acetylation of initial secondary cellulose acetate with acetic anhydride in the presence of potassium acetate as a catalyst.

Secondary cellulose acetate containing 44.2% bound acetic acid was prepared by hydrolyzing the original secondary cellulose acetate under very mild conditions. These samples had degrees of polymerization of 370,370 and 333, respectively.

To study the reaction speed of acetylation and deacetylation depending on reaction conditions the latter were conducted in solution, with acetic acid at 4.8%, 2.8% and 0.5% water content at different temperatures (337.4; 367.2; 377.4; 388⁰ K) and with a small amount of catalyst sulfuric acid (0.1%) at lower temperatures (Z19. 327.8, 338.7⁰ K). It was found that during the interaction of secondary cellulose acetate with acetic acid with different water content, two reversible processes occur simultaneously, namely, hydrolysis of acetyl groups and reacetylation of hydroxyls until equilibrium is reached.

The rate constants of these reactions are different and are determined by the nature of the hydroxyl groups. In the interaction between acetic acid containing 2-3% water and incompletely substituted acetyl cellulose, the degree of substitution (equal to 2.3-2.4 at first) slightly increases and then gradually decreases. In more saponified cellulose acetates with a bound acetic acid content of 44.2%, the initial rise in the degree of substitution is sharper. This is explained by the fact that in the beginning the content of free primary hydroxyl groups in the initial samples exceeded their equilibrium amount for a given acetic acid-water mixture and, therefore, there was rapid acetylation of the excessive amount of primary hydroxyls. At the same time, the content of free secondary hydroxyl groups was initially below the equilibrium value, so that deacetylation occurred in these positions, but at a slower rate than the initial acetylation of primary groups. When the acetylation resulted in the primary hydroxyl content approaching the equilibrium value, continued hydrolysis at the secondary positions became the main reaction.

The results of the experiments are shown in Fig. 25 and 26.

The introduction of a catalyst does not affect the final result of the processes, but accelerates them. It was found that these reactions, as well as the reaction of esters of low molecular weight alcohols, are bimolecular and can be described using the equation of reaction kinetics of the second order. As a result, the rate constant for acetylation with acetic acid and deacetylation of primary and secondary hydroxyls was determined. According to Hiller's data, the rate of acetylation of primary hydroxyls is 1.7-1.8 times greater than that of secondary hydroxyls, and the rate of detachment of acetyl groups replacing primary hydroxyls is 60-70 times greater than that of acetyl groups replacing secondary hydroxyls. These data are in good agreement with the data of Malim and his collaborators [2]. The author confirmed his conclusions by tritylation method. There are only a few reports on cellulose acetates hydrolyzed below the solubility limit in acetone in the literature. These are the patents of Fordyce [3,4], the works of Howlett and Martin [5] and Malm and co-workers /6/.

Howlett and Martin performed deep saponification of acetylcellulose in homogeneous (in solution) as well as in heterogeneous (in fiber) media. The results obtained in both cases were compared. Different industrial brands of secondary cellulose acetates with 53% of bound acetic acid were the objects of the study, which are shown in Table 8.

Figure 25 shows curves of heterogeneous saponification of these fibers by alkali solutions. The figure shows that there are

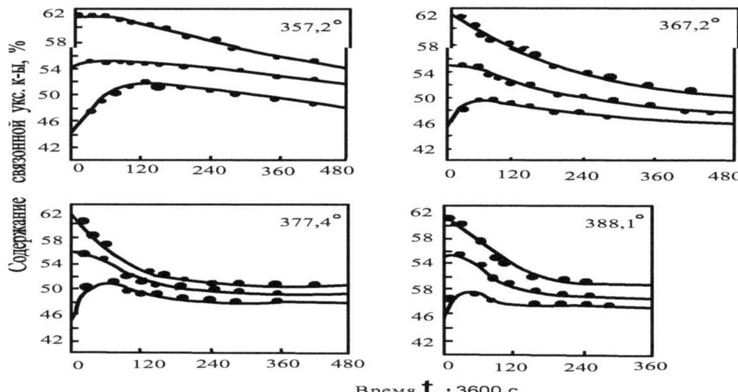

Fig.25. Homogeneous reaction of cellulose acetate samples with 97.2% acetic acid at different temperatures (without a catalyst).

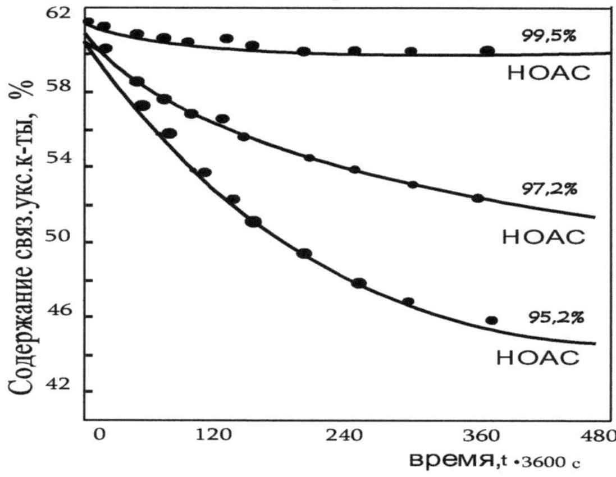

Fig.26. Homogeneous reaction of cellulose acetate with acetic acid of various concentrations at 367.2°К.

90

Table 8.

**Content of bound acetic acid depending on the grade of secondary
cellulose acetates.**

Samples,stamps	Content of bound acetic acid in %
"Celanese" fiber	52,6
"Zansil" - "-.	52,7
"Rhodiaceta" - "-	53,4
"Seraceta" - "-	53,8

Differences in saponification rate and that "Rhodiaceta" fiber is saponified most slowly compared to the others. This is explained by different supramolecular structure of "Rhodioceta" and other fibers, which is obtained in the process of their drawing. This conclusion is also confirmed by X-ray diffraction. Analysis of X-ray diffractograms showed that "Rhodioceta" is actually a more oriented product.

In the case of a homogeneous reaction, all differences between the acetylcellulose varieties disappear. If the indicated acetylcellulose fibers are dissolved in acetone and a certain amount of 0.1N aqueous NaOH solution is added to this solution, no precipitation of acetylcellulose occurs and the reaction can be carried out homogeneously to a considerable degree of saponification. Saponification of acetylcellulose by alkali in acetone solution can be described by the second order reaction equation:

$$K=\frac{2,303}{t(a-b)}*\lg\frac{b}{a}\left(\frac{a-x}{b-x}\right)$$

Where "a" and "b" are the initial concentrations of reagents in g/mol per liter

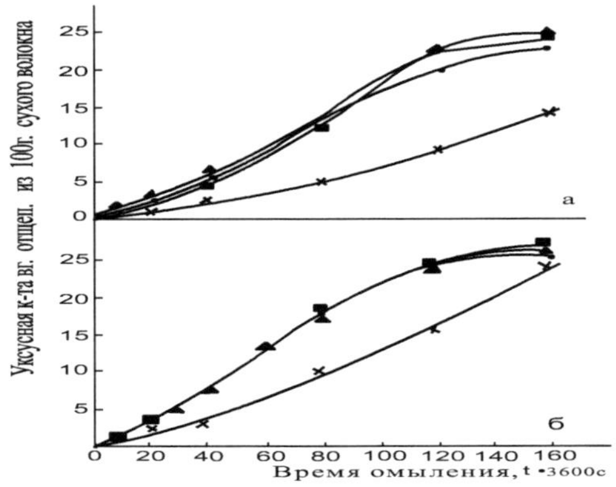

Fig.27. Amount of acetic acid released as a function of saponification time at 298⁰ K.
A-saponification with 0.05 N NaOH solution,B-saponification with 0.05 N KOH solution

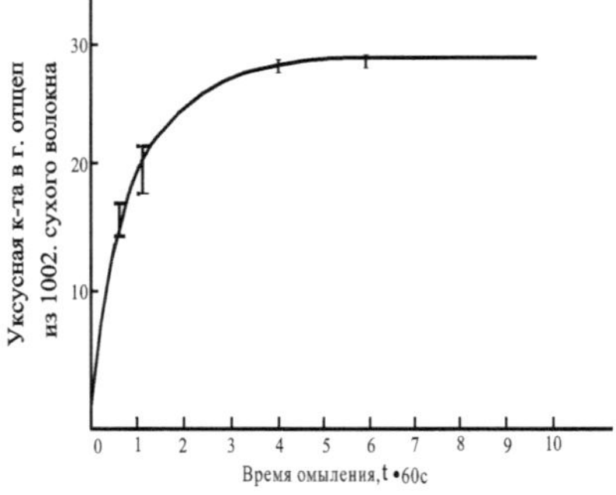

Fig.28. Saponification of acetate fibers in a solution (0.05 NaOH or KOH).

92

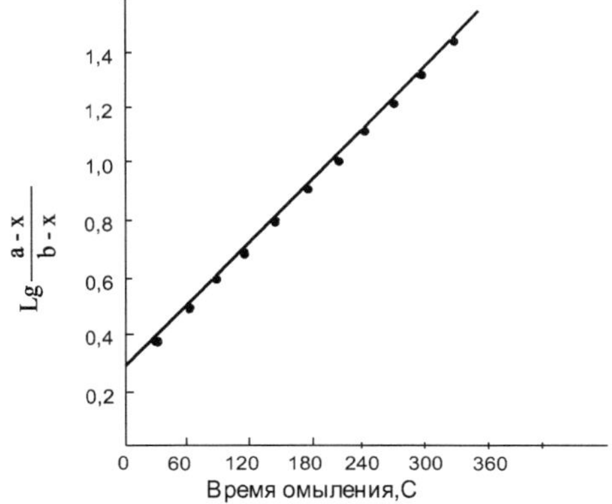

Fig.29. Dependence $\lg\dfrac{a-x}{b-x}$ of the saponification time.

"X"-concentration in g/mol in a liter of any of the two components used after time (sec)

K is the rate constant in liters/gmol sec.

Table 9 shows data on saponification of different acetyl cellulose grades with water-acetone solutions of NaOH and KON, and in Fig. 28 these results are expressed graphically as a kinetic curve.

Table 9.

Saponification of various grades of acetyl cellulose in a homogeneous medium.

Acetate fiber	Used alkali	Amount of acetic acid detached from 100 g of dry fiber after different time periods (in 0.6–10 sec)					
		1/2	1	2	∠	6	10
Celanese	NaaON	46,4	21,6	25,7	29,6	30,0	30,7
	CON	17,9	21,6	27,0	3C,1	30,8	30,8
Zansil	NaaON	16,8	21,6	25,7	2S,6	30,0	30,7
	CON	17,9	20,2	27,0	3C,0	30,8	30,8
Rhodiaceta	NaaON	14,9	18,0	25,0	2S,7	30,4	30,8
	CON	16,1	20,6	26,4	2S,7	30,9	30,9
Seraceta	NaaON	15,4	21,1	27,2	2S,5	30,4	30,7
	CON	14,9	21,6	26,3	2S,3	30,3	30,7

As can be seen from this table and the graph, the saponification rate of all the mentioned grades of acetylcellulose in the solutions was the same, and caustic alkalis NaOH and KOH also had the same effect. Figure 29 shows the values $\lg \frac{a-x}{b-x}$ as a function of time at 298^0 K. From this figure it can be seen that the points satisfactorily lie on the same straight line. Thus, the reaction does meet the conditions of a bimolecular reaction and the slope of this straight line can give the rate constant of this reaction.

It should be noted that the data obtained by the authors show that in the case of saponification with alkaline solutions there is no difference in the reaction rate of primary and secondary acetyl groups.

It is more accurate to say that these differences may not be so great as to be reflected in the change in the rate constant of the total process.

It is interesting to compare the rates of cellulose acetate cleavage in acidic and alkaline media. The work /6/ was devoted to this question. The effect of pH and temperature on the saponification rate of cellulose acetates was studied in the region of pH from 2 to 10 and temperature from 296 to 368^0 K. It was shown that the saponification rate of cellulose acetates is different depending on the pH of the medium as well as the temperature. Acetyl cellulose containing 30.8% acetyl was the object of the study. During saponification, the acetyl content decreased to about 34% in all cases. Therefore, the products were soluble in acetone and other organic solvents. Saponification of acetyl cellulose was performed heterogeneously, in buffer solutions with a certain pH value.

Since the amount of water in the experiments remained practically constant and the solutions themselves were buffered, the cleavage reaction can be regarded as a reaction of the first order. Figure 30 shows the change in the rate of hydrolysis depending on the pH value of the medium.

This figure shows that the saponification rate of acetates as a function of pH has a minimum. The minimum reaction rate occurs in the pH range of 4.5-4.7. The curves also allow a comparison of the saponification rate in acidic and alkaline media. This comparison shows that the rate of cleavage of ester bonds in acetylcellulose in an alkaline environment is several orders of magnitude higher than in an acidic environment.

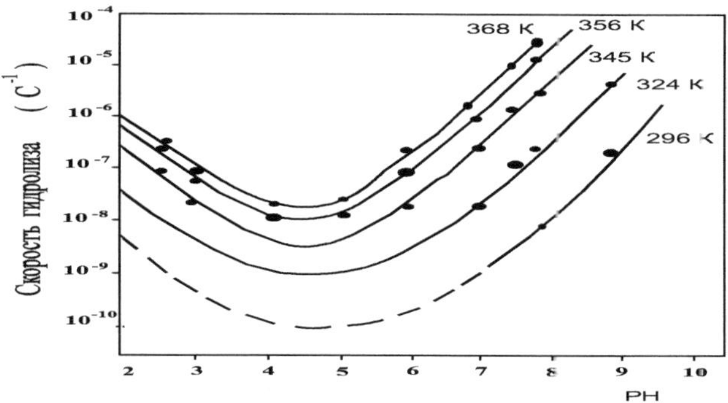

Fig.30. Hydrolysis rate of cellulose acetates as a function of pH value.

Concluding this section on washing and hydrolysis of acetylcellulose to different degrees of substitution it is necessary to note further the known properties of deep-hydrolyzed acetylcellulose preparations. Apart from the works [7-10], there is only one work in the literature in this respect by Malm et al. [2] describing the properties of deeply hydrolyzed cellulose acetates mainly in application.

Methods of direct preparation of soluble partially substituted cellulose acetates have been recently described in the literature [11,12]. The essence of these experiments was that cellulose was treated with an excess, relative to cellulose, of concentrated sulfuric acid dissolved in dioxane and acetic anhydride, also taken in large excess relative to cellulose (600-700%).

It is known that sulfuric acid forms a complex compound with dioxane, β-chloroethyl ether and other compounds that sulfate cellulose. It can be assumed that in this case, too, acetone-mixed cellulose ethers are formed in the beginning, i.e., also cellulose ethers in which part of the hydroxyl groups are substituted on acetyl radicals and another part on sulfogroups.

The formation of acetosulfates was observed by Malm and coworkers during acetylation of cellulose in the presence of large amounts of sulfuric acid and Na sulfate [13].

The authors of [11] also suggest that part of the hydroxyl groups can be esterified by the sulfuric acid dioxanate complex to form very unstable ester bonds, which are easily destroyed by water.

As a result, precipitation of such mixed cellulose ether in water or in organic solvents produces not completely substituted cellulose acetates. Thus, in this case we can also talk not about direct acetylation of cellulose, but sequential reactions of obtaining the tri ether and its subsequent saponification.

Thus, information about the properties of these products is very limited. Their properties in theoretical respect are especially poorly investigated. Information about water-soluble cellulose acetates is absent in the world literature.

Water-soluble acetylcellulose.

Currently, the most developed method of obtaining water-soluble acetylcellulose (WAC) is based on hydrolysis of triacetylcellulose under homogeneous conditions.

To obtain HPAC the method /7/ was adopted, which allowed acetylation and subsequent hydrolysis to be performed continuously and in the same apparatus. For acetylation a cotton linter was used with the characteristics given in Table 10.

Table 10.

Characteristics of the initial sample used for acetylation.

Source sample	Content (%)				Average SP (according to the viscosity of the copper-ammonia solution)	Low molecular weight fractions by p-value in 10% NaOH solution	Transparency 7% solution of acetyl cellulose syrup (mm)
	α-Cellulose	Pentosanov	Resins	Ashes			
Cotton Liner	99,2	0,5	0,06	0,07	1380	1,4	290

For each experiment, 1 part of carefully prepared, lump-free, linter with a moisture content of 7-8% and 2.4 parts of glacial acetic acid were taken. After the ice-cold acetic acid was added to the pulp, the mixture was stirred for one hour at 311^0 K. A solution consisting of 4 parts of glacial acetic acid and 0.09 parts of concentrated sulfuric acid (d=1.84) was then added. Mixing was continued at the same temperature for another 45-60 s. The mixture was then cooled to about 291°K. (All of these processes were necessary to activate the cellulose). At this temperature another 2.7 parts of 98% acetic anhydride and 6.12% (of the weight of the cellulose) sulfuric acid were added. The temperature was controlled and slowly raised to 305-308°K over a time interval of (2.0-2.5)-3600s, while stirring was continued. After this time, the reaction mass was transformed into a viscous transparent solution (syrup) free of fibers. A mixture of one part water and two parts acetic acid was added to the syrup thus obtained. The mixture was added carefully, drop by drop, for one hour, since this operation is associated with the release of heat due to the hydration of excess acetic anhydride. This step ended the stage of preparation of triacetylcellulose and its solution in acetic acid. After that, hydrolysis of triacetylcellulose was performed to obtain a water-soluble product. In this stage, the temperature of the solution was raised to 313^0 K, after which it was kept in the thermostat for a certain time with continuous stirring.

In the hydrolysis process, as noted above, two reactions take place: hydrolytic cleavage of the ester bond and reesterification. Both of these reactions depend on the relative amount of water available in the mixture. Thus, in order for the hydrolysis process to take place, the amount of water in the solution must increase during this process. In addition, the solubility of the hydrolyzed products changes as the hydrolysis process proceeds. The more hydrolyzed the product, the more water must be in the mixture to dissolve it. Therefore, during the hydrolysis process water was added to the initial solution of acetyl cellulose and its amount was determined by titrating the individual samples of the solution with water. Thus, the hydrolysis process was carried out as follows.

After incubating the solution at the above temperature for 16 hours, a sample of the solution was taken in such an amount that it contained approximately 125 mg of cellulose acetate. This solution was diluted in a beaker with 125 ml of acetone, and then titrated with water from a burette. The end of the titration was the moment of appearance of opalescence in the solution due to precipitation of acetylcellulose, This titration allows to determine to what extent hydrolysis has taken place and how much water can be added without fear of precipitation of acetylcellulose from the solution. After 16 hours of the hydrolysis reaction, 45 mL of water was added so that its concentration in the solution was 13%.

At time 48-3600 s, the concentration of water in the mixture was increased to 36%. Water was then added every 24-3600 s. throughout the hydrolysis time. The final concentration of water in the mixture was 73-75%. During the second half of the hydrolysis reaction, when the content

of acetyl groups in the ether decreased to ~30%, titration with water became impossible due to the increased solubility of acetylcellulose in it. Therefore, methyl ethyl ketone (MEK) titration was used to control the depth of hydrolysis. The titration was performed as in the case of titration with water from the burette until opalescence appeared. A sample of 5 g solution was taken for titration. This amount of solution was diluted with 100 ml of 50% alcohol and subjected to titration. During the specified hydrolysis time, the content of acetyl groups dropped to 16-18%. After the end of hydrolysis, which was established by a number for methyl ethyl ketone of $10 \div 15$ ml, the solution was poured into acetone. The resulting white fibrous precipitate was washed five times with a precipitant, the last two washes with acetone in which a small amount of potassium acetate (0.004%) was dissolved to bind the residual amount of sulfuric acid.
 After washing, the sludge was air dried.

In the obtained cellulose acetate determined the content of bound acetic acid by alkaline saponification and water solubility /4/.

To study the kinetics of deep hydrolysis experiments were carried out at different temperatures, namely at 313, 323 and 333°K. According to the hydrolysis stages, solution samples were taken, cellulose acetates with different hydrolysis depths were precipitated (with water, alcohol or acetone) and the content of bound acetic acid was also determined. When the content of bound acetic acid in the ether was higher than 30%, the ether was precipitated in water, less than 30% in acetone or alcohol.

Fig. 31 shows curves expressing the change in acetic acid content in cellulose acetate during deep hydrolysis at 313°K. Fig. 32 shows curves of changes in the content of bound acetic acid at 323 and 333° K.

These figures show that the content of bound acetic acid decreases to 16-18% at 313°K in 280-3600 s, at 323°K in 1203600 s, and at 333°K already in 603600 s.

As can be seen from Fig. 31, the curve expressing the dependence of the content of bound acetic acid on time is characterized by two sections- the first reflecting the relatively rapid course of the hydrolysis reaction, and the second, the flatter one associated with a relatively slow process.

As noted above, the reaction of hydrolysis of cellulose acetate at a small concentration of water in the mixture, considered as a reaction of the second order, which is reversible and reaches equilibrium. This reaction is complex and consists of two mutually related processes: the reaction of primary hydroxyl groups and acetate groups, which proceeds quickly, and the reaction of secondary hydroxyl groups and acetate groups, which proceeds more slowly.

However, in order to obtain the conditions necessary for deep hydrolysis and to maintain the solubility of the hydrolyzed product, a constant addition of water to the mixture was made. Water was added gradually so that no precipitation of the cellulose ether occurred. At the point when the acetic acid content dropped to 35-40%, the amount of water added was 300% of the weight of the acetylcellulose, and its concentration in the mixture was about 30%.

Subsequently, the amount of water increased even more. This allowed us to consider the reaction of deep (from 35-40 to 16-18% bound acetic acid) hydrolysis of cellulose acetate as a reaction proceeding almost in excess of the second reagent (water).

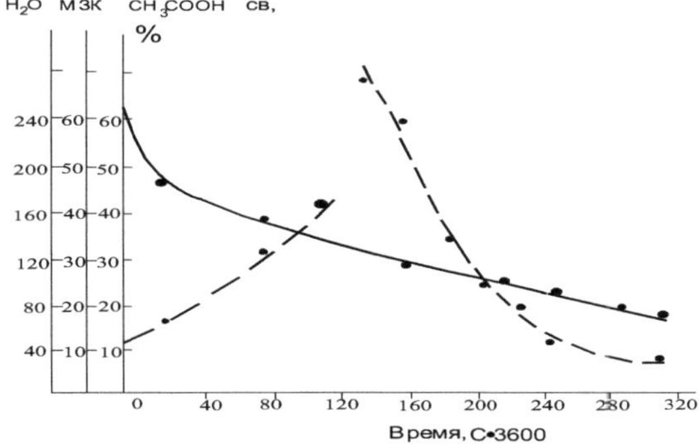

Fig.31. Changes in the bound acetic acid content of the water resistance value and the methyl ethyl ketone stability value with the hydrolysis time at 313⁰ K. 1- H₂ O, 2-MEC, 3-CH₃ COOH.

Since the reaction occurred when water was added to the mixture, it did not reach equilibrium and was practically irreversible. Therefore, the kinetics of the deep hydrolysis reaction was calculated using the 1st order reaction equation.

Table 11. shows the results of calculations of the rate constant of the reaction of deep hydrolysis, starting from a time point that corresponded to the content of 35-40% bound acetic acid. The calculation was made according to the formula

$$K=-\frac{1}{t}\ln\frac{Co-X}{Co}$$

where: Co-concentration (mol) of bound acetic acid in the initial cellulose acetate;

99

X in the hydrolyzed acetate at time t.

As can be seen from the data in the table, the constancy of the values in the above interval is satisfactory. Calculation of K for the first hours of hydrolysis at 313°K using the 1st order reaction equation, results in values that differ from each other. There are two reasons for this. The 1st order reaction equation is not applicable, first, because of the relatively small amount of water in the mixture. Second, as already mentioned, the primary and secondary acetate groups hydrolyze at markedly different rates.

Table 11.

Kinetics of deep hydrolysis of cellulose acetate.

τ (-3600c)	CH UNON$_{cb}$ %	K From^{-1} x 3600	Csr FROM^{-1} -3600
at 313° K			
120	33,30	0,004600	
144	33,80	0,004000	
168	28,50	0,004107	
192	25,95	0,004062	$(4,15\pm0,10)\cdot10^{-3}$
216	23,25	0,004030	
240	20,70	0,004123	
264	18,15	0,004091	
288	16,00	0,004003	
at 323° K			
65	30,05	0,0099	
90	20,50	0,1184	$(10,91\pm0,8)\cdot10^{-3}$
120	17,00	0,0110	
at 333° K			
22	37,50	0,01990	
40	24,05	0,02098	$(19,16\pm1,2)\cdot10^{-3}$
60	17,50	0,01861	

Therefore, a change in the concentration of primary acetate groups during the first hours of hydrolysis will lead to a change in the reaction rate coefficient calculated from the change in the average content of acetate groups in the product, i.e., from the value related to the sum of primary and secondary groups. All, the phenomena noted for the reaction at 313^0 K will occur more rapidly at higher temperatures. Therefore, as can also be seen from the data in the table, the K values at 323^0 K are constant (within the error of the experiment) at 60-3600 s of hydrolysis time, and at 333^0 K-already at 22-3600 s.

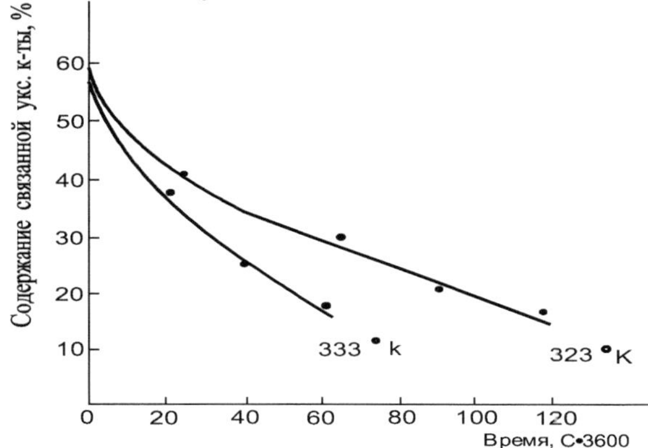

Fig.32. Change of bound acetic acid content in cellulose triacetate as a function of temperature and hydrolysis time.

The effect of temperature on the reaction rate is shown in Fig. 33 as a dependence of ln K on the inverse of the absolute temperature T.

The logarithm of the rate constant ln K is linearly related to the inverse temperature (1/T):

$$\ln K = \frac{A}{T} + B$$

This Arrhenius equation, where A and B are individual constants characteristic of the reaction, allows us to determine its activation energy (E= -AB), which turned out to be 69.196 kJ/mol.

Temperature coefficient of the reaction of deep hydrolysis:

$$\gamma_n = \frac{K_t + n \cdot 10}{K_t} = 2 / 2,5$$

Simultaneously with the hydrolysis of the ester, i.e., the detachment of acetyl groups, the glucoside bond in the cellulose acetate macromolecule is cleaved.

It should be noted that both esterification, cellulose itself and hydrolysis of the resulting cellulose ether is accompanied by significant destruction of the cellulose chain. This simultaneous degradation must also be controlled so that the polymer can further be useful for its application in the production of fibers, plastics and films. Therefore, it was of interest to determine the size of the degradation and the rate constant of this process during the hydrolysis reaction.

To determine the degree of polymerization of cellulose acetate, cellulose was regenerated by mild saponification with 0.5N sodium methylate solution in anhydrous methanol.

The regenerated cellulose was further used to determine the degree of polymerization. The degree of polymerization of the regenerated cellulose was determined by the viscosity of its solution in iron-vinosodium complex (IVNC)-/(C H O$_{4663}$)-Fe/_Na$_6$ [15].

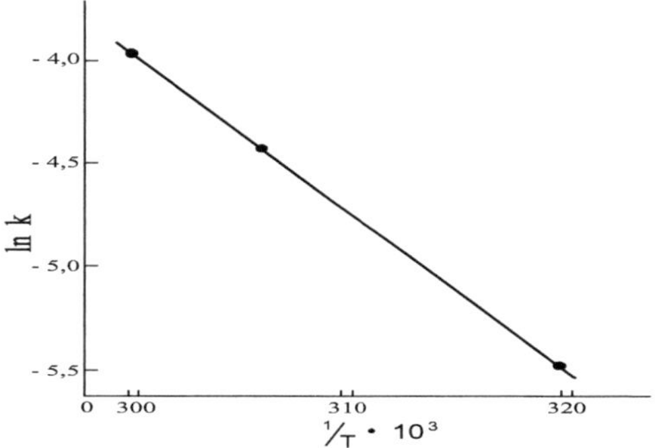

Fig.33. Dependence of the logarithm of the hydrolysis rate coefficient
of cellulose acetate from the inverse value of the absolute temperature.

Cellulose solutions in this complex are practically insensitive to oxygen, stable and transparent. The iron-nitrium-sodium complex was prepared by the method proposed by Valtasaari [16].

The characteristic viscosity was determined by the formula [17.p. 292].

$$|\eta| = \frac{\eta_{уд} - К_1 \cdot \eta_{уд}}{1 + К_2 \cdot \eta_{уд}}$$

where: C- the concentration of the solution, g/l;

$К_1$ и $К_2$ -constants, $К_1 = 0,25$; $К_2 = 0,1$

The degree of polymerization was determined by the formula:

$$СП = \frac{|\eta|}{Km}$$

102

where: Km= 8.14-10^{-4}

Changes in the degree of polymerization of cellulose acetate at 323°K with the hydrolysis time are shown in Fig. 3.4. Since, as it was found [18,19], the macromolecule destruction during homogeneous hydrolysis has a random character, it was assumed that the distribution arising in this case is such that

$$\frac{M_w}{M_n} = \frac{P_w}{p_n} \cong 2 \quad (1)$$

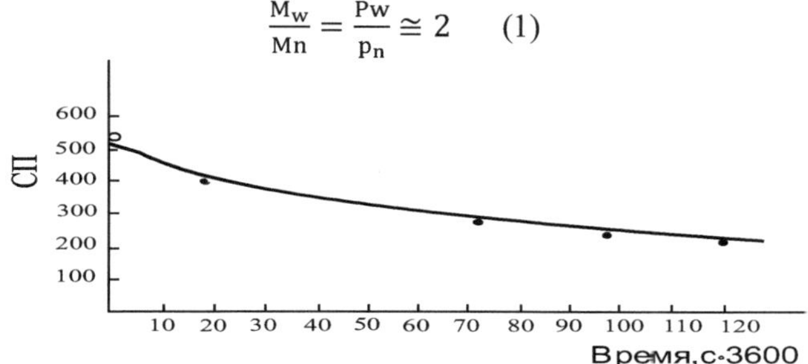

Fig.34. Changes in the degree of polymerization of cellulose acetate at 323⁰ K with the duration of hydrolysis.

This specificity is characteristic of homogeneous destruction reactions. The kinetics of destruction was determined using the 1st order equation:

$$K = \frac{1}{\tau}\ln\left(\frac{C_0-x}{C_0}\right) = -\frac{1}{\tau}\ln\left(1 - \frac{x}{C_0}\right) \quad (2)$$

where: Co- concentration of the component (in this case, the concentration of β-glycosidic bonds) at zero time;

X - concentration of the component by time τ;

Since, with relatively shallow hydrolysis, the fraction of broken bonds $\frac{x}{C_0}$ is much less than unity, the equation is converted* to the form: $K = \frac{\alpha}{\tau}$ (3)

where $\alpha = \frac{x}{C_0}$ -is the fraction of broken ties by the time τ. The proportion of broken bonds, according to [20], can be judged by the drop in the average number degree of polymerization:

$$\alpha = \frac{1}{p_{n\tau}} - \frac{1}{P_{n_0}} \quad (4)$$

*ln(1-2) $\cong -2$ if $\alpha \geq 0,1$,

where : $p_{n\tau}$ - Numerical average SP for the time τ;

P_{n_0} - is the average numeric SP at the beginning of hydrolysis.

103

Combining (1), (3) and (4) and taking into account that average viscosity degree of polymerization P_{Vo} at its definition according to Staudinger equation coincides with average weight degree of polymerization [21,p.142-146], we obtain:

$$k = \frac{1}{\tau}\left(\frac{2}{P_{v\tau}} - \frac{2}{P_{vo}}\right) \quad (5)$$

The results of the calculation of the rate constant of hydrolysis at 323^0 K are given in Table 12.

In order to reduce the hydrolysis time, secondary acetylcellulose produced by industry and having the following characteristics was used as a feedstock for obtaining water-soluble acetylcellulose: bound acetic acid content 54.3% ($Y=236$), degree of polymerization (SP) =280 [22].

Table 12.

The rate constant of cellulose acetate degradation during hydrolysis at 323^0 K.

№ n/a	Hydrolys is time, x-3600 s	$\lvert\eta\rvert$	SP	SP= $\frac{\text{СП}}{2}$	K, c^{-1}- 3600	K, c^{-1}- 3600
1.	0	0,423	510	255	-	
2.	24	0,354	425	212	$3{,}7\text{-}10^{-5}$	
3.	72	0,246	300	150	$3{,}8\text{-}10^{-5}$	$(3{,}80\pm0{,}16)\cdot10^{-5}$
4.	96	0,212	260	130	$4{,}0\text{-}10^{-5}$	
5.	120	0,187	230	115	$4{,}0\text{-}10^{-5}$	

Secondary acetylcellulose was dissolved in 75% acetic acid solution. After formation of 12.5% viscous and transparent solution of secondary acetylcellulose, hydrolysis was performed in the same way as in the case of hydrolysis of triacetylcellulose. However, in this case we used not sulfuric acid but chloric acid ($HClO_4$) as a catalyst in the amount of 15% of acetylcellulose weight.

The course of hydrolysis was also monitored by titrating separate samples of MEK solution and water.

Fig. 35 shows curves characterizing changes in the content of bound acetic acid during hydrolysis at different temperatures for tri- and diacetates of cellulose.

The kinetics of deep hydrolysis of secondary acetylcellulose was studied and the reaction activation energy (E) was determined, which turned out to be 66.88 kJ/mol (Table 3.4).

Solubility diagrams of cellulose acetate hydrolysis products in various organic solvents were studied (Fig. 35).

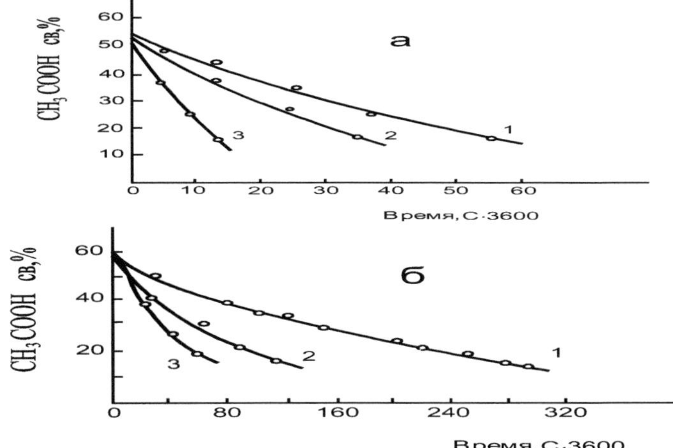

Figure 35. Changes in the content of bound acetic acid:
a - in secondary cellulose acetate: b - in cellulose triacetate.
1- at 313, 2-323, 3-333°K.

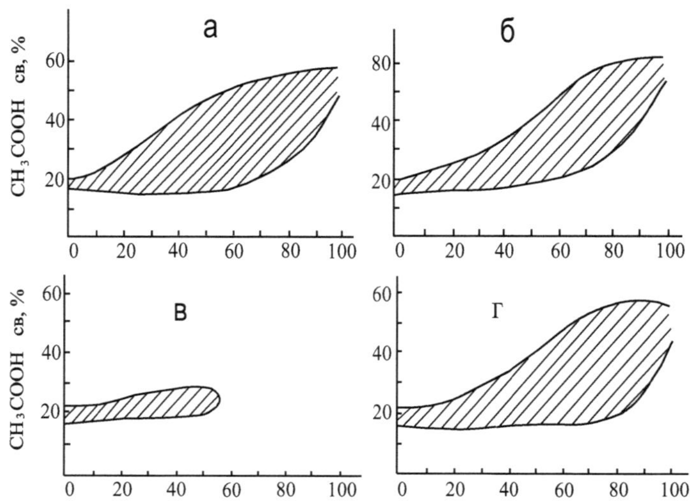

Fig.36. Solubility diagrams of hydrolysis products in various mixed solvents depending on the content of bound acetic acid: a- acetic acid + water; b-acetone + water; c-alcohol + water; d- diaxane + water.

Table 13.

Kinetics of cellulose acetate hydrolysis.

Temper a tour,°K	Hydrol ysis time. x-3600s	CH_3 UNCO sb, %	!	K	K_{cp}
				FROM^{-1} -3600	
	0	54,3	2 3 6	-	
313	14	44,3	1 7 3	0,024 124	
	26	36,0	1 3 0	0,023 968	$(23,77+0,6)$ -10^{-3}

106

Temperature, °K	Hydrolysis time x-3600s	|η|	SP	SP =SP /2nV	K,c⁻¹ - 3600
	38	26,0	85	0,024 580	
	57	18,8	58	0,022 476	
	5	46,9	188	0,045 220	
323	15	37,5	140	0,041 182	$(42.78 \pm 1.4) \cdot 10^{-3}$
	25	26,8	89	0,043 342	
	36	18,3	57	0,041 399	
333	5	35,3	124	0,128 231	
	10	24,8	89	0,082 217	$(111,95 + 1,7) \cdot 10^{-3}$
	13	18,2	57	0,125 425	

The degrees of polymerization and rate constants of secondary cellulose acetate degradation during hydrolysis were also determined /11/. The results of these studies are shown in Fig. 37 and Table 14.

Table 14.

Degradation rate constants of secondary cellulose acetate during hydrolysis.

Temperature, °K	Hydrolys is time. x-3600s	|η|	SP	SP =SP /2nV	K,c⁻¹ - 3600	BYcp , from⁻¹ - 3600
	0	-	28 0	140	-	
313	8	1,1 0	26 0	130	10,10- 10⁻⁵	

	35	0,90	210	105	$9,95 \cdot 10^{-5}$	$10,37 \cdot 10^{-5}$
	50	0,80	190	95	$11,06 \cdot 10^{-5}$	
	5	1,00	244	122	$21,00 \cdot 10^{-5}$	
323	24	0,78	185	93	$22,10 \cdot 10^{-5}$	$22,06 \cdot 10^{-5}$
	36	0,66	153	77	$23,09 \cdot 10^{-5}$	

In the literature there are no data on Km and α values in the Mark-Hauvinck equation to determine the molecular weights for many polymers, especially water-soluble ones. In this regard, it was of some interest to separate the BPAC into separate fractions and determine its molecular weight by osmotic and viscometric methods [11, 12]. Fractionation of water-soluble acetylcellulose was performed by its precipitation from a 2% aqueous solution by adding acetone. Thirteen fractions were isolated (Table 15).

Table 15.

Fractionation of water-soluble acetylcellulose.

faction number	Amount of acetone added	Fractional yield		$|\eta|$	Total mass fraction of the fraction
		Г	%		
I.	360	0,01791	5,61	-	93,49
II.	250	0,1689	5,29	1,95	87,88
III.	100	0,1554	4,87	1,74	82,59
IV.	50	0,1784	5,59	1,70	77,72
V.	50	0,1800	5,64	1,67	72,13
VI.	25	0,3238	10,14	1,64	66,49
VII.	75	0,4109	12,88	1,56	56,35
VIII.	25	0,4089	12,81	1,50	43,47
IX.	95	0,1631	5,11	1,44	30,46
X.	68	0,1789	5,60	1,40	23,35
XI.	119	0,2986	9,37	1,30	19,75

| XII. | 219 | 0,1876 | 5,88 | 0,80 | 10,38 |
| XIII. | 213 | 0,1586 | 4,50 | 0,70 | 4,50 |

When determining the molecular weight of polymers by the osmotic method, the choice of membrane is of great importance. The most suitable membranes are those based on polyvinyl alcohol [28], polychlorotrifluoroethylene, and polyvinyl butyral [29]. Cellophane films are also used as a membrane [30]. Cellophane film pre-treated with 10% caustic soda solution proved to be quite suitable for determining the osmotic pressure of dimethylformamide solution of water-soluble acetylcellulose.

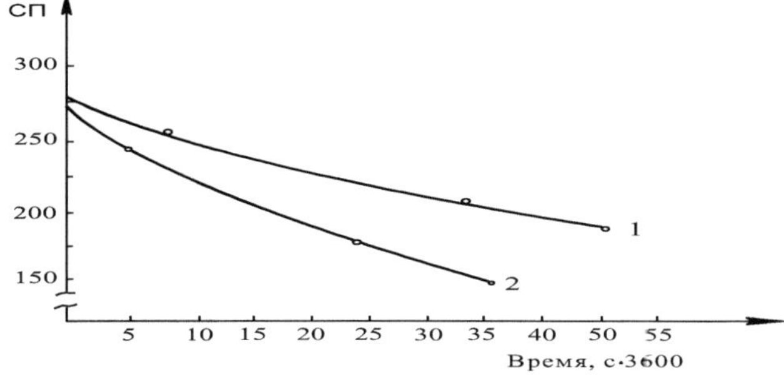

Fig.37. Changes in the degree of polymerization of secondary cellulose acetate with the duration of hydrolysis: 1-at 313°K; 2- at 323°K.

To quantitatively characterize the treated membrane, the specific permeability constant G for dimethylformamide was determined (calculated from the data obtained) [24].

After determining the membrane specific permeability constant, the osmotic pressure of the solution of fractions (II-IX) of water-soluble acetylcellulose in dimethylformamide at 398°K was measured. The results obtained were plotted as a dependence of P/C_2 on C_2 and extrapolated to zero concentration (Fig. 38).

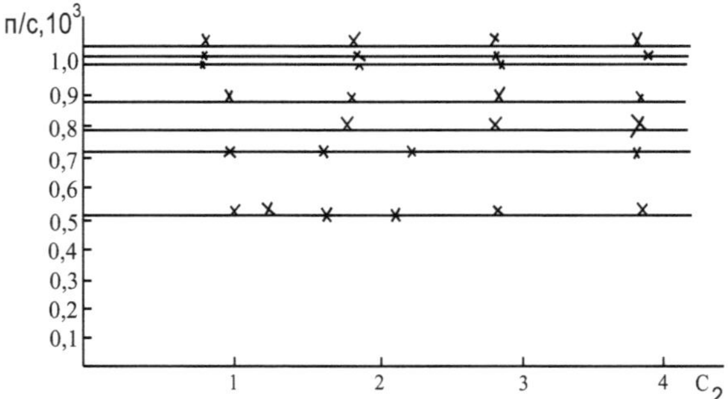

Fig.38. Dependence of osmotic pressure on concentration dimethylformamide solutions of water-soluble acetylcellulose fractions.

From the data of the reduced osmotic pressure (P/C_2)-$C_2 \to 0$, the average molecular mass of the sample under study, the second virial coefficient constant B_2, and the Huggins interaction constant M were calculated. From the graphical relationship between the logarithm of characteristic viscosity $\lg |\eta|$ and the logarithm of molecular mass $\lg M_n$ determined by osmotic method, Km and α values were calculated. Based on the calculated values of Km and α, the average viscosity molecular weight of the fractions (Table 16) and the MMR of water-soluble acetylcellulose were calculated, the values of which are shown in Table 16 and Fig. 39.

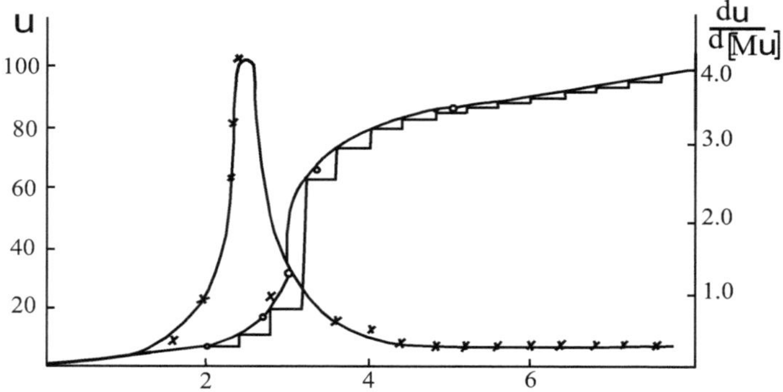

Fig.39. Molecular weight distribution of water-soluble acetylcellulose.

Table 16.

Molecular weights of water-soluble acetylcellulose fractions.

№ fractions	η	K_1	$(n/C)_{2C_0}$	Mn	lg||η|	lg Mn	Mv	Mv/ Mn
II	1,95	0,2882	0,000 51	47 91 0	0,2 900	4.6 804	507 80	1,06
III	1,74	0,2750	0,000 52	33 91 0	0,2 405	4.5 303	369 20	1,08
IV	1,70	0,2750	0,000 79	30 73 0	0,2 304	4.4 875	345 80	1,12
VI	1,64	0,2700	0,000 88	27 55 0	0,2 148	4,4 401	312 70	1,13
VII	1,56	0,2960	0,001 00	24 43 0	0,1 931	4,3 879	271 90	1,11
IX	1,44	0,2740	0,001 30	18 80 0	0,1 584	4,2 742	217 30	1,16
X	1,40	0,2860	0,001 50	16 30 0	0,1 461	4,2 119	200 80	1,22
		0,001 10	0,001 10	22 21 0	Unfractionated water-soluble acetylcellulose			

Note: $|\eta|=0.04074 * M^{0,357}$ in dimethylformamide
at T= 298°K. For all fractions $\alpha =$
0.357;
Km = 0.04074; In $*10_2^{-4}$ =1.50; M
=0.4998

Table 16 shows that HPAC is characterized by unusually low values of α in the Mark-Hauvinck equation. This suggests that its molecules in

111

solution represent dense balls, practically not permeable to the solvent. Small variations in the Mv/Mn ratio in the fractions indicate a high homogeneity in molecular weight of HPAC. The obtained values of K and α allow for the determination of the molecular weight of HPACs.

Determination of the chemical composition by infrared spectroscopy.

To take IR spectra, fibrous samples taken according to the hydrolysis stages and dried in the air were subjected to rapid grinding to a powdery state in an electromagnetic vibrating mill, then mixed with KBτ and pressed into tablets [31]. IR spectra were taken with a Nippon-Bunio spectrophotometer, model DS-301, on NaCl prisms between 4000 and 660 cm⁻¹ [28].

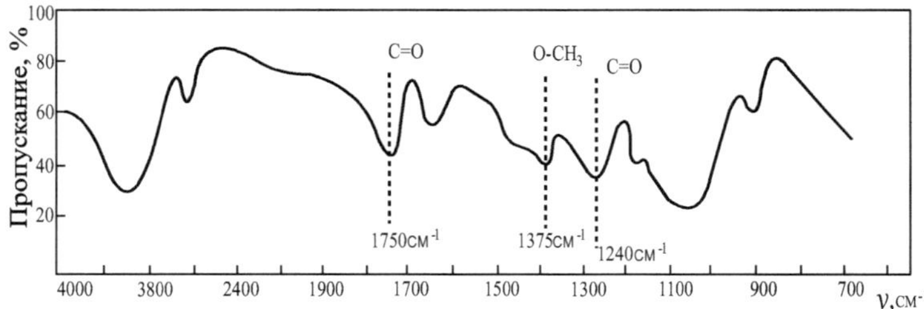

Fig.40. Infrared spectrogram of water-soluble acetylcellulose (bound acetic acid content 17.50%).

Fig. 40. shows the IR spectrum of water-soluble acetylcellulose.

The analytical bands characterizing this product are also shown here. Such bands are the "OH" valence vibration band (3200-3600 cm⁻¹) and various bands related to vibrations of bonds in acetate groups (C=0, C-CH₃ , C-0). The assignment of these bands to the vibrations of certain groups and bonds was made on the basis of the works of O'Connor [31], Huggins [32], and other authors [33].

Changes in optical density in the region of analytical bands of cellulose acetate by hydrolysis stages were measured. The values of the optical densities of the indicated bands were determined from the transmittance as $\lg\frac{I_0}{J}$ using a baseline, which was drawn according to the Huggins technique. Its construction is clear from Fig. 41.

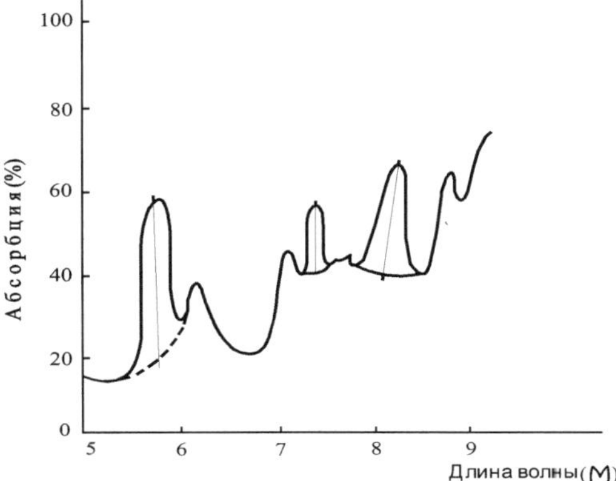

Figure 41: Drawing the baseline for the analytical bands of cellulose acetate.

In order to avoid the influence of sample concentration, the ratio of the optical density of the bands in the region of 1750, 1375, and 1240 cm^{-1} to the optical density of the "OH" band was calculated $-\lg \frac{J_0}{J_R} / \lg \frac{J_0}{J_{OH}}$

Table 3.8 and Fig. 42 show the data characterizing the change in the above ratio depending on the content of bound acetic acid in the products of hydrolysis of cellulose acetate.

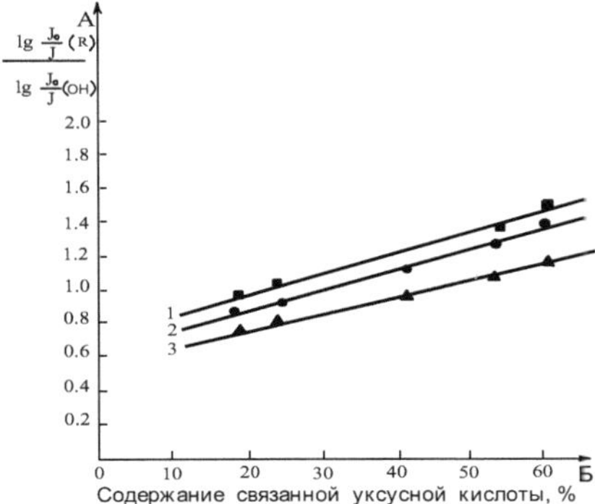

Figure 42: Dependence between the ratio of optical densities of analytical bands and the degree of hydrolysis of cellulose acetate.

1.-(-C=0); 2.-(-C-SN$_3$); 3.-(-0-C).

As can be seen from these data, there is a good correlation between the relative optical density of these bands and the degree of substitution. The straight-line dependence of these values allows a quantitative determination of the degree of substitution. The analysis of IR spectrograms also shows that HPAC does not contain any functional groups other than hirdoxyl and acetyl groups in its composition.

Table 17.

Value of the ratio of optical densities of analytical bands depending on the degree of hydrolysis.

Content of bound acetic acid, %	$\dfrac{\lg\frac{I_0}{J}(R_1)}{\lg\frac{I_0}{J}(OH)}$	$\dfrac{\lg\frac{I_0}{J}(R_2)}{\lg\frac{I_0}{J}(OH)}$	$\dfrac{\lg\frac{I_0}{J}(R_3)}{\lg\frac{I_0}{J}(OH)}$
59,95	1,62	1,49	1,24
51,78	1,46	1,36	1,12
40,95	1,19	1,16	1,00
22,81	1,04	0,96	0,82
17,74	1,00	0,86	0,76

Determination of the position of acetyl groups.

As indicated above, the rate of detachment of acetyl groups replacing hydroxyls at C_6 , is 60-70 times higher than the rate of detachment of secondary acetyl groups (C_2 and C_3). Therefore, under conditions of deep hydrolysis, it should be noted that water-soluble acetylcellulose will have substituents mainly only at C_2 and C_3 .

To prove this point, the tritylation reaction of HPACs was performed /34/. It is known that the reaction rate of tritylation of primary hydroxyl groups is 10-13 times higher than that of the C_2 and C_3 positions [35]. Tritylation was performed with triphenylchloromethane-$(C_6 H)_{53} \cdot C \cdot Cl$.

Triphenylchloromethane (TFCHM) was obtained by the method described in [26, p.256] and purified by recrystallization.

The TFCM purified in this way (Tpl = $383 \div 385^0$ K) was used for esterification reactions. The esterification was carried out according to the Gelferich method [36, p.258].

To determine the degree of substitution by triphenylmethyl radicals we analyzed the elemental composition of the obtained products. The results of this analysis showed that the carbon content was: C=73.7%, H=6.01%.

The amount of carbon was used to determine the degree of substitution of the trityl ether according to the formula:

$$C3 = \frac{162 \cdot C + x(42 \cdot C - 2400) - 7200}{22800 - 242 \cdot C}$$

where: C-percent carbon content;

x is the degree of substitution of water-soluble acetylcellulose;

C3-triphenylmethyl radicals calculated using this formula was 1.03.

Theoretically, if the interaction with the primary hydroxyl groups is exceptional, if the reaction is complete, and finally, if all of the primary hydroxyl groups are unsubstituted, the degree of substitution should be equal to 1. The first two conditions are unlikely to be fully satisfied. Nevertheless, the obtained value of the degree of substitution of triphenylmethyl radicals equal to 1.03 indicates that in water-soluble acetylcellulose most of the primary hydroxyl groups are free.

Thus, the acetyl groups in water-soluble acetylcellulose can be assumed to be statistically randomly distributed in the C_2 and C_3 positions.

To prove the statistically normal distribution of the substituted glucose links, cross-fractionation of water-soluble acetylcellulose was performed in [37,38].

Based on the results obtained, integral and differential composite distribution curves were constructed and chemical heterogeneity parameters were calculated.

The results of the study showed that the curve of the substitution degree distribution is unimodal and characterizes a statistically normal distribution (Fig. 43).

The same authors first attempted to study the distribution of acetate groups of water-soluble acetylcellulose by high-discharge NMR spectroscopy. However, the obtained spectra were poorly resolvable, since the instrument frequency was only 100 MHz, and although signals inherent to the two acetyl groups were observed in the spectrum, their quantification from the spectra was difficult.

Physical structure of water-soluble cellulose acetates.

The physical structure of water-soluble acetylcellulose (meaning "amorphous" or "crystalline" state) should have a significant influence on the properties of this product, in particular on its solubility. This structure was investigated by X-ray diffraction [10].

The X-ray structure of acetyl cellulose samples was traced as its hydrolysis and the detachment of acetyl groups proceeded [8].

Figure 44 shows diffractograms of cellulose acetate samples with different degrees of hydrolysis.

As can be seen from the figure, the initial acetylcellulose is characterized by an amorphous structure (lower curve in Fig. 44).

All intermediate radiographs and radiographs of the final product containing 16-18% bound acetic acid and soluble in water are also radiographs of an amorphous substance.

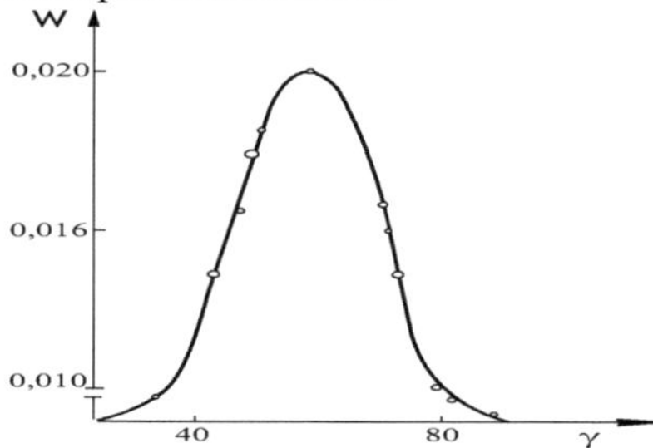

Fig.43. Differential curve of composite distribution according to the degree of substitution of water-soluble acetylcellulose.

It can be noted that the two weakly blurred maximums of the X-ray at angles 2θ, characteristic of triacetate, as the acetate groups are hydrolyzed shift towards the angles corresponding to the maxima characteristic of hydrate cellulose, however, in general, as already mentioned, there is no significant increase in intensity.

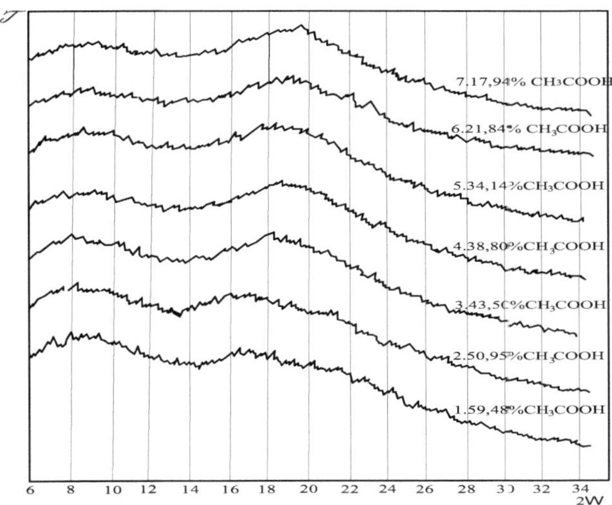

Fig.44. X-ray diffraction patterns of the products in the stages of hydrolysis of cellulose acetate.

The formation of the supramolecular structure of water-soluble acetylcellulose was also traced using an electron microscope. It was shown that the initial sample of triacetylcellulose is characterized by a homogeneous structure without highly ordered supramolecular formations. As a consequence, its fragmentation (ultrasonic or mechanical method of dispersion) results in broad layered aggregates (Fig. 45a).

In contrast, in aqueous solutions of acetylcellulose ($\gamma = 50$) are detected (by the "thermal attachment" method) by tiny anisodiametric particles about 200-400 A wide^0 and 0.1 MKM long (Fig.45-6).

Further electron microscopic studies showed that when a solution of water-soluble acetylcellulose slowly dries on a solid substrate, fibrillar formations also arise (Fig. 45-c).

It is easy to see that they are quite imperfect, very thin, and tend to be mutually ordered. Thus, the appearance of a significant number of hydroxyl groups in the chains of acetylcellulose macromolecules led to its structural reorganization, characterized by the provision of fibrils and the emergence, therefore, of some new, higher level, order.

However, as follows from the X-ray diffractometric data, the formed fibrillar aggregates of macromolecules are still amorphous.

117

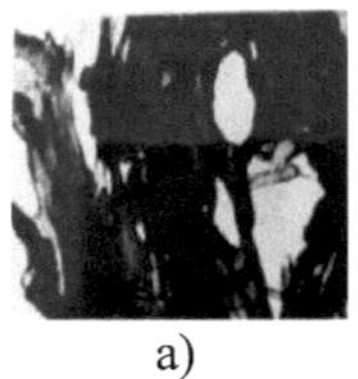 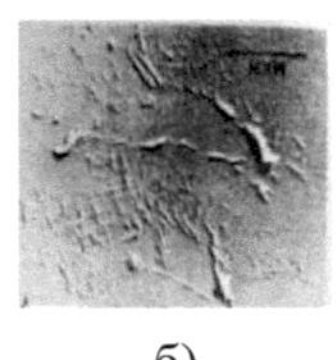 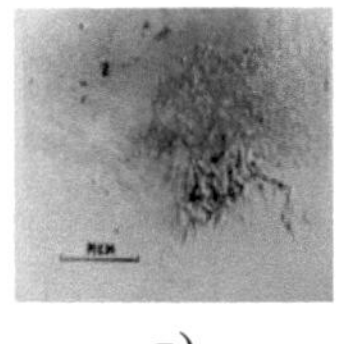

а)   б)   в)

Fig.45. Electron micrographs of products of hydrolysis of cellulose acetate.
a- triacetyl cellulose, b-water-soluble acetyl cellulose, and c-after drying on a solid substrate.

Further complete detachment of acetate groups regenerates cellulose, giving a diffractogram with more sharply pronounced maxima than that of water-soluble acetate, as shown in Fig. 46.

Thus, when all acetyl groups are detached, the crystallinity of the sample increases. These data agree well with the results of Renby and Noe [39], who by hydrolysis of water-soluble cellulose acetate in aqueous buffer solution at pH=9-10 and at different temperatures obtained precipitates having an X-ray pattern of highly crystalline cellulose.

In connection with these data on the structure of the products obtained by the stages of deep hydrolysis of cellulose acetate, it is interesting to note the work of Conrade and Chrisley [40].

The authors also used X-ray diffractography to study the effect of the degree of substitution, but not during saponification, but during heterogeneous acetylation on the structure of cellulose acetate. It was shown that the crystallinity of the partially substituted products decreased with increasing degree of substitution. This is very clearly shown in Fig. 47-a.

The figure shows that the characteristic maxima of native cellulose 14.6; 16.3 and 22.6° (respectively in 101, 102, 002 planes) at angle 20 decrease in intensity with the process of acetylation, but all reflexes retain their position to the degree of substitution 1.59 (29.2% As).

The complete disappearance of crystallinity upon further acetylation can be seen in Figure 3,17-6, which shows a series of diffractograms of cellulose acetates up to a substitution degree of 2.93.

Comparing the two figures we can come to the following conclusion that during acetylation of cotton cellulose the crystallinity of the structure is more or less maintained up to substitution degree 2.73, and starting from substitution degree 2.73 and up to 3.0 the structure is completely amorphous. Comparing with the obtained data it can be noted that in contrast to the heterogeneous acetylation products, water-soluble acetates are characterized by an amorphous structure at a much lower degree of substitution.

The presence of some acetate groups prevents the formation of a crystal structure in water-soluble acetates, and free hydroxyl groups fix the resulting structure. The formation of hydrogen bonds is one of the reasons, which probably limits the phase transformation of cellulose by fixing, as already mentioned, any structure.

Therefore, it seemed very important to obtain information about the state of hydrogen bonds in water-soluble cellulose acetates.

The mechanism of formation of inter- and intra-molecular bonds was described in detail in chapter 1 of the book.

In connection with the previously discussed X-ray diffraction structure of water-soluble cellulose acetates, however, it should be noted that their amorphous structure favors the formation of many types of hydrogen bonds, while in the crystalline state certain types of bonds should be assumed. A significant number of works have been devoted to the study of hydrogen bonds in cellulose ethers by infrared spectroscopy [41-44]. It was shown that there is a certain correlation between the character of the valence vibration band of hydroxyl groups in the IR spectrum of cellulose ether and the properties of this ether.

Thus, Baun, Holiday, Trotter [41], Gerbo [42], Zhabankov [43], Mironov et al. [44] showed that the viscosity of acetates and other cellulose ethers depends on the homogeneity of substitution of these cellulose ethers. It turned out that the latter characteristic is also related to the shape of the valence vibration band of hydroxyl groups in the IR spectrum of these derivatives.

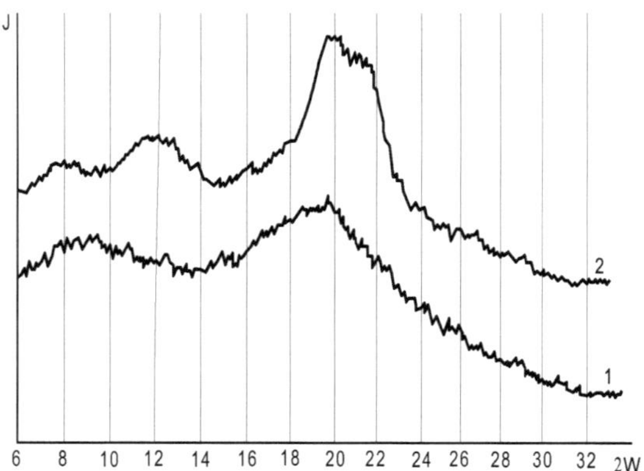

Fig.46. X-ray radiographs of water-soluble acetyl cellulose (1) and fully hydrolyzed cellulose acetate - regenerated cellulose (2).

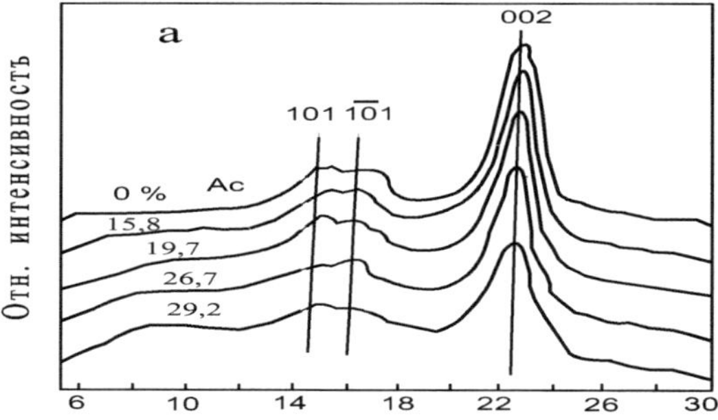

120

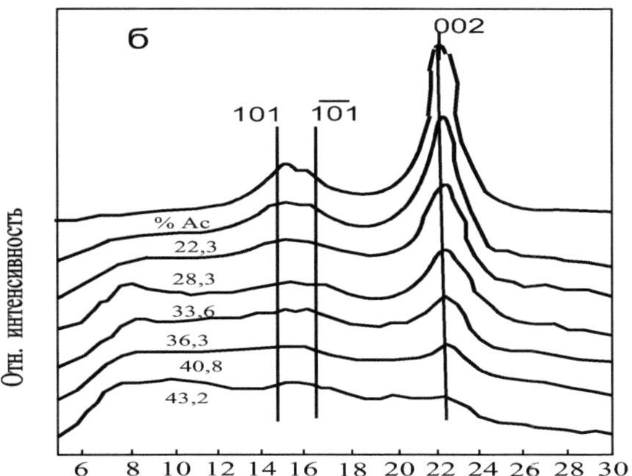

Fig.47. X-ray diffractograms of partially substituted acetates.

Brown suggested using the symmetry index of the hydroxyl group band to characterize the homogeneity of the cellulose ether, which is the ratio of the half-width of the hydroxyl group valence band on the low frequency side to the half-width on the high frequency side, as seen in Fig. 48.

In a heterogeneously substituted product, i.e., a product in which unsubstituted hydroxyls are arranged in groups, there may be associates of hydroxyl groups with sharply different hydrogen bond strengths. In a "homogeneous" product hydroxyl groups are randomly placed along the chain, the distribution of associates with different hydrogen bond strengths will obey the law of normal distribution.

Assuming that the curve bounding the contour of the OH-group absorption band is a distribution curve, we should expect that a statistical normal distribution yields a symmetric distribution curve, while a deviation from the normal distribution yields a non-symmetric curve.

This provision was taken as the basis for the spectroscopic study of water-soluble cellulose acetates. Measurement of the symmetry index of the valence band by hydrolysis stages in obtaining the water-soluble product yielded the results, which are shown in Table 18.

As can be seen from the data in the table, the symmetry index of the OH-group vibration band during deep hydrolysis changes little and is close to unity and, thus, characterizes the symmetric absorption band. In contrast, during heterogeneous acetylation, as was shown by Hurtubize

[45], this index changes considerably from 1.6 to 0.9 in the full range of substitution degrees, as can be seen in Fig. 49.

These data also fit well with those of Gerbo [42] and Mironov [44], who also showed that the character of the symmetry index change depending on the distribution of OH-groups in relatively highly substituted cellulose acetates is related to the homogeneity of substitution.

Thus, we can assume that the symmetry index really characterizes the uniformity of the distribution of hydroxyl groups, because during homogeneous hydrolysis we should note exactly this distribution.

Fig.48. Determination of the symmetry index of the valence oscillation band of the OH-groups.

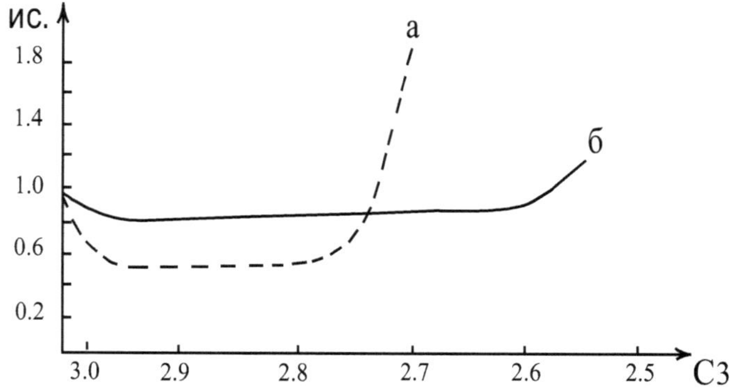

Figure 49: Changes in the symmetry index during heterogeneous acetylation (a) and homogeneous acetylation (b).

Table 18.

**Characteristics of the hydroxyl group band in
depending on the degree of hydrolysis of cellulose acetate.**

Content of bound acetic acid, %	Symmetry index in homogeneous hydrolysis	Maximum position ν_{OH} See^{-1}
59,95	1,00	3452
51,78	1,03	3455
40,95	1,08	3450
34,00	1,12	3420
22,81	1,12	3442
17,74	1,16	3440

Table 18 shows that water-soluble cellulose acetates have a maximum ν position$_{OH}$ at 3440 cm^{-1} . Other researchers [41] have shown that regenerated cellulose is also characterized by the OH band at 3443 cm^{-1} , i.e., weaker hydrogen bonds than those of natural cellulose. In this connection, the average energy of hydrogen bonds depending on the degree of substitution was determined.

The average energy of hydrogen bonds was calculated using the formula [46].

$$E = \frac{1}{K} - \frac{\nu_0 - \nu}{\nu} = -\frac{1}{K} \cdot \frac{\Delta\nu}{\nu_0} \text{ , kJ}$$

where: K is a constant equal to $6.69 \cdot 10^{-2}$ to J^{-1}

ν, ν_0 - frequency of unbiased and biased bands.

The results are shown in Table 19.

As can be seen from the data in the table, the hydrogen bonding energy of water-soluble acetylcellulose is slightly lower than the hydrogen bonding energy of regenerated cellulose.

Table 19.

**Change in the average energy of hydrogen bonding in the product
deep hydrolysis of triacetylcellulose depending on the degree of
substitution.**

Sample name	Degree of substitution, γ	E, kJ/mol

123

Initial acetyl cellulose	297	9,82
Partial hydrolysis product	237	11,07
-"- -"-	210	11,45
-"- -"-	130	12,16
-"- -"-	85	13,58
Water-soluble acetylcellulose	58	15,38
Regenerated pulp	0	16,09
Source pulp	0	17,89

It should be said, however, that for the statistical normal distribution of hydrogen bonds in cellulose, it is obviously not only the random distribution of hydroxyl groups in the cellulose ether macromolecule that is most important, but also the amorphous structure of this cellulose ether.

However, these two factors are mutually related.

Effect of functional groups on the solubility of water-soluble acetylcellulose.

Water-soluble cellulose ethers, including acetyl cellulose, have at least two types of functional groups. In the case of acetyl cellulose they are hydroxyl group and acetyl groups. The water solubility of cellulose ethers is related to the interaction of its functional groups with the solvent. However, this issue is far from being adequately studied. The interaction of hydroxyl groups and water molecules is beyond doubt. However, the interaction of other functional groups with water is not so clear. The role of these functional groups in ensuring the solubility of cellulose ether in water may also lie in the fact that they are some "spacers" preventing the interaction of hydroxyl groups in acetylcellulose and thus ensuring the interaction of these groups with water dipoles.

The structure factor affecting solubility has already been discussed above.

It was of great interest to clarify the role of acetyl groups in the dissolution of these products in water. This question was solved by comparing the properties of deeply hydrolyzed cellulose acetates with the properties of propionate cellulose hydrolysis products.

This cellulose ether differs from cellulose acetates only in that it has a radical length in the ether increased by one methylene group - CH_2 .

Comparison of the properties of acetylcellulose hydrolysis products with

Products with different contents of bound propionic groups were obtained by deep hydrolysis of cellulose tripropionate in a homogeneous medium /47/.

It is known that cellulose tripropionate is more difficult to obtain compared to triacetyl cellulose. This is explained by the fact that the reactivity of organic acids used to obtain cellulose esters decreases with increasing molecular weight of the acid, probably due to the influence of steric factors.

In order to increase the reactivity of cellulose in the esterification process, activation of cellulose with ethylenediamine followed by its displacement by repeated washing first with acetic, then propionic acid /47/ was carried out. The activated cellulose was esterified with a mixture consisting of six weight parts of propionic acid anhydride and 0.06 weight parts of sulfuric acid catalyst (6% of the weight of the cellulose).

The esterification was performed at $+288^0$ K. The main part of the reaction proceeds for 1-3600 s. By the end of this time the reaction product is almost completely dissolved in the reaction mixture. The mixture was left further at room temperature (4-4.5)-3600 s to complete the reaction and dissolve completely.

To the thus obtained viscous transparent solution of cellulose tripropionate, water (in excess) was added to destroy the unreacted propionic anhydride.

To accelerate the hydration of propionic anhydride the solution after adding water was heated at 353°C for 2x3500 s. The available sulfuric acid was then neutralized by adding propionate-Na. Then a certain amount of concentrated hydrochloric acid was added to the solution. This acid was used as a catalyst in the hydrolysis process, as preliminary experiments have shown that the rate of hydrolysis of cellulose propionate in the mixture propionic acid - water in the presence of $H_2 SO_4$ is much lower than the rate of hydrolysis (under similar conditions) of cellulose acetate in the mixture acetic acid - water. Thus, if the hydrolysis of cellulose triacetate product with $\tau = 55$ was obtained at 313^0 K for 250-3600 s, while the hydrolysis of cellulose tripropionate required more than 500-3600 s. hydrolysis time.

In the experiment with HCl catalyst, cellulose propionate underwent hydrolysis, which contained 65.5% of bound propionic acid (theoretically, the three-substituted product should have 67.2% of bound propionic acid).

The initial concentration of cellulose tripropionate in the solution was 10%, the HCl concentration was 0.74% and the propionic acid concentration was 86%.

Hydrolysis was carried out by heating the solution at 313^0 K under continuous stirring [47].

The hydrolysis of cellulose tripropionate changes its solubility in the reaction mixture. To maintain constant solubility and hydrolysis water was added to the solution at certain intervals. The course of hydrolysis was monitored by titrating separate samples of the solution with methyl ethyl ketone and water (the same way as was done in acetyl cellulose hydrolysis) and by determining the bound propionic acid in the reaction products. For this purpose, solution samples were taken and cellulose propionates with different depths of hydrolysis were precipitated (with water or acetone). The content of bound propionic acid in them was determined by the alkaline saponification method.

To determine the degree of polymerization, celluloses were regenerated from their respective esters by mild saponification with 0.5 N sodium methylate solution in anhydrous methanol in the same way as in the case of cellulose acetates.

Regenerated cellulose was further used to determine the degree of polymerization. The degree of polymerization of cellulose was determined by the viscosity of its solution in the iron sodium complex (IVNC)-[C H 0_{436})-Fe] - Na$_6$.

Fig. 50 shows curves expressing changes in the content of bound propionic acid in propionate cellulose during deep hydrolysis at 313^0 K and changes in stability of solutions with respect to water and methyl ethyl ketone.

Data on the change in the content of bound propionic acid as a function of time show that when using as a catalyst hydrochloric acid (8.6% by weight of tripropionate cellulose) can hydrolyze propionate cellulose to the content of bound propionic acid ~20% ($\approx$50), for 200-300 seconds, ie for the same time as the cellulose acetate when hydrolyzing the latter with a catalyst - sulfuric acid.

Kinetics of the hydrolysis of tripropionate cellulose as well as acetyl cellulose can be calculated starting from a certain time of hydrolysis (corresponding to the content of bound propionic acid–40%) formally by the 1st order reaction equation. The calculation was made according to the change in the amount of bound propionic acid as a function of time. The data are presented in Table 20.

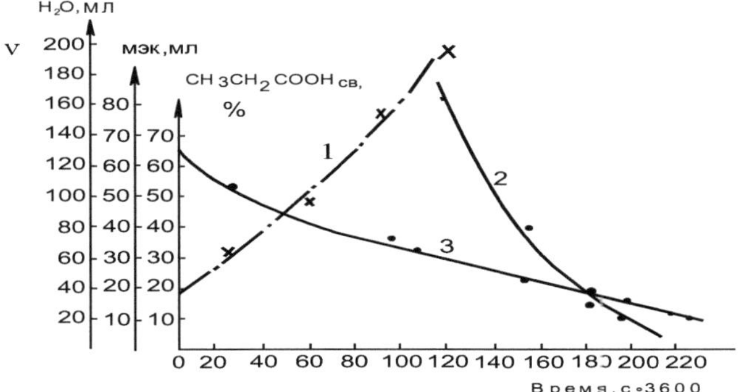

Fig.50. Variation of the bound propionic acid content of the water resistance values and the methyl ethyl ketone stability value with the hydrolysis time at 313°K. 1-N₂ O;   2- MEC; 3- CH₃ CH₂ COOH;

These data show the constancy of K value. Simultaneously with the hydrolysis of the ester, i.e. the detachment of propionate groups, there is a cleavage of the glucoside bond in the macromolecule of cellulose propionate. It should be noted that the formation of cellulose propionate already goes with a significant destruction of the cellulose molecule. In this case, this destruction proceeds to a greater extent than in acetylation of cellulose. Thus, in our case, the original cellulose, which had a degree of polymerization of 1400, had a degree of polymerization of 170 after esterification. The same phenomenon was also noted by other authors. So, in [48] Iwao Terasaki studying the reaction of cellulose with propionic acid showed significant destruction of cellulose and reduction of the degree of polymerization in the final product to 100.

Table 20.

Kinetics of deep hydrolysis of cellulose propionate at 313⁰ K.

τ -3600 c	CH₃ CH₂ COOH sv, %	By, c⁻¹ -3600 s	By, cp⁻¹ -3600 s
80	40,13	0,007181	
104	34,66	0,007608	
134	29,29	0,008136	
155	26,31	0,007532	$(7,84\pm0,83)\cdot10^{-3}$
184	21,86	0,007806	

127

201	19,37	0,008543	
218	17,07	0,008367	
227	16,15	0,007549	

It was of interest to determine the size of the degradation and the rate constant of this process during the hydrolysis reaction. The change in the degree of polymerization of cellulose propionate at 323°K with the hydrolysis time is presented in Table 21.

Table 21.

Rate constants of cellulose propionate degradation in the process of hydrolysis at 323°K.

№/№ P/S.	Time,hydr., τ -3600s	$|\eta|$	SP	K, c^{-1} - 3600 s	$To_{cp,}$ from^{-1} -3600
1.	0	0,136	167	-	
2.	24	0,114	140	$9,3 \cdot 10^{-5}$	$(9,49 \pm 0,09) \cdot 10^{-5}$
3.	82	0,082	101	$9,5 \cdot 10^{-5}$	
4.	130	0,066	81	$9,5 \cdot 10^{-5}$	

To determine the rate constant of the depolymerization reaction, the same assumption was made as in the acid hydrolysis of acetylcellulose.

As shown above, as the acetyl groups in the cellulose triacetate are detached in solution, the resulting products become soluble in mixed solvents containing water in increasing proportions, then in pure water, and finally they stop dissolving as well as pure cellulose.

The physical meaning of this process is probably as follows. Acetate groups at a high degree of substitution of hydroxyl groups of cellulose prevent, due to its hydrophobic nature, dissolution of ether in water.

When these groups are detached, the hydrophobic "pockmark" of these groups along the chain molecule in cellulose becomes sparser. Water molecules become able to penetrate to the free hydroxyls of the partially substituted cellulose ether and interact energetically with them to form hydrate shells, allowing dissolution first in mixtures containing water, and then in pure water. At the same time, the remaining ether groups still serve as "spacers" which prevent the cellulose chains from coming so close together, where strong hydrogen bonds between hydroxyl groups of neighboring cellulose chains arise. These remaining ester groups must be distributed statistically along the macromolecule. When a large number of these strands are removed, hydrogen bonds between the hydroxyls on separate sections of the long chain molecules arise, which act as strong cross-links that form a cross-linked structure. A product such as cellulose is not soluble in pure water.

As the results of experiments have shown, cellulose tripropionate during its mild hydrolysis in solution behaves in the beginning quite similarly to cellulose acetate. This is clearly seen in Fig 50, which shows the change in water resistance of cellulose propionate solutions with hydrolysis.

However, when a certain degree of hydrolysis is reached, further detachment of propionate groups does not lead to dissolution of the products in mixtures richer in water than the mixtures in which the product was soluble up to this point of hydrolysis. Hydrolyzed cellulose propionates containing 15% or less of bound propionic acid become insoluble in none of the mixtures containing water. It is not possible to obtain a product soluble in water. The behavior of cellulose propionates in this sense is opposite to that of cellulose acetates, which, at 16-18% of bound acetic acid become well soluble in water. The above phenomenon may be explained by the following circumstances. The propionic acid residue bound to cellulose has a more hydrophobic character. This general definition can be specified, however, when considering the nature of chemical bonds of propionic acid and acetic acid residues in the corresponding cellulose ethers.

In two connections:

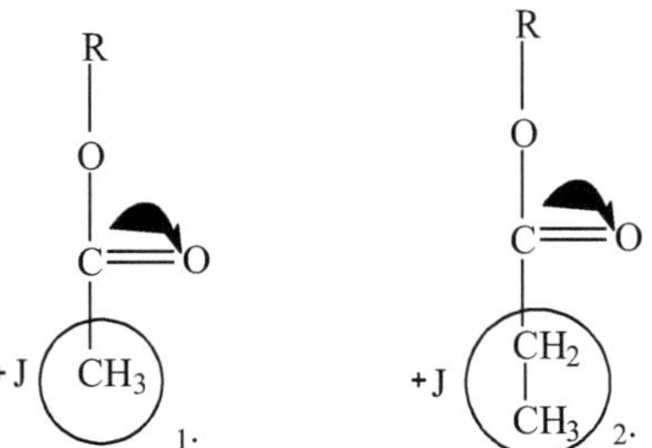

the reactivity of the carbonyl group, based on its polarity, will be different. This difference will result from the fact that the polarization π-bond $\diagup\!\!\!\diagdown C = O$ is more pronounced in compound 1 than in compound 2 because the +J effect of the end group CH -CH$_{23}$ is greater than that of the -CH group$_3$. As we know, the induction effect of substituents can be used

129

as the induction constant of substituents δ^*. If δ^* for methyl radicals is 0.00, then for ethyl radicals it is 0.10. Negative values of δ^* correspond to +J - effect of substituents /49, p.77-181/.

As a result, in the case of compound 1, there can be a chemical addition of a nucleophilic reagent to the carbonyl group of the carbonyl group

water with the formation of hydrates:

$$\ce{>C=O + H-\overset{-}{O}-H <=> >C<^{-OH}_{OH}}$$

or the formation of hydrogen bonds of the oxygen atom with water dipoles:

$$\ce{>C=O\cdots\cdots\cdots\cdots H-OH}$$

The formation of such hydrates, probably due to steric reasons, takes place in acetate cellulose only at a certain density of acetate radicals on the cellulose chain, i.e. at a certain degree of acetyl cellulose hydrolysis.

In the case of cellulose propionate formation of hydrates at the acid radical obviously does not occur. This product at any degree of hydrolysis is insoluble in pure water, but deeply hydrolyzed cellulose propionate is soluble in water with a small addition (10%) of acetone or alcohol. Obviously, therefore, in the disassembled case of two hydrolyzed acyl cellulose derivatives, solubility is determined not only by physical factors (amorphous structure, uniform distribution of functional groups), but also by the need for solvation of both hydroxyl and ether groups.

Considering the data on the value of the degree of polymerization and propionates of cellulose and comparing them with the value of the degree of polymerization of HPAC, which is 250, also suggests that the molecular weight of cellulose ether is not a factor in this case, determining the solubility in water.

Solubility diagram of cellulose propionate with different contents of bound acid in mixtures acetone + water, alcohol + water, propyl alcohol water and propionic acid + water are shown in Fig. 51 (a-d). Figs. 51. a and b show two dissolution regions at room temperature and when heated to 343-373°K, since the solubility of hydrolyzed cellulose propionates increases with increasing temperature.

To confirm the assumptions made about the hydration of the acetic acid residue in cellulose acetate, IR spectra of a concentrated solution of acetylcellulose in heavy water were taken.(D_2O)

The spectra of acetylcellulose solutions in heavy water were investigated because the analytical bands of acetylcellulose cannot be studied in H_2O due to their overlap with the water bands.

Fig. 52 shows the IR spectrum of D_2O, and Fig. 53 shows the IR spectrum of a 10% solution of HPAC in D_2O. Comparison of these spectra, as well as the spectrum of HPAC in the dry state (see Fig. 40 above) shows that the 1750 cm^{-1} band relating to the valence vibrations of -C=O, which is present in the original acetylcellulose, shifts in the interaction with heavy water towards lower frequencies - to 1725 cm^{-1} .

This magnitude of displacement is significant and indicates the interaction of the acetate group with heavy water molecules, with the formation of hydrate forms discussed above.

Similar results were obtained later in the work of G.A. Petropavlovskii et al. [50] in a study of the solubility of cellulose, methyl and ethyl cellulose ethers, where it was also shown that the dissolution of partially substituted cellulose ethers in water occurs during hydrophilic interaction of the introduced alkoxyl radical with water. This interaction was directly proved by shifting the corresponding absorption bands in the IR spectrum during the interaction with D_2O.

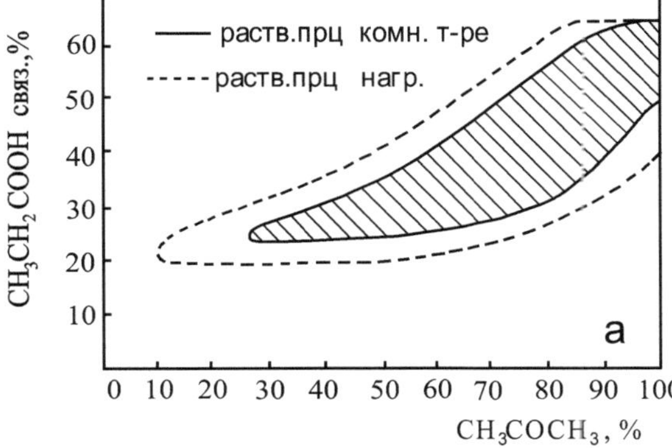

131

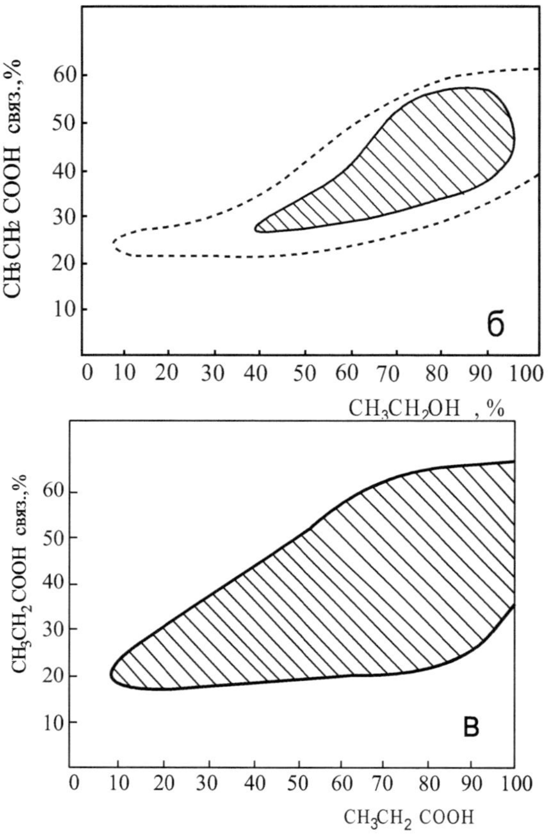

CH₃CH₂ COOH связ.,%
60
50
40
30
20
10
б
0 10 20 30 40 50 60 70 80 90 100
CH₃CH₂OH , %
CH₃CH₂COOH связ.,%
60
50
40
30
20
10
В
0 10 20 30 40 50 60 70 80 90 100
CH₃CH₂ COOH

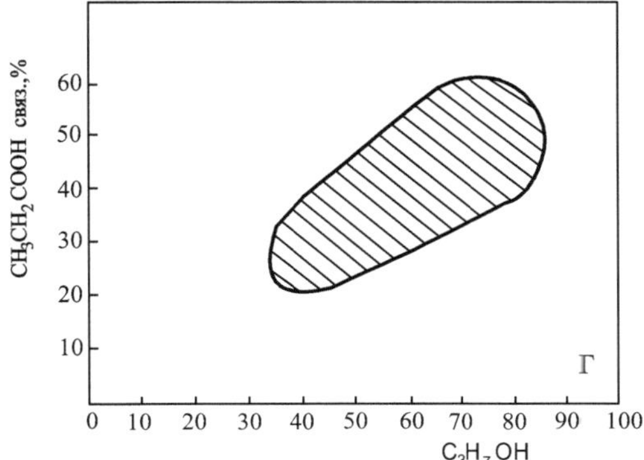

Fig.51 Solubility diagrams of hydrolysis products in various mixed solvents depending on propionic acid content: a-acetone + water; c-propionic acid + water; b- ethyl alcohol + water; d - propionic alcohol + water.

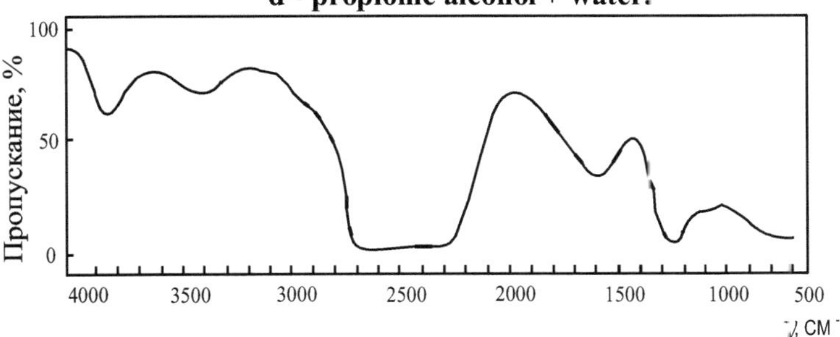

Fig.52. IR spectrum of heavy water (D₂ O)

Fig.53. IR spectrum of 10% water-soluble acetylcellulose solution in D O.$_2$

Properties of water-soluble cellulose acetates.

As can be seen from the above, water-soluble cellulose acetates, unlike other water-soluble cellulose ethers, are compounds in which the ester bond can be easily cleaved under mild conditions. Saponification of acetyl cellulose with alkali solutions makes it possible to obtain regenerated cellulose whose macromolecule degree of polymerization is close to that of the original product. In view of this, it was of interest to study exactly those properties of acetylcellulose, which are associated with this difference from other water-soluble cellulose ethers.

In particular, it is of great interest to study the possibility of regulating the structure of films obtained from water-soluble acetylcellulose by influencing the solvent and changing the content of ester groups, while keeping the molecular weight and molecular mass distribution constant.

Another important point is the ability to perform chemical reactions with water-soluble reagents. In this case, water-soluble acetylcellulose can model cellulose in such homogeneous reactions that cannot be performed due to the insolubility of cellulose (for example, "cross-linking" reactions).

In this connection, the properties of various films obtained from water-soluble acetylcellulose and the chemical interaction of acetylcellulose (in the form of films) with a bifunctional reagent were considered.

It is well known that all important mechanical properties of polymers, especially tensile strength, elongation at rupture, impact strength, elasticity, etc., depend largely on the average molecular weight or average degree of polymerization [57]. With increasing degree of polymerization, starting from a certain value, there is a sharp increase in mechanical characteristics until reaching some high values of molecular weight, above which the increase in strength becomes insignificant. Each polymer has, for example, its critical SP value, below which it has no mechanical strength.

For polymers, the absolute values of the critical degree of polymerization vary; polyamides have some strength already at SP=40, cellulose at SP=60, and many vinyl polymers SP=100. In fact, all polymers used in practice are in the SP range of 200-2000, which usually corresponds to molecular weight of 20,000-200,000. HPACs, as shown above, have a polymerization degree of 200-250.

But in addition to the molecular weight, the properties of films and fibers are greatly influenced by their structural organization, which may be different under different molding conditions. Although there are a large number of works [52-54], in which an attempt was made to establish the relationship between the supramolecular structure of the polymer and the mechanical properties of the fiber, and significant progress has been made in the formation of high-strength fibers, this issue is still unresolved completely. In a number of cases, polymer films with an amorphous supramolecular structure have been shown [53-54] to have higher mechanical properties than crystalline films. This can be explained by the fact that in partially crystallized polymers, large internal stresses arise under mechanical action, which lead to crack growth and material failure. In an amorphous polymer, these stresses are distributed more everly. Of course, we are talking here about non-oriented fibers and films.

In this case, the question of the relationship between radiographic structure and properties is addressed for the first time by a series of successive structure transformations without changing important indicators such as molecular weight or molecular weight distribution.

The study was first performed on HPACs in the form of films, which also allowed us to determine the effect of functional groups, stereoregularity of the structure, and the influence of the solvent on the structure and strength of the films.

The HPAC had a degree of substitution of 0.5 (γ =50) and a degree of polymerization of SP=200-250. The HPAC films were obtained from a 5% aqueous solution at room temperature.

X-ray diffractogram of such a film is shown in Fig.54. upper curve. It is quite clear that the diffractogram characterizes the amorphous structure, having two broad blurred maxima, at the same 2θ angles as for the film of ordinary cellophane. It should be noted here that the acetate and hydroxyl groups present in water-soluble acetyl cellulose violate the stereoregularity of the macromolecules, so that this product obviously cannot form a highly ordered system. Such an amorphized film has the following strength: tensile strength 98.0-107.0 MPa and elongation at break of 15-20%.

As can be seen from the relatively low degree of polymerization (close to the critical value) the film has a fairly high mechanical strength. A film of water-soluble acetyl cellulose can be transformed under mild conditions into a film of regenerated cellulose. This operation can be performed in both aqueous and nonaqueous media. Saponification in aqueous medium was accomplished as follows. The HPAC film was incubated in a desiccator under aqueous ammonia solution vapor for 6x600s and then transferred to a slightly alkaline buffer solution (pH=9), where it was incubated for 6x36000s at room temperature. After that, the film was washed with distilled water to neutral reaction and air dried. Saponification of the film in a non-aqueous medium was carried out by treating the film with a solution of Na-methylate in methyl alcohol at room temperature. Completeness of saponification was checked by determining the content of bound acetic acid by alkaline saponification as well as by IR spectroscopy. Under the above conditions of saponification, the acetyl groups are completely detached. A diffractogram of the hydrate cellulose film regenerated in the presence of water is shown in the same figure with curve 2.

As can be seen, this diffractogram has changed little compared to the original one, however, the strength of such films changes.

Their tensile strength becomes 107.0-127.4 MPa and elongation at break 24-27%. Since the molecular weight did not change during this process, nor did the X-ray patterns change significantly, the change in mechanical properties should be attributed to a change in the chemical structure of the product. The 50 acetyl groups that accounted for an average of 100 glucose residues were detached, and hydroxyl groups replaced them. It should be assumed that this increased the intermolecular interaction between the chains and thus led to an increase in strength.

Hence, the role of polar groups in the polymer macromolecule, which carry out dipole-dipole interaction, is clear. The role of these groups (hydroxyl groups in this case) is also to fix a certain mutual arrangement of macromolecules, which is evident from the fact that the detachment of acetyl groups in aqueous medium does not lead to a change in the radiographic structure. Only many hours (6-600 s) heating at water boiling point of this saponified film leads to a different radiographic structure, namely, the structure of cellulose II, which can be clearly seen from curve 3.

At the same time, the strength also changed quite naturally. Tensile strength 137.0-157.0 MPa, and elongation slightly decreased 18-21% as the ordering of the system increased. However, these changes in strength are not large.

It is especially important to note that the elasticity of the film changes little as the ordering of the system increases. This fact is due to the fact that the intermolecular forces due to the interaction of hydroxyl groups played a role here as well. This fact emphasizes the importance of the prior creation of a certain network of intermolecular forces in order to obtain an advantageous supramolecular structure. The amorphous structure with respect to elongation at rupture, is the most advantageous, as can be seen from curve 4 and the mechanical strength data corresponding to this type of structure. Saponification in a non-aqueous medium allows to obtain from HPAC a film characterized by an even more amorphous structure (curve 4) and having a tensile strength of 163.0-186.0 MPa at 30-35% elongation, i.e. the highest strength that could be obtained in this case.

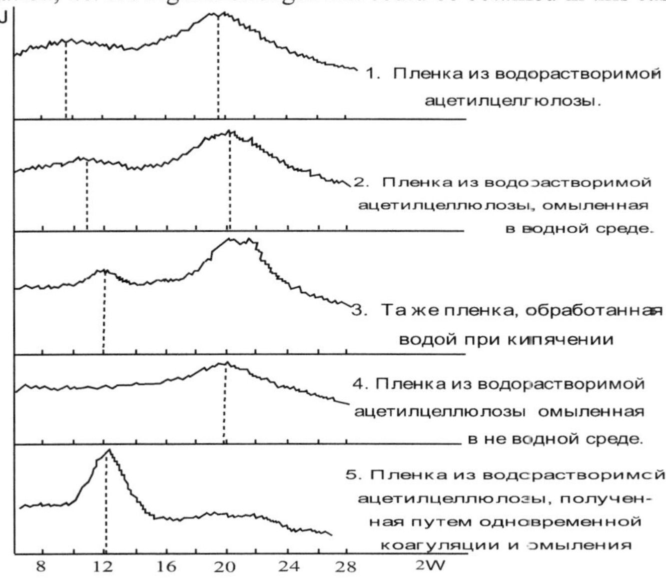

Fig.54. X-ray diffractogram of films obtained by different saponification methods.

We cannot say that this structure is always preferable, but it can be argued that the creation of such a structure and a certain density of strong intermolecular bonds in it, is the necessary first step of forming a desirable (to obtain well-stretchable films) supramolecular structure.

This is confirmed by the data cited in [55,56], which showed that the introduction of small amounts of bridging bonds (by crosslinking) can increase the orientation ability of the film, and thereby dramatically increase its drawing strength by reducing chain slippage and premature fracture by the weakest bonds.

A clear example is the experience of obtaining a film from HPAC (from its aqueous solution) by simultaneous coagulation and saponification. Such a film was obtained as follows. An aqueous solution of acetylcellulose was applied in an even thin layer to a glass plate which was placed in a desiccator over vaporous aqueous ammonia solution. This first caused gelatinization of the solution and then syneresis of the gel. After that, the gel film thus formed was washed with water to neutral reaction and dried. The structure of the resulting film is shown in diffractogram 5. If we compare this diffractogram with the diffractogram of conventional hydrate-cellulose films obtained by the wet forming method, we can see that they are very similar.

This diffractogram has an anomalously high maximum at 12^0 20, obviously related to the planar orientation in these films.

The mechanical strength of such films is the lowest, especially the elongation at break of 3-5%.

This is explained by the fact that the formation of such a structure occurred when there was no previously created network of intermolecular forces. This led to the creation of large supramolecular formations with the structure of cellulose II having relatively low perfection of crystallinity.

3.2.1. Formation of fibers based on water-soluble cellulose acetates.

At present, the chemical industry, along with large-tonnage production of fibers for textile purposes, develops small-tonnage, but very important, production of fibers for special purposes and, above all, for medical purposes.

We must assume that the studied properties of water-soluble cellulose acetates suggest that it is appropriate to use acetylcellulose fibers for the latter purposes.

Therefore, the properties of HPAC fibers used as suture, tamponage materials and polymeric bases for drugs with antimicrobial properties were studied. Along with this, the general conditions of molding water-soluble fibers from aqueous HPAC solutions were studied.

The stability of polymer solutions to the action of precipitators is an important factor in the manufacture of films and fiber molding [57,58].

The stability of polymer solutions is determined by the amount of precipitate required for polymer precipitation to begin. This technique is reduced to the titration of a certain volume of solution with a precipitant. Based on the titration results, the precipitation number was calculated according to the known formula /59/.

$$K = \frac{y}{y_0 + y} \cdot 100$$

where: Y,Wo - amount of precipitate and solution taken for titration, ml.

Stability of aqueous solutions of HPAC to active isopropyl alcohol is 54.5, to ethyl alcohol 63.0, acetone 64.0, methyl alcohol-65, 5, dioxane 82.0, and hexane, heptane, acetic acid, oxalic acid, tartaric acid do not cause coagulation.

These data show that organic nonsolvents act weakly on water-soluble acetylcellulose solutions, while isopropyl, ethyl, methyl alcohol and acetone are relatively strong nonsolvents. The most effective precipitant is isopropyl alcohol.

Studies of the effect of water on the precipitating ability of isopropyl alcohol showed that the precipitating ability of isopropyl alcohol decreases with increasing amount of water. This seems to be explained by the fact that HPAC has a large number of free hydroxyl groups ($\overline{Y}$=250) and as the amount of water in the precipitant increases, the solubility of the polymer increases, leading to an increase in the number of precipitations.

In /60,61/ the resistance of aqueous and alkaline solutions of PVS and CMC to the action of various inorganic substances, in particular, various mineral acids and salts, was studied. In this regard, the resistance of aqueous solutions of HPAC to the action of some salts was studied; the following results were obtained /62/:

Salts	Concentratio n,g/l	Deposition rate
$MgSO_4$	70	55,0
$MgSO_4$	100	33,0
$MgSO_4$	150	21,0
$MgSO_4$	200	11,7
$CuSO_4$	150	33,6
$CuSO_4$	200	20,7
$CH_3 COONa$	250	20,7

NaCl	200	37,2
Na$_2$ CO$_3$	200	25,7
Ca(CH$_3$ COO)$_2$	220	25,0
CaCl$_2$	150+200	not coagul.
Al$_2$ (SO)$_{43}$	70+100	not coagul.

From this we can see that not all salts have the same precipitation effect, and in terms of precipitation ability they can be arranged as follows:

$MgSO_4 > CH_3 COONa > Ca(CH_3 COO)_2 > Na_2 CO_3 > CuSO_4 > NaCl$

The precipitating effect of salts, apparently, is explained by the fact that when they are introduced, the coagulation of HPAC solutions is associated with the dehydration of its molecules, due to the binding of the solvent precipitant. As the concentration of $MgSO_4$ increases the precipitation number decreases. The effects of organic and inorganic precipitators on water-acetone solutions of HPAC were also studied. The number of precipitation with increasing acetone content in the solvent (water) decreases with the action of organic precipitators and increases with the action of inorganic precipitators (Fig. 55). This can be explained by the fact that when acetone is added to the solvent, the efficiency of water use increases due to the strengthening of the hydrophobic nature of the precipitant. In salt precipitation due to solvation by acetone molecules of acetyl radicals in HPAC, polymer solubility in this system is improved, which leads to an increase in the number of precipitation.

Thus, the obtained results made it possible to establish some regularities in the coagulation process of HPACs and proceed to further studies directly in the conditions of fiber molding.

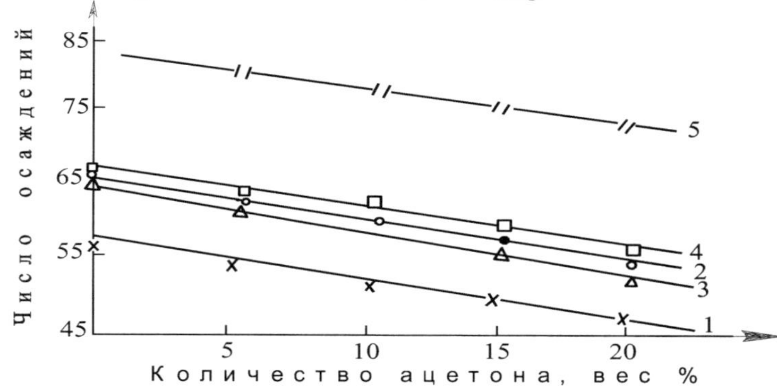

Fig.55. Stability of water-acetone solutions of HPACs to the action of organic precipitators.1-isopropyl alcohol; 2-ethyl alcohol; 3-acetone; 4-methyl alcohol; 5-dioxane.

But before proceeding to the formation of fibers and films, the properties of diluted and concentrated HPAC solutions were investigated.

Study of the properties of diluted and concentrated water-soluble acetylcellulose solutions.

Diluted solutions. It is known that the main factors determining the properties of diluted polymer solutions, including cellulose ethers, are the degree of polymerization, degree of esterification, temperature and thermodynamic "quality" of the solvent [63p.420-423]. Two solvents - water and dimethylformamide (DMF) have practical value for obtaining spinning solutions of HPAC, therefore viscosity properties of diluted (from 0.0625 to 0.50 g/100 ml) solutions of HPAC in water and DMF /64/ were investigated.

The characteristic viscosity in aqueous and DMF solutions was determined in a dilution viscometer at 298-343^0 K. The measurement results are shown in Table 22.

As can be seen from Fig. 57, the value of /η/ decreases with changes in temperature and solvent quality. It is known [65.p.233] that the characteristic viscosity is related to the size of macromolecules by the relation

$$|\eta| = \Phi \cdot \frac{(h^{-2})^{3/2}}{M}$$

Where: F is the Flory constant, the value is 2.22×10^{21} , if /η/ is measured in 100 ml/g, and the mean square distance between the ends of the macromolecules $(h^{-2})^{3/2}$- in centimeters

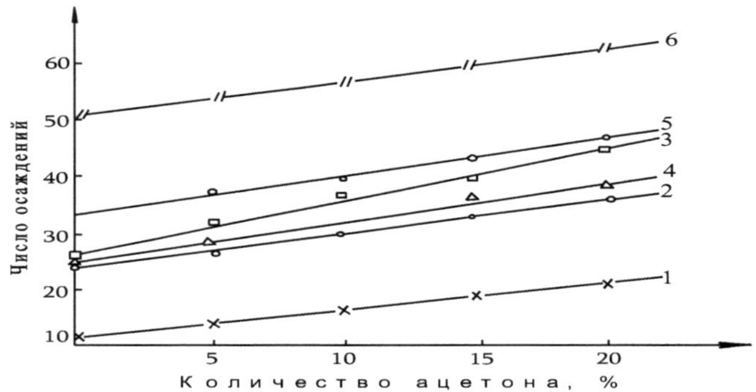

1.MgSO₄ ; 2.CH₃ COONa ; 3.Ca(CH₃ COO)₂; 4.Na₂ CO₃ ; 5.CuSO₄ ;
6.NaCl.

**Fig.56. Stability of water-acetone solutions of HPAC to the
action of inorganic precipitators.**

Table 22.

**Characteristic viscosity /η/ and Huggins' constant K' of water-
soluble acetylcellulose solutions.**

Temperature, K	/η/	K'	Temperature, K	/η/	K'
solvent - water			solvent - DMF		
298	1,45	0,60	298	1,28	0,57
303	1,34	0,58	303	1,21	0,51
313	1,24	0,55	313	1,12	0,44
323	1,05	0,51	323	1,02	0,44
333	0,89	0,37	333	0,99	0,32
343	0,79	0,31	343	0,95	0,28

Measurements of macromolecule sizes in solution are aimed at finding
out their dependence on the thermodynamic interaction of the polymer
with the solvent as well as on the temperature.

142

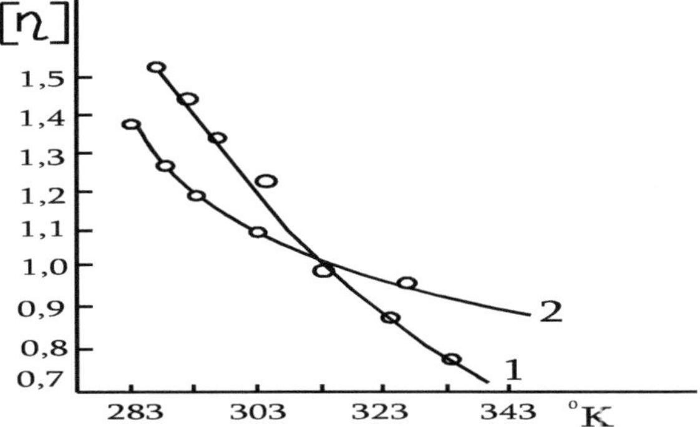

Fig. 57. Temperature dependence of the characteristic viscosity
of HPAC in aqueous solution (1) and in DMF solution (2).

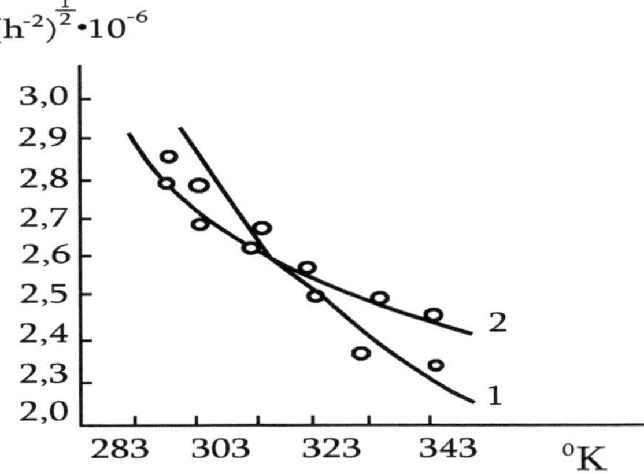

Fig. 58. Dependence of the RMS size of the HPAC molecular
ball of the molecule on temperature in aqueous solution (1) and
in DMF solution (2).

As can be seen from Fig. 58, with a change in the "quality" of the
solvent and temperature of the solution the size of the molecular ball of
HPACs changes. At the same time, the change of conformations seems to
be limited by the comparative rigidity of the cellulose chain.

To assess the intermolecular interaction in solutions, we calculated
the temperature coefficient of viscosity, which in turn can be used to

143

calculate the apparent activation energy of the viscous flow ΔE according to the equation similar to the Clapeyron-Clausius equation

$$\frac{d\ln\eta}{d(d_{\mathrm{T}})} = \frac{\Delta E}{R}$$

The value ΔE can be easily calculated graphically using the dependence $\ln\eta = f\,(1/T)$. As can be seen from the data given above, in particular Fig. 59 and Table 22, the viscosity of the solutions decreases with increasing temperature, i.e., the temperature coefficient of viscosity is positive.

As can be seen from Fig. 59, the value ΔE for HPAC solutions increases almost linearly with increasing polymer concentration. In addition, the change ΔE depends on both polymer concentration and temperature. As the temperature of the solution increases, the intensity of the intermolecular interaction is characterized by larger values, hence ΔE increases. This intensification of intermolecular contacts is due to the partial unfolding of macromolecules.

As shown in Table 22. with increasing temperature in almost all cases there is a decrease in the Huggins constant K', which indicates an increase in the equilibrium flexibility of macromolecules, which was calculated by the formula /63, p.414/.

$$K = \frac{\eta_{y\text{д}}\,/\,c - c[\eta]}{c[\eta]}$$

The K values calculated by this formula correlate well with the results of calculating the $(h^{-2})^{1/2}$. It is known that the viscosity of solutions changes over time due to structure formation, which is associated with the release in the form of the second phase of the fraction, insoluble in a given medium and at a given temperature. This in some cases leads to general gelation of the solution /68/. The change in the viscosity of diluted HPAC solutions depending on the storage time was investigated. As can be seen from Fig. 60. the value of the specific viscosity of diluted solutions practically does not change during 14-15 days, after which a slight decrease in viscosity, apparently due to degradation processes or other side effects, is observed. Thus, viscosity properties of diluted solutions of HPAC are determined (at given values) by Y) concentration, temperature and thermodynamic quality of the solvent.

Concentrated HPAC **solutions** in polar solvents (water, DMF, DMSO and ethylene glycol) are true, thermodynamically equilibrium structured systems with a well-defined anomaly of viscous properties manifested in non-Newtonian flow.

It is known that the structure of concentrated solutions due to the emergence of contacts between the fluctuating packs of solvated macromolecules to a certain extent and the flexibility of polymer chains. The properties of fibers and films based on cellulose ethers depend largely on the initial structure and characteristics of their solutions. Information on the rheological properties of water-soluble solutions of simple and complex cellulose ethers is almost absent in the literature /66/.

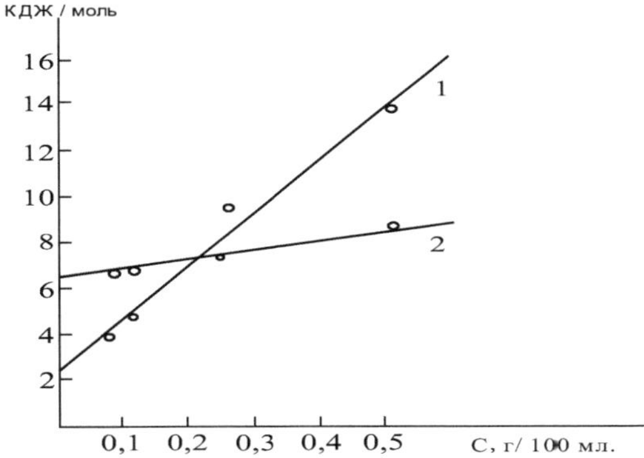

Fig. 59. Concentration dependence of the apparent activation energy of the viscosity flow of HPAC solutions in water (1) and in DMF (2).

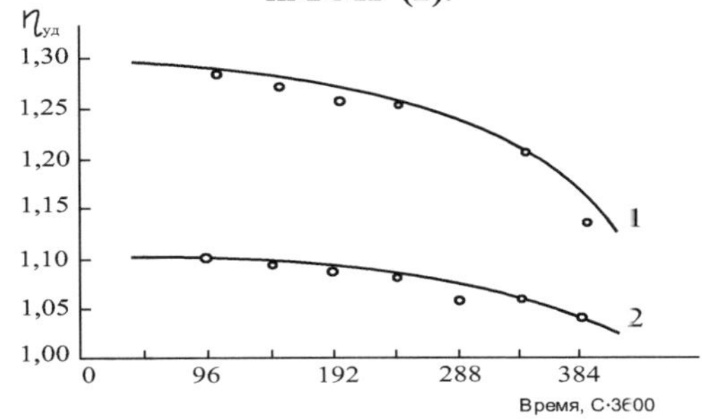

Fig.60. Changes in specific viscosity of HPAC solutions in water (1) and in DMF (2) depending on storage time.

145

In this regard, the rheological properties of HPACs were studied. The above-mentioned solvents were used as solvents: water, DMF as well as DMSO and ethylene glycol [67].

The viscous flow process of concentrated HPAC solutions was studied in the range of shear stresses $4 \cdot 10^{-1} - 1,1 \cdot 10^{-3}$Pa in the concentration range of 8-16% (weight) on the device "Rheostest-2" as a function of temperature, as well as the storage time of the solutions.

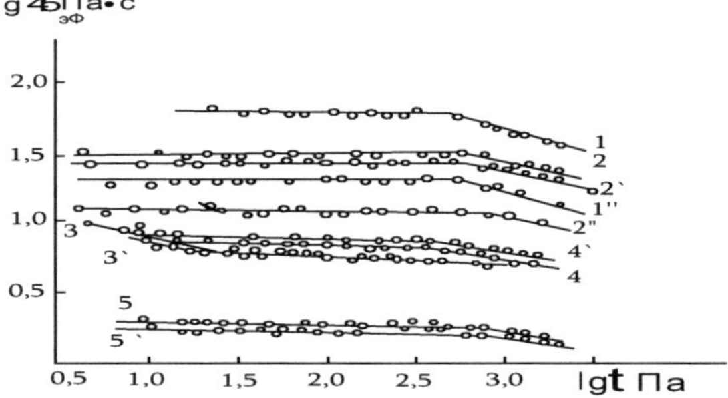

Fig.61. Flow curves (dependence of the logarithm of the effective viscosity on the logarithm of the BPAC shear stress) depending on the type of solvent, concentration and storage time of solutions.

T-ra (o_K): 1,2,2', 3,4,4', 5,5'-293;
1", 2" - 313
R-lys: 1,1*4,4'-H_2 0; 2,2', 2"-DMSO;3,3' 5,5'-DMF
HPAC k-value (%): 1,1", 2,2",3,3'-16; 4,4',5,5'-12.
2'-5'-flow curves were taken 10 days after preparation of the solutions

The values of the activation energy of the viscous flow were calculated (ΔE) and "viscous" volume (V^*) average kinetic unit sizes. The calculation (V^*) was performed using the Eiring formula [68]

$$V = \frac{2K \cdot \left[\frac{E}{R} - (\ln \eta - \ln A\tau) \cdot T \right]}{\tau}$$

η-viscosity, $A\tau$ - pre-exponential multiplier, which has the meaning of viscosity at infinitely high temperature: T-temperature, in K^O

τ - shear stress,
R-universal gas constant,
K is Boltzmann's constant.

146

HPAC forms viscous and transparent solutions in water. With time, as can be seen from Fig. 61 (curves 4.4.), there is some, though very slight, increase in the viscosity of aqueous solutions. The viscosity of HPAC solutions in DMSO and DMF kept for some time at room temperature decreases slightly (Fig. 61, curves 5.5^1, - 2.2^1). Dissolution of HPACs in EG occurs at temperatures above 343°K. The initial viscosity of 8% solutions in EG has a value of the order of 100.0 Pa-C and higher. When shear stresses are applied to them, there is a fairly sharp decrease in viscosity (Fig.62, curve 1), in one hour after stopping the device and measuring the same solution, the flow curve takes the form corresponding to curve 1-a (Fig. 62), when the solution is measured again immediately after stopping the device-2a. After a day there is a partial restoration of the solution structure subjected to the action of shear stress (curve 2, Fig.62). Higher viscosity of HPAC solutions in EG compared to solutions of this polymer in DMF, DMSO is apparently explained by higher affinity of the solvent and HPAC, formation of stronger intermolecular bonds. Decrease in temperature of HPAC solution in EG below 343 k leads to the formation of a strong gel.

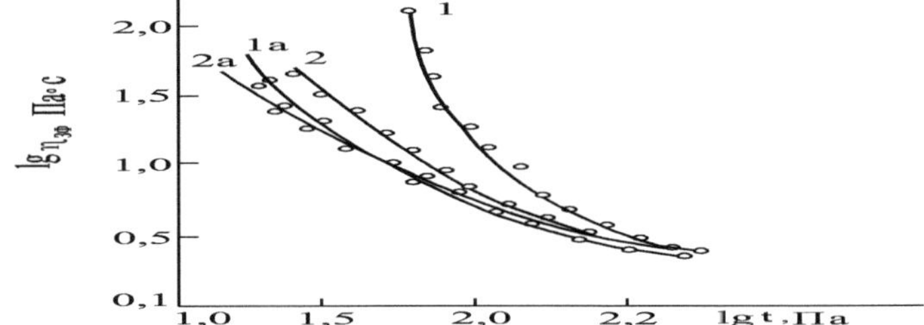

Fig. 62. Flow curves of HPAC in EG at 353°K and polymer concentration of 8 wt.%.

Solution changes: 1 - 24-3600s after preparation; 1a - 1-3600s after stopping the instrument; 2 - again after 24-3600s; 2a - 45-3600s after stopping the instrument.

147

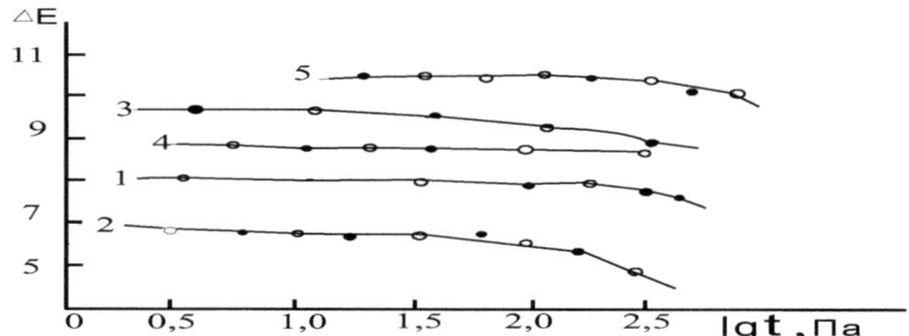

Fig. 63. Dependence of apparent activation energy of viscous flow of HPAC solutions on shear stress at different polymer concentrations and in different solvents.

Concentration (wt.(%): 1,2-8; 3,4 - 12; 5-16;

Solvents: 1,3,5 - water; 2 - DMF; 4 - DMSO.

The aqueous HPAC solutions, having higher viscosities as compared to the equi-concentrated dimethyl sulfoxide solutions, also have higher values of viscous flow activation energy, which is a quantitative measure of intermolecular interaction (Fig. 63 and see Table 23). However, in spite of this, the strength of the spatial structure of HPAC solutions in DMSO and DMF appears to be not much higher than that in water (Fig. 61). It may be due to the fact that macromolecules in water are largely unrolled, but the bonds are weaker than in DMP and DMSO where macromolecules are more coiled. With the increase τ the size of the kinetic units also decreases (see Table 23). The viscous volume increases with increasing temperature (see table). This is explained by the fact that as the temperature rises, macromolecules unfold, intermolecular interactions become easier and associates and bundles form more easily in the solution. However, the application of stress to these systems leads to gradual destruction of structural formations of supramolecular order.

The above experimental data show that the values ΔE and V^* increase with increasing solution concentration (Table 23 and Figure 63). Increasing the polymer concentration in the solutions increases the probability of intermolecular contacts and the formation of stronger supramolecular formations with less mobility and flexibility.

The value V^*, which is a measure of the mobility of structural elements and their sizes, is of great importance in aqueous solutions and less in dimethioformamide solutions, i.e. $V^*H\,O_2 > V^*ДМСО > V^*DMF$. This means that less mobile and larger DMF-sized structural elements (associates, bundles) are formed in water than in DMSO and DMF.

148

In connection with the above, we can assume that in aqueous HPAC solutions, the formation of complex supramolecular structures is easier than in dimethylsulfoxide and dimethylformamide solutions, so in aqueous solutions, a solid structural network will rather arise, and the solution will rather go to the gel-like state.

Earlier the results of researches of stability of 0.5% aqueous, dimethylformamide and water-acetone solutions of HPAC to action of organic and inorganic nonsolvents were given. It was found that isopropyl alcohol and acetone are stronger precipitators for the above mentioned HPAC solutions. By comparing the precipitation ability of aqueous solutions of some salts and acids from aqueous and aqueous-acetone solutions, it was shown that some salts also have the highest precipitation ability. However, due to the difficulty of cleaning the fiber from the remaining salts, the use of inorganic salts as precipitation baths seems inexpedient.

For further selection of optimal molding conditions we studied: the influence of spinning solution temperature, composition and temperature of sedimentation bath, nozzle and plasticizing drawings on physical and mechanical properties of water-soluble fiber. These series are

Table 23.

The value of the apparent activation energy of the viscous flow /ΔE/ at $\tau = 100$ Pa and "viscous volume" / $V*10^{16}$, cm³ / at different shear stresses /τ/.

Solvent	Concentration, %	$T,^0 K$	ΔE_{100} кДж/моль	V*-shear stresses τ, Pa						
				3.16	10.0	131.6	100.0	177.8	316.8	1000.0
Water	80	293	33,44	263	88	28	8,8	4,9	2,6	-
		313		284	94	30	9,5	5,3	2,9	-
		333		303	101	32	10,2	5,6	3,1	-
		353		318	106	34	10,7	5,9	3,1	-
	12	293	38.03	342	101	30	9,3	-	2,9	-
		313		366	107	33	9,9	-	3,2	-
	16	293	42.21	-	-	28	9,1	5,1	3,5	0,84
		313		-	-	38	12,3	6,9	3,9	1,15
		333		-	-	39	12,6	7,1	4,0	1,17
		353		-	-	40	12,9	7,3	4,1	1,19
DMSO	8	293	33,44	260	85	30	5,5	5,0	2,5	-
		313		271	91	32	8,5	5,2	2,9	-
		333		300	97	35	9,9	5,5	3,0	-
	12	293	36,37	-	88	28	8,8	4,6	2,0	-

		313		-	94	30	9,4	4,9	3,0	-
		333		8,7	99	32	9,9	5,3	3,2	-
		353		-	106	34	10,6	5,6	3,4	-
	16	293	38,87	-	106	33	10,9	-	3,5	-
		313		-	109	34	11,1	-	3,6	-
		333		-	111	35	11,4	-	3,7	-
		353		-	113	36	11,5	-	3,8	-
DMF	8	293	26,75	250	74	24	7,6	4,2	2,2	-
		313		268	79	25	8,2	4,5	2,4	-
		333		284	84	27	8,6	4,8	2,6	-
	12	293	31,77	262	84	26	8,0	-	2,5	-
		313		278	88	28	9,0	-	2,7	-
		333		296	95	30	9,6	-	2,9	-
		353		315	101	31	10,1	-	3,0	-
	16	293	33,02	-	91	25	7,8	-	2,3	-
		313		-	98	25	8,4	-	2,5	-
		333		-	104	29	8,9	-	2,5	-
		353		-	110	30	9,5	-	2,8	-

studies are essentially the initial stage in the study of the conditions of fiber molding on the basis of HPAC, as well as other its derivatives, and are carried out for the first time [69-71]. Fibers were molded from 10-12% solutions of HPAC. The molding results are shown in Tables 24-27.

Table 24.

Change of physical and mechanical properties of fiber depending on polymer concentration and spinning solution temperature.

Indicators	Temperature,^{0} K				
	288	293	298	303	308
10% solution					
Strength, gs/tex	5,4	4,8	3,8	3,2	3,2
Elongation, %	13,0	13,6	15,2	14,8	15,4
12% solution					
Strength, gs/tex	5,1	5,6	6,2	6,0	5,4
Elongation, %	13,0	14,3	13,4	14,4	14,8

Fibers from 10% solutions at 288-293°K formed satisfactorily, and from 12% solutions at 293-303°K. As can be seen from the table, the optimum temperature value for the 10% spinning solution is 288-293°K, and for the 12% solution 293-303°K.

This optimum temperature is apparently explained by its influence on the viscosity of the spinning solution, due to more intensive interaction of macromolecules in concentrated solutions. In accordance with the known data [72], as the temperature of the spinning solution increases, the diffusion rate of the precipitant deep into the fiber increases only up to a certain limit, since the formation of a dense surface gel layer delays the diffusion. The resulting finished fiber has an irregular, defective structure. When the spinning solution is heated above 303-308°K, the strength of the fiber decreases significantly. Obviously, the difference in temperature of the spinning solution and the settling bath during the fiber formation process causes a disturbance of the uniformity of the gel fiber mesh structure.

The formation of structural heterogeneities in the fiber (cracks, large supramolecular structures) can be prevented, first, by appropriate selection of conditions of polymer deposition from the molding solution, as well as by plasticization stretching [73].

As mentioned above, fiber formation in the deposition bath is a diffusion process, the rate of which changes as the amount of residual solvent in the fiber decreases. The rate of diffusion of the precipitating medium into the forming gel obviously depends, firstly, on compaction and ordering of its structure during fiber formation, and secondly, on the composition of the precipitation bath. In this regard, in order to soften the precipitation capacity of the precipitation bath, a small amount of water polymer solvent was added to the composition of the precipitation bath. The obtained data are shown in Table 25.

The molding process in organic baths and baths with 5% water content (Table 25) proceeds satisfactorily.

Table 25.

Influence of precipitation bath composition on physical and mechanical properties of water-soluble fiber.

№	Ratio of precipitation bath components, weight %			Number of depositions	Physical and mechanical parameters of the molded c-n	
	isopropanol	acetone	water		Strength, gs/tex	Elongation, %
1.	100	-	-	54.5	7.4	12
2.	-	100	-	64.0	7.3	10.8
3.	95	-	5	57.0	5.4	13.4

151

| 4. | - | - | 10 | 58.5 | does not coagulate |

It should be noted that the deposition conditions become harsher especially in anhydrous acetone and isopropyl alcohol, which leads to a decrease in elongation of the formed fibers in such deposition baths. Forming fibers in an acetone bath causes the elementary fibers to stick together during drying.

The possibility of adhesion of fibers is explained by the strong volatility of the precipitation bath, i.e. acetone and the specific physical and chemical properties of water-soluble fiber, i.e. good solubility in water.

In the process of evaporation of volatile acetone, the residual solvent water remaining in the fiber causes intensive swelling and partial dissolution of freshly formed fiber. With its subsequent volatilization, the elementary fibers and threads on the spool are glued together.

When molding under milder conditions, sticking of fibers also occurs. This is apparently due to the fact that the presence of water in the settling bath (5-10%) also leads to a deterioration in the coagulation ability of the bath (the number of settling increases) and partial dissolution of the fibers, and thus, the strength of the forming gel is reduced. Forming of fibers in the sedimentation bath, containing isopropyl alcohol, differs from the previous one by more uniform course of diffusion processes and sticking of threads during drying is excluded. This is apparently explained by the fact that isopropyl alcohol has a higher boiling point ($356°K$) than acetone and is close to water.

This leads to a more uniform evaporation of water solvent and isopropyl alcohol precipitant. The temperature of sedimentation bath was changed in a relatively narrow range (283-298 $°K$) due to strong evaporation of isopropyl alcohol and acetone. With further increase in the temperature of the deposition bath conditions of fiber deposition become more severe, which leads to a simultaneous decrease in both strength and elongation of the fiber.

The influence of the spinning draft was studied by changing the feed of the spinning solution from the live section of the die. The results of the influence of the spinning draft (in the range from 0 to 200%) on the physical and mechanical properties of the fibers are shown in Table 26.

Table 26.

Effect of filter drawing on the physical and mechanical properties of water-soluble fibers.

Indicators	Filter extraction, %				
	50	70	100	150	200
Strength, CH/tex	4,8	8,5	6,6	6,2	5,0
Elongation, %	16,6	15,0	14,8	13,2	12,0
Density, g/cm³	1,4070	1,410	1,410	1,4120	1,4035
SU	47,5	50,0	48,5	53,0	57,5
Water sorption at 65% vol.	11,2	14,0	14,5	15,3	-

As can be seen from the table, the increase in fiber strength after nozzle drawing is insignificant. This phenomenon can obviously be explained by the fact that the wet forming process results in a weakly oriented fiber structure as a result of the nozzle drawing.

The effect of die drawing on strength is significant before fiber structuring, after that it is negligible. Therefore, it is necessary to draw in a highly plasticized state. It is known that the fiber, torn off after the deposition bath, does not have the necessary properties for operation, due in large part to the features arising during molding heterogeneity of structures and the presence of micro- and macro-cracks, pores, channels, etc. These defects are removed in the future after the plasticizing drawing, as in this case there is a "healing" of micro-and macro-defects in the structure. In this regard, drawing of freshly formed fiber, in order to further strengthen it, is a necessary stage.

The results obtained on the effect of plasticizing drawing on the physical and mechanical properties of the fiber are shown in Table 27.

Table 27.

Effect of plasticizing drawing on the physical and mechanical properties of water-soluble fibers.

Indicators	Plasticizing extract, %					
	0	30	35	50	70	100
Strength, gs/tex	5,6	6,2	6,5	8,8	7,5	7,2
Elongation, %	16,0	15,2	13,0	15,0	14,2	12,0
Density, g/cm³	1,4120	1,4450	1,4615	1,4650	1,4670	1,4115
SU	42,5	-	47,5	45,5	54,0	57,5
Double ray refraction	-	0,013	-	0,015	0,026	0,035

As can be seen from Table 26 and 27, with increase of drawing and plasticizing, increase of fiber strength passes through the maximum. Based on the data obtained, the following conditions of forming water-soluble fibers based on partially substituted cellulose acetate can be selected:

Concentration of the spinning solution	10-12%
Spinning liquor temperature	293-295^0 K
Precipitation bath temperature	293^0 K
Filter extractor	70%
Plasticizing hood	50%

The fibers obtained under these molding conditions had a strength of 10.0-11.0 cN/tex and an elongation of 13-15%.

Study of the properties of fibers by different physical and chemical methods.

At present, it is an established fact that the main physicochemical and mechanical characteristics of polymeric fibers, including cellulose fibers, are largely determined by their macro- and microstructure [74.75].

The structure of the surface layers of the fibers plays an important role [76].

Table 28.

Types of VRAC Expert fibers.

System	Sample No.	Spinning solution		Temperature of spinning solution ,^{0} K	Filter extraction in %	Plastic extraction in %	Composition of the siege bath	Plastic bath composition
		Syrup in %	Solution in %					
I.	1	10	-	293-295	0	0	Isopropyl alcohol	
	2	-//-	-	-//-	50	-		
	3	-//-	-	-//-	100	-	-//-	-//-
	4	-//-	-	-//-	200	-	-//-	-//-
II.	5	-//-	-	-//-	0	50	-//-	-//-
III.	6	-	12	-//-	0	0	-//-	-//-
	7	-	water	-//-	50	-	-//-	-//-
IV.	8	-	r	-//-	100	-	-//-	-//-
	9	-	-//-	-//-	200	-	-//-	-//-

V.	10	-	-//-	-//-	0	50	-//-	-//-
VI.	11	-	-//-	288	150	35	-//-	-//-
	12	-	-//-	293	-//-	-//-	-//-	-//-
	13	-	-//-	298	-//-	-//-	-//-	-//-
	14	-	-//-	303	-//-	-//-	-//-	-//-
VII.	15	-	-//-	293-295	-//-	-//-	Acetonisopro pyl alcohol	
VIII.	16	-	-//-	-//-	150	50		
	17	-	10	-//-	0	0		
	18	-	DMF	-//-	50	-	-//-	-//-
	19	-	r	-//-	100	-//-	-//-	-//-
	20	-	-//-	-//-	200	-//-	-//-	-//-
IX.	21	-	-//-	-//-	0	50	-//-	-//-
	22	-	-//-	-//-	-	100	-//-	-//-
	23	-	-//-	-//-	-	200	-//-	-//-
X.	24	-	-//-	-//-	70	50	-//-	-//-
XI.	25	-	10 waters.	-//-	-//-	-//-	-//-	-//-

Many works [77,78] are devoted to electron microscopic study of supramolecular structures during formation of viscose fibers, but less so for di- and triacetate fibers [79-81]. However, there are no works on study of supramolecular structures by electron microscopy of water-soluble fibers produced by wet method in the world literature.

Electron microscopic studies of the surface structure of water-soluble acetylcellulose fibers depending on the conditions of production were of some interest.

10-12% aqueous and dimethylformamide solutions were prepared for obtaining experimental fibers from HPACs. In addition, for comparison, experimental molding of fibers from syrup (hydrolysis solution), excluding the stages of precipitation and washing, was carried out. Isopropyl alcohol and acetone (each separately) were used as a precipitation bath.

Table 28 shows the forming conditions for these fibers, which were grouped into X I systems.

Table 29 shows the characteristic physical and mechanical parameters for some types of fibers.

The following instruments and methods were used for researches: light microscopy for comparative estimation of fibers section (MBI-6) - fibers microcuts were studied; electronic microscopy by means of two-stage (carbon-styrene) replica method [82] the microstructure of fibers surface layers was studied. A device of "Tesla" type (direct electron-optical magnification of about 10 thousand times) was used in the work. Microscopic researches showed that cross sections of HPAC fibers with zero extraction value have different forms: rounded and bean-shaped (Fig. 64 a). The diameter of the fibers with rounded cross sections is in the range of 10-20 microns. The short part of the "bean-shaped" sections is 20 microns, the wide part is 40-60 microns.

Table 29.

Physical and mechanical parameters of the experimental HPAC fibers.

System	Sample No.	Strength, cN/tex	Elongation, %	Density, g/cm^3	SU	Water sorption at 65% vol.	Double ray refraction
III	6	5,6	13,4	1,4120	42,5	11,2	-
	7	4,8	13,4	1,4070	47,5	12,0	0,010
	8	5,6	14,6	1,4100	48,5	14,5	0,013
IV	10	8,8	10,2	1,4050	45,5	12,6	0,014
V	13	6,2	13,5	-	-	-	-
VIII	17	6,4	12,9	-	-	-	-
	19	7,0	12,0	-	-	-	-
	20	8,9	11,4	-	-	-	-
IX	21	9,6	11,0	-	-	-	-
	22	9,0	12,6	-	-	-	-
X	24	9,8	12,2	-	-	-	-

+ Molding conditions are given in Table 28.

With increasing drawing, the cross sections mainly acquire a rounded shape, and the diameter of the fiber decreases. Thus, cross sections with diameters from 10 to 20 μm are observed for HPAC fiber with 200% filter drawing (Fig. 64.b). This circumstance suggests that increase of nozzle drawing contributes to obtaining relatively homogeneous in shape of fiber sections, but still different in diameter.

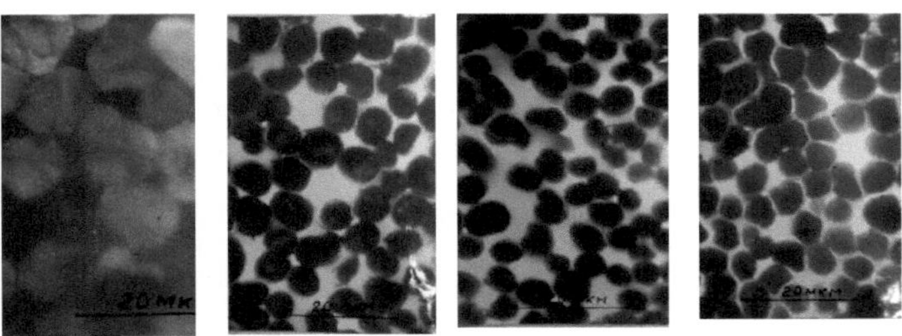

A   Б   В   Г

Fig.64. Cross-sections of HPAC fibers obtained under different forming conditions.

It should be noted that insignificant application of plasticizing drawing does not lead to significant changes in the shape of fibers in comparison with zero drawing, regardless of the type of spinning mass. Cross-sections of HRAC fibers, molded at temperatures 283,288 and 293°C, have different forms (elliptic, spherical), and they become most homogeneous at 298°C (fig.64.c). The latter have a predominantly rounded appearance with a diameter of about 30 μm. Individual oblong sections of fibers have dimensions in width of the order of 15 μm and length of 30 μm.

The shape and size of the cross sections of the fibers, as shown by experiments, also depend on the composition of the precipitation bath. More homogeneous fibers are obtained when the precipitation bath contains isopropyl alcohol (Fig. 64. d).

The results of electron-microscopic studies of the surface of the fibers of system 1, molded from 10% HPAC syrup with different degrees of elongation, are shown in Fig. 64.a-c.

These pictures show that fibers with zero degree of drawing have relatively homogeneous isotropic surface microstructure (Fig.65.a), in some places there are transversely arranged (in relation to the wave axis) ordered elements of the structure. With a small degree of die drawing

157

(50%) some anisotropic pattern appears. Globular aggregates detected on the surface layers are predominantly located along the fiber axis. Thus, for the fiber surface with 100% filament drawing (Fig. 65.b), a clear direction of the fiber axis and the character of the location of the globular aggregates along the fiber axis can be observed. Along with this, as can be seen from the electron micrography, thin elements of fibrillar-type structure arise. In the case of HPAC fibers with 200% filament draw, a fine fibrillar structure is observed on the surface (Fig. 65.c). Obviously, with a considerable degree of nozzle drawing the HPAC globules start to straighten and are oriented along the fiber axis.

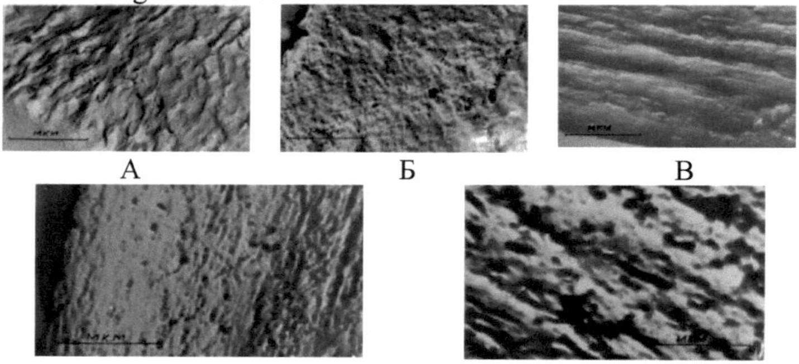

А　　　　Б　　　　В

Г　　　　Д

Fig.65. Electron microphotographs of the surface of HPAC fibers obtained under different conditions.

For the fiber type in question, there are no globular aggregates on the surface. The above-mentioned patterns of fiber surface structure may undergo significant changes, if additional plasticizing drawing is applied to them. So, if the fiber surface (system 1) with 50% filter drawing was characterized by isotropic globular structure without any signs of orientation of the latter, in case of application of 50% plasticizing drawing a noticeable anisotropy effect is revealed (Fig.65.d), although the globular nature of the surface structure is preserved. One can also see the oriented arrangement of the particles (at the same time some aggregation) along the fiber axis.

It is interesting to note that for HPAC fibers molded from 12% aqueous solution with 50% plasticizing drawing (0% nozzle drawing) the fiber surface has a heterogeneous fibrillar structure, although the elements of the structure are not large in size and they are aggregated in the direction of drawing. Under the electron microscope, their "ribbon-like" associations are clearly visible. It can be assumed that in contrast to HPAC syrup in aqueous solution in obtaining the fibers there are favorable conditions for the emergence of anisodiametric formations. By this and, apparently, it is possible to explain occurrence of heterogeneous "ribbon structure" of fibers surface (sample 10 of system IV) in contrast to fibers from syrup (samples 2 and 5). It is established, that in case of the specified fibers without drawing, in contrast to fibers, formed from syrup (Fig.66 a), on the surface there are globular elements of HPAC with predominant aggregation along the fiber axis (Fig.66 e).

The size of the globules ranges from 500 to 1000°A. The width of linear aggregates consisting of globules is about 900-1000°A.

With a certain degree of elongation (up to 50%) there is a destruction of large linear structures and the formation of smaller ones with a width of about 500°A. Here one can also notice areas with chaotic, distribution of globules without any signs of aggregation.

A further increase in the degree (up to 100%) promotes the unfolding of globular elements into relatively short elements, although the optical heterogeneity of the fiber surface microstructure remains. For the fibers having 200% filament drawing, an excellent picture is observed on the surface, from which one can clearly judge the presence of a whole set of supramolecular formations: globules, anisodiameter particles of curved shape, as well as fibrillar aggregates. All of them have an orientation along the fiber axis.

Some peculiarities of surface microstructure have been found for the fibers of the system, i.e. for the fibers molded from 12% aqueous solution with 150% filtering and 35% plasticizing, but at different temperatures of spinning mass (288,293,298 and 303⁰ K) and it has been established, that even at a small difference of temperature the fibers with different microstructure are formed. So, at temperature 288°K on the surface of fibers mainly globular aggregates and their various associations are observed, at 293°K - scaly pattern, but polymorphism is observed in the nature of the structures. For these two types of fibers also noticed that in the arrangement of the structural elements on the surface of the fibers a certain order was not detected, i.e. no orientation effect is felt, although the fibers are molded at the die drawing-150, and plasticizing-35%. Further studies showed that at a spinning mass temperature of 298°K, the effect of orientation of linear structural elements along the fiber axis is clearly detectable on the fiber surface. However, the degree of packing of

HPAC macromolecules in these supramolecular formations is apparently not high enough; they cannot be identified as the finest fibrillar formations. Similar structures appear in the surface layers of such fibers, which are molded from the solution at 303^0 K.

It was interesting to compare the surface structure of two more types of fibers: both fibers were molded from 12% aqueous solution, at 293-295^0 K, die drawing-150% and plasticizing bath also almost the same, but very different from each other: in the first case when obtaining the fiber in the deposition bath was acetone, and in the plasticizing bath - a mixture of acetone and isopropyl alcohol; in the second case, the composition of deposition and plasticizing baths was the same isopropyl alcohol.

It should be noted that the first types of fibers are characterized by a rather loose packing of the fibrillar structure on the surface and observed optically empty places-defects in the structure. The second types are most characterized by rather dense packing of structural elements, but there is a polymorphism of the supramolecular structure.

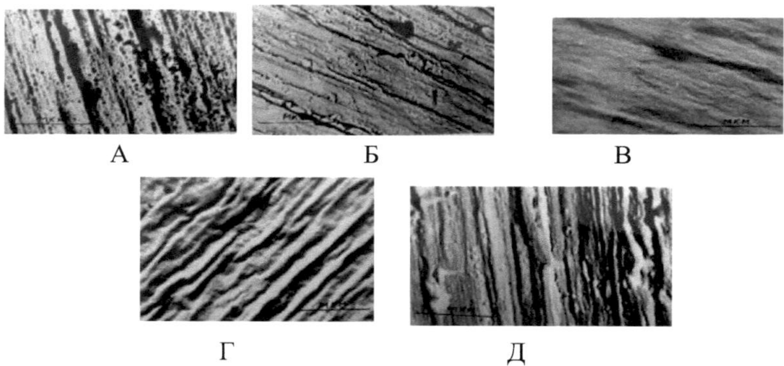

А   Б   В

Г   Д

Fig.66. Electron microphotographs of the surface of HPAC fibers obtained under different conditions.

The results of electron-microscopic studies of HPAC fibers of U system show that they have approximately the same surface layer microstructure regardless of different degrees of nozzle drawing (Fig.66.b). Some differences observed between them can be attributed to the fact that at low degrees of drawing relatively large globular formations of HPACs appear, and at higher degrees of drawing the smallest globules

The characteristic patterns of fibers surface were also found for the fibers of system IX, to which different plasticizing drawings were applied: 50, 70 and 100% (Fig.66.c). It was found that plasticizing drawing of - 50% results in fibers with optically homogeneous surface structure, where clearly marked structural elements are absent, in some places only longitudinal strips - fiber folds are visible (Fig.66.c). With increasing degree of plasticizing elongation at first there are a large number of longitudinal bands (at 70%), then certain linear type of structural elements, although the degree of their ordering is not high enough (Fig.66. d).

The structure of two more types of fibers was also investigated: in the first case, molded from 10% DMF solution with die drawing 70% and plasticizing 50%; in the second case, fibers obtained from 10% aqueous HPAC solution, in the same degrees of die and plasticizing drawing, electron-microphotographs of their surface structure are shown in Fig. 66. In both cases on the surface of the fibers can be seen not homogeneous both in shape and size, as well as in the degree of orientation of the elements of the structure.

It should be noted that, in the second case, the structural formations have a strongly swollen appearance and, in addition, they are not optically homogeneous. So, from the above data on the microstructure of various experimental fibers the following conclusions can be made, namely, depending on technological conditions of HPAC fiber molding one can observe different patterns of their surface structure. Apparently, the most ordered are the samples, which were significantly oriented due to nozzle and plasticizing drawings and molded from syrup or from dimethylformamide solution, but not from water solution. In the latter case both the structure is imperfect and the physical and mechanical properties are low.

In order to find out the interrelation of different forming conditions (temperature of spinning solution, composition and temperature of sedimentation bath, nozzle and plasticizing drawings) with structural and physical-mechanical parameters of the formed fibers, a series of studies by various physical and chemical methods was also carried out.

For the study, we used the method of determining the DLP, the acoustic method of determining the orientation of the fibers, as well as the methods of isometric heating and three-mechanical [75]. The obtained data are presented in Table 30.

Table 30 shows that the fiber orientation slightly increases with increasing die drawing, accompanied by an increase in strength. The exception is the fiber with VF=50%. Such a drawing results in a low-oriented fiber with a lower transition temperature to the plastic state (see Table 30). As for plasticizing drawing, the change of properties is more noticeable. As the plasticizing drawing increases from 0 to 100%, the strength, total molecular orientation, and anisotropy determined by the optical method increase. The IN curves (Fig. 67) for HPAC fibers molded from DMF solutions are similar in form to the curves for HDC fibers with one maximum in the temperature range 338-358°K, though the absolute values of the maximums for HDC fibers are significantly higher than those for HPAC. A significant part of the stresses (especially the low-temperature region of the maximum) is due to the removal of moisture from the fiber, as a consequence, its contracting, which leads to higher stresses. Indeed, rather high values of sorption (from 11.2 to 15.3 at 65% relative humidity) for these fibers confirm such an assumption. Nevertheless, these fibers also possess intrinsic stresses, since the full decrease of stresses in the IN curve occurs at a temperature above 473°K, then the process of removal of sorbed moisture from the fiber should end earlier [84]. Usually with increase of drawing and corresponding increase of orientation for all types of fibers [84,85] a considerable increase of internal stresses is observed. As the die drawing increases, there is a decrease in $\Delta\sigma$ (the difference between the maximum stress and the initial load) in the IN curves.

However, in the case of HPAC, the opposite phenomenon is found. This may be due to the predominant influence of sorption, decreased due to density increase, during drawing of HPAC fibers over stress increase due to orientation growth. At the same time, "natural" internal stresses during drawing of strongly swollen fibers practically do not increase, since the drawing process goes on the mode of plastic deformation. No clear correlation between the temperature of the maximum on the SI curve and the value of drawing was observed. However, the fibers subjected to plastic drawing have a higher maximum temperature.

Table 30.

Physical and mechanical parameters of water-soluble fibers.

№/ №P/S.	Extract, %		$\sigma*10^5$ n/m^2	T,^{0}K max at IN	T$_c^0$K	Strength, SN/tex	DLP	Speed of sound, km/sec
From DMF solution								
1	WF=0	GP=0	23,5	337	471	5,6	0,017	3,23
2	WF=0	GP=0	20,0	337	465	5,7	0,010	2,94
3	WF=100	GP=0	8,3	344	473	6,0	0,017	3,57
4	WF=200	GP=0	5,3	337	488	8,1	0,018	3,57
5	WF=0	GP=50	4,0	347	474	7,4	0,017	3,85
6	WF=0	GP=70	10,2	347	483	8,0	0,020	3,92
7	WF=0	GP=100	4,4	351	489	9,2	0,019	3,77
8	WF=70	GP=50	18,3	337	483	10,6	0,020	4,17
From an aqueous solution								
9	WF=0	GP=0	75	349	457	5,6	0,08	2,33
10	WF=50	GP=0	105	345	490	4,8	0,010	2,38
11	WF=100	GP=0	63	341	481	6,8	0,013	2,56
12	WF=200	GP=0	77	343	475	–	0,015	2,86
13	WF=0	GP=50	80	353	–	8,8	0,014	2,44
14	WF=70	GP=70	95	348	–	8,2	0,026	2,56
15	WF=70	GP=50	70	353	–	9,8	0,012	3,60

This fact indicates that with plasticizing stretching the total number of intermolecular bonds increases, the proportion of overstressed bonds decreases and, as a consequence, the temperature at which macromolecules or their associates can move relative to each other increases.

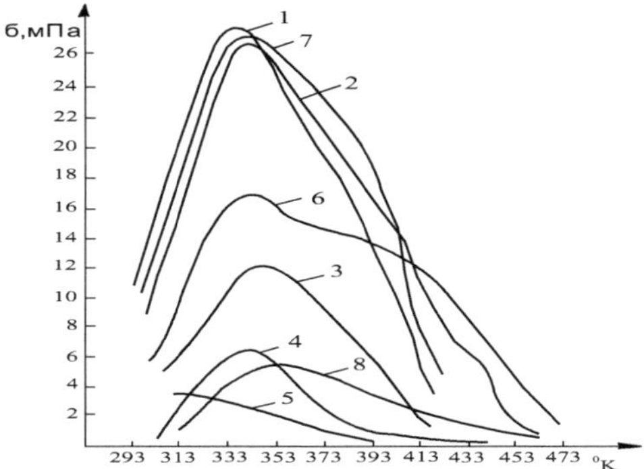

Fig.67. Isometric heating curves of HPAC fibers molded from DMF solutions depending on molding conditions.

1	WF=0	GP=0	5	WF=0	GP=50%
2	WF=50%	GP=0	6	WF=0	GP=70%
3	WF=100%	GP=0	7	WF=70%	GP=50%
4	WF=200%	GP=0	8	WF=100%	GP=100%

Figure 68 shows thermomechanical curves, which resemble the TMC for hydrate-cellulose fibers (HCC). Here there is also a region with negative strains, but the glass transition temperature of these fibers is much lower than that of HCV. This can be explained by weaker intermolecular interaction of HPAC fibers, which is proved by lower values of hydrogen bonding energy of HPAC in comparison to HCV (see Table 19). Tc slightly increases with increase of the filter drawing. With increasing plasticizing draft, Tc increases more markedly. This fact is in accordance with the fact of increasing temperature maximum on the curves, which coincides with the literature data for other types of fibers, such as triacetate fibers molded wet from acetic acid syrups and fiber based on PVS (85,86).

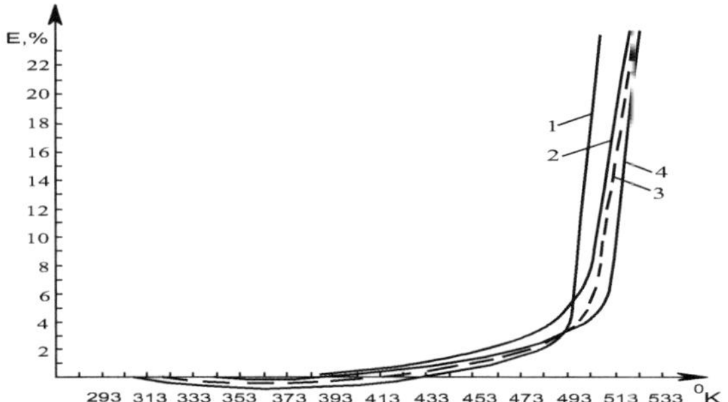

Fig.68. Thermomechanical curves of HPAC fibers depending on molding conditions.

1. WF=0   GP=0
2. WF=50%   GP=0
3. WF=0   GP=50%
4. WF=70%   GP=50%

It should be noted, that receiving of fibers from dimethylformamide solutions with high drawings, for example, VF=200% or VP=100% is connected with some difficulties, i.e. in the process of drawing high breakage takes place. In contrast to fibers molded from dimethylformamide solutions, filver drawing has no significant effect on the strength of fibers molded from aqueous solutions. At the same time, the fiber DLP increases slightly, as well as the ultrasonic speed.

Plasticizing draw noticeably increases the strength of fibers molded from aqueous solutions, their anisotropy and general molecular orientation. In general, absolute values of DLP indicate insignificant anisotropy of HPAC fibers, as well as acoustic characteristics indicate low orientation of these fibers, although in each series of increasing of nozzle and plasticizing drawing the rate of ultrasonic passage through the fiber increases monotonically. IN curves of fibers from aqueous solutions (Fig.69) also have one maximum in the region of 343-353°K. There is no clear correlation between the value of internal stresses and the value of drawing of the fibers (see Table). However, high stresses at low sorption correspond to fibers with low strength and low orientation (VF=50% VP=0%).

Fibers with high strengths also have high ultrasonic velocity values. However, fibers with 100% and 200% nozzle drawings have the same acoustic parameters and VLP, but they have low strengths. This phenomenon can be explained by the fact that the nozzle drawing results in a heterogeneous structure, which is characterized by the presence of defects and pores. After plasticizing drawing, the pores disappear and the micro- and macro-cracks partially heal. Therefore, plasticizing drawings lead to a hardening of the fiber without significant improvement in orientation

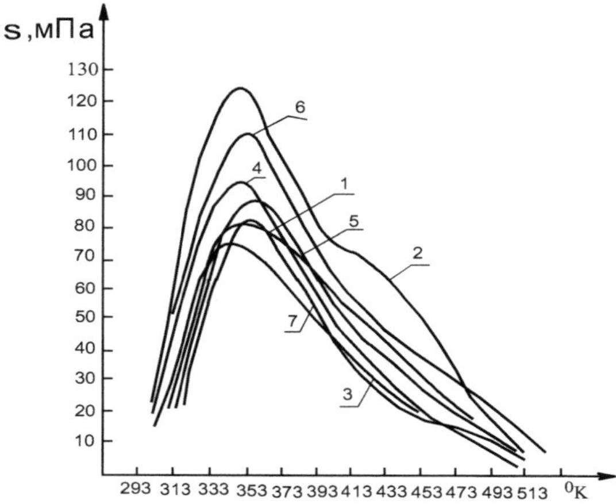

Fig.69. Isometric heating curves of HPAC fibers molded from aqueous solutions depending on molding conditions.

1	WF=0	GP=0	5	WF=0	GP=50%
2	WF=50%	GP=0	6	WF=0	GP=70%
3	WF=100%	GP=0	7	WF=70%	GP=50%
4	WF=200%	GP=0			

parameters, apparently, due to the healing of structural defects.

TMK have the same shape as for the fibers from dimethylformamide solutions, the transition temperature to the viscous state lies in the region of 473-493°K.

The absence of a highly elastic plateau in this case can be explained by the fact that when the fiber is heated, water is removed and the transition to the glassy state shifts towards higher temperatures. Thus, the glass transition temperatures given in the table correspond to the transition to the highly elastic state of the dehydrated polymer, which by this time is

HPAC, i.e. the very method of determining the glass transition temperature excludes the possibility of finding a highly elastic plateau in such a hydrophilic fiber.

3.2.2 Modification of the properties of water-soluble acetylcellulose by chemical cross-linking reactions.

Interaction of N-methylene compounds with cellulose is widely used to impart crease-resistant properties to cellulose fabrics [87-90]. It is believed that such interaction proceeds with formation of transverse bridges between cellulose chains.

It should be emphasized that, in most cases, the criterion for the formation of bridging bonds is the insolubility of the cellulose treated in this way in its characteristic solvents.

However, this criterion in the case of heterogeneous treatment seems most doubtful, since due to the small swelling of cellulose in water, the bifunctional reagent is distributed only in the most accessible areas of the cellulose fiber. It is likely that part of this reagent capable of polymerization under the conditions of this reaction fills the capillary space of the same fiber in the form of homopolymer, which can be a mechanical obstacle to the penetration of the solvent to the cellulose macromolecules.

This assumption by no means excludes that there may also be a reaction with cellulose on the surface of the microfibrils, but this is not true "cross-linking" in the sense of forming one giant three-dimensional molecule.

At heterogeneous cross-linking the physical structure of cellulose fibers also has a great influence. The dependence of the number of cross-links on the type of cellulose (on the physical structure) during treatment with dimethylthiourea has been studied [91,92].

The results are shown in Table 31.

Table 31.

Dependence of the number of cross-links on the type of cellulose when cross-linking with a 10% DMTM solution.

Fiber	Content, %		Formaldehyde at the AGZ, mol.	Nitrogen at the AGZ, moth	Number of transverse links per 100 AGZs
	formaldehyde	nitrogen			
Rami	2,18	1,85	0,1600	0,2280	4,60

Linen	2,80	1,85	0,1600	0,2280	4,60
Cotton	2,85	1,90	0,1650	0,2370	4,70
Polynosis	4,30	2,85	0,2580	0,3680	7,40
Viscose technical yarn	4,65	3,10	0,2800	0,4000	8,00
Viscose textile yarn	4,95	3,30	0,3000	0,4300	8,60
Viscose textile staple	5,55	3,70	0,3440	0,4880	10,00

The table shows that when modifying DMTM under the same conditions, nitrogen content and number of cross bonds in hydratecellulose fibers are much higher than in natural ones (cotton, linen and ramie). This is due to the different molecular packing density of cellulose fibers. The maximum number of cross-links is observed in viscose textile staple fibers, viscose textile yarn and viscose technical yarn, because they have a smaller share of highly ordered areas; therefore, there are more favorable conditions for penetration of the crosslinking agent inside the fibers and intensive process of cross-linking formation.

It is natural to assume that homogeneous cross-linking should differ from heterogeneous cross-linking both in terms of the speed of the process and the cross-linking results.

Therefore, the homogeneous mixing of the crosslinking agent and cellulose is of great interest, which can be carried out using the example of water-soluble dimethylurea (or dimethylylethylurea) and partially substituted acetylcellulose. In addition, the latter can serve as a convenient model compound for obtaining "cross-linked" hydrate cellulose with a uniform distribution of bridging bonds.

Printed works in the field of cross-linking of water-soluble cellulose ethers are mainly represented by the works of G.A. Petropavlovsky and his collaborators (93-95/ and the author, performed jointly with him [87,96].

The object of the study was HPAC with a degree of substitution of 0.55 (γ=55) and a polymerization degree of 200-250. The crosslinking reagent used was dimethylethylurea (DMM) synthesized according to [97] and a ZnSo catalyst$_4$. Cross-linking was performed as follows. An aqueous solution of DMM or DMEM with the catalyst was added to an aqueous solution (concentration-5%.) of HPAC. The mixture was mixed thoroughly, then an even layer of the solution was applied to the glass, which gave a good film when drying. The film was then heated for a certain time at 423°K and subjected to various tests.

When selecting the most suitable conditions for the interaction of dimethylurea and HPAC, the questions of the effect of temperature and time of heat treatment, the amount of dimethylurea and the catalyst on such properties as solubility and swelling of the "crosslinked" film were solved.

When DMM and HPAC solutions are mixed and the solvent subsequently dries during film formation, there is no segregation or separation of the solid components, so we can assume a truly homogeneous distribution of DMM and HPAC molecules.

Adding up to 2% ZnSO$_4$ by weight of HPAC also does not cause any changes in mixing homogeneity. Thus, the interaction of DMM and HPAC in the case of their homogeneous mixing can be estimated in this case by the solubility of products obtained after heat treatment, as well as by the amount of firmly bound DMM not removed during extraction of the crosslinked film with water. It is clear that not all the amount of DMM introduced into the film forms monomolecular bridging bonds between cellulose macromolecules, some DMM molecules can be attached by reaction with cellulose only one hydroxyl group of DMM, some DMM molecules can be associated already in aqueous solution, so cross bridging can consist of dimeric and polymeric DMM molecules.

Fig. 70 shows the dependence of solubility at room temperature in water of cellulose acetate film with DM (16% of the weight of acetylcellulose, which is 1 g/m DM in 4 g/m acetylcellulose) on the heating time at 423°K.

Another cellulose acetate film containing 4% DMM and 1.1% ZnSO$_4$ (catalyst) was used to study the effect of temperature on film solubility when boiled in water.

Table 32. shows the results of these experiments.

Table 32.

Influence of temperature and processing time on the solubility of cross-linked acetylcellulose films.

№/ №	Film composition	Temperature	Processing time	Solubility in water by

n/a		processing, °К	τ -60 c	boiling in % of film weight
1		393	5	100
2	HPAC+4% DMM+1,1% ZnSo₄	393	15	92
3		393	30	41
4		423	15	42

Based on these preliminary experiments (Fig. 70 and Table 32), it was found that the most rapid change in solubility is observed under reaction conditions at 423°К. At this temperature, the minimum solubility is reached in 15-60 s.

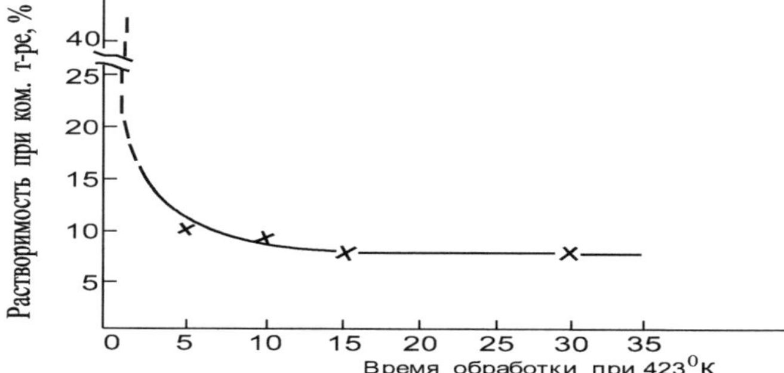

Fig.70. Effect of time of heat treatment of acetylcellulose film with 16% DMM on its solubility in water at room temperature.

The reaction can also take place at a lower temperature, but the time to reach the same degree of solubility increases considerably. Therefore, all the results discussed below refer to experiments performed under the conditions of terioblasting at 423°K for 15-60 s.

Next, the effect of the amount of dimethylurea on the swelling and solubility of cross-linked films at a constant DMM/ZnSO ratio₄ =7, 3/I was studied.

The results of the experiments are shown graphically in Fig.71.

As can be seen from the figure, the minimum solubility in the copper-ammonia reagent is almost reached already at the introduction of 8% DMM from the weight of HPAC. Swelling continues to decrease beyond this limit, indicating an increase in the number of cross-links. It should be noted that the minimum solubility of such films, which contained 16% DMM and 2.2% ZnSO$_4$ by weight of HPAC, was about 20% in the copper-ammonia reagent. Films

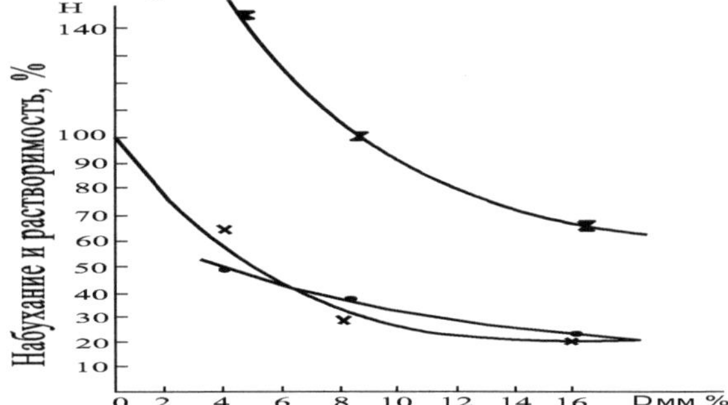

Fig.71. Dependence of swelling in water (o-o-o-o-), swelling in 6% NaOH (x-o-x) and solubility in copper-ammonia solution (x-o-x) at room temperature on the amount of DMM.

had low swelling in water, and the swelling in a 6% NaOH solution was about 70%. It should be noted that changes in the swelling and solubility of the films also occur in the interaction of HPAC with DMM in the absence of the catalyst, but these changes are much smaller than in the first case, as shown in Table 33.

Table 33.

Effect of DDM and ZnSO$_4$ on the solubility of copper-ammonium reactant films after heat treatment.

Number of DMMs of the weight of the HPAC,%	Quantity of ZnSO$_4$ of the weight of the HPAC,%	Solubility in copper ammonia reagent, (%)
-	-	100
16	-	58
16	2,2	20,6
8	-	76,3
8	0,55	44,7
8	1,1	31,0

The data in this table suggest that the amount of ZnSO$_4$ also affects the change in solubility. At the molar ratio of DMM: ZnSo$_4$> 4 the solubility increases significantly, which can be seen in Fig.72.

As mentioned above, obviously not all amounts of DMM react with HPAC. If we exclude the possibility of the formation of a significant amount of homopolymer in the homogeneous mixing of DMM and HPAC, the unreacted DMM can be extracted from the film with boiling water, in which it is well soluble.

Table 34 presents data on the nitrogen content in the cross-linked films after extraction with water under boiling for 3600 s with a change of solvent.

Table 34.

Nitrogen content in cross-linked films after treatment with boiling water.

Raw film properties	N content (%)	Bound DMM from the weight of the film, (%)	Effectiveness of using DMM (% of the initial amount of DMM)
VRATZ-89.8%, DMM-8% $ZnSO_4$ -2.2%	1.36	5.8	75
VRATZ-84.0%, DMM-16%	0.63	2.7	34

The data in Table 34. show that DMM is for the most part firmly retained and not extracted by hot water. Its amount (in agreement with the data on swelling and solubility) is significantly greater if the catalyst ZnSO is present$_4$. For example, when introducing 16% of DMM to the weight of HPAC, but without a catalyst, its content of the initial amount was only 34%, while when introducing a catalyst, it is already about 80%. These data show that in the presence of the catalyst $ZnSO_4$ HPAC and DMM really enter into chemical bonding with each other, as both components, soluble in water, lose this property after the reaction. It is interesting to note that the catalyst $ZnSO_4$ also participates in the reaction. This conclusion can be inferred from the fact that the $ZnSO_4$ introduced into the film together with the DMM remains in the film after extraction with hot water in the same amount as this salt was introduced at the beginning. The amount of $ZnSO_4$ in the film after hot water extraction was determined on the basis of sulfur content determined by Schoniger /98/. These data are given in table 35.

172

There are references in the literature to the formation of Zn^{2+} coordination complexes with DMM and the participation of these complexes in the formation of bridging bonds /99,100/.

Table 35.

ZnSO content₄ in cross-linked HPAC films

Introduction to the film compositions in % of VRP		Detected after extraction with water in % of VRAC	
DMM	$ZnSO_4$	S	$ZnSO_4$
4	0.55	0.11	0.55
8	1.10	0.23	1.16

To confirm the chemical interaction of water-soluble acetylcellulose with DMM, IR spectra of the initial reagents and reaction products were studied. Fig. 73 shows the IR spectra of HPAC and HPAC film mixed homogeneously with DMM. The latter spectrum has a number of complex absorption bands, which in some cases are the result of superposition of absorption bands in the spectra of DMM and HPAC. Figure 75 shows the IR spectrum of pure DMM.

Analysis of the absorption bands in the spectrum of the HPAC film together with DMM shows that, compared to pure HPAC, there are two characteristic absorption maxima in the region of 1670 cm^{-1} and 1560 cm^{-1}, which can be attributed to the C=0 and H-N valence vibrations [101] of the DMM bond.

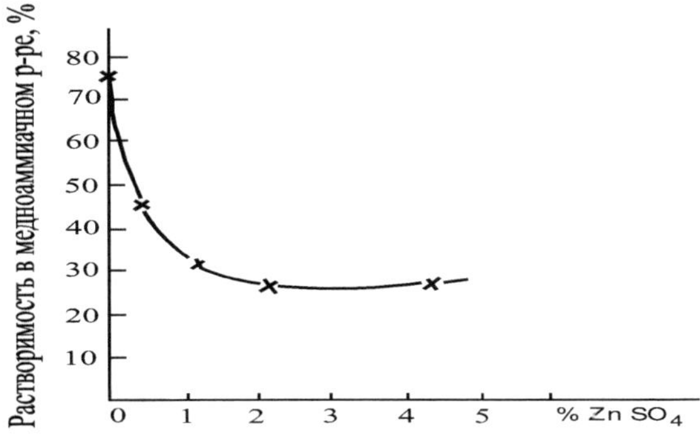

Fig.72. Effect of the amount of catalyst *(ZnSO₄)* on the

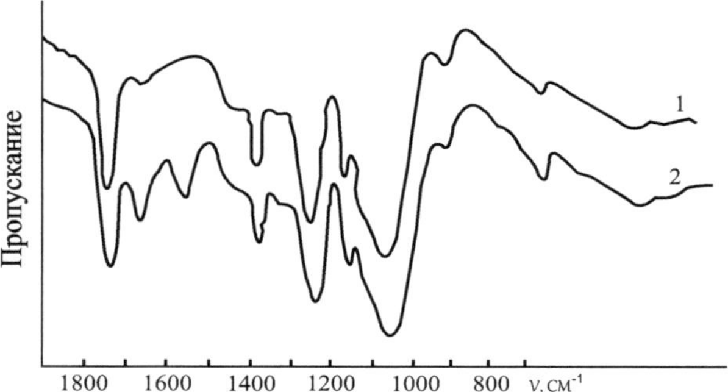

Figure 73. IR spectra of HPAC (1) and mixture of HPAC with dimethylolurea (2).

solubility of crosslinked films in copper-ammonia reagent.

It should be noted that, compared to the spectra of pure DMM, the position of the absorption band maxima is shifted to 1560 cm^{-1} for N-H and from 1640 cm^{-1} to 1670 cm^{-1} for C=0 vibrations. Such a shift of the indicated bands apparently occurred due to the dissociation of DMM molecules during its transition to the dissolved state and is preserved when the "solid" solution of DMM in HPAC (in the film) is obtained. It should be noted that a similar band shift is observed during melting of crystalline DMM. When heating the HPAC film with DMM in the presence of the catalyst at 423°K for 15-60 s, there is no change in the position of the above bands. However, in this case, other changes in the spectrum can be observed.

Table 36 shows quantitative characteristics of changes in some of the absorption bands in the spectra studied. The band of valent vibrations of hydroxyl groups is significantly shifted to the region of low frequencies. This is observed in both heated and unheated films and can be explained by the position of N-H valence vibrations and the shift resulting from the formation of new hydrogen bonds between OH groups of cellulose and DMM. At the same time, the band of valence vibrations of OH groups in the IR spectra of films changes in intensity. The intensity of this band in our case was conveniently expressed by referring its area to the absorption band area C=0 in the acetyl group just like the 1245 cm band^{-1} (C-0 valence vibrations) does not change during all treatments and can thus be an internal standard.

Table 36.

Variation of some absorption bands in IR spectra and HPAC+DM films

Quantity (% of VRAC)		Thermo processing	processing of cross-linked film	ν he	$\nu c = 0$	$\frac{Д1750}{Д1380}$	$\frac{Д1245}{Д1380}$	$\frac{Д1560}{Д1380}$	$\frac{Д1670}{Д1380}$	$\frac{Д1430}{Д1380}$	$S_{\nu_0 H}$ / $S_{\nu c = 0}$
DMM	ZnSO										
-	-	+	-	3460	1745	1.2	1.8	-	0.14	0.47	8.43
16	-	-	-	3410	1745	1.1	1.8	0.8	0.8	0.54	10.8
16	-	+	-	3420	1740	1.2	1.8	0.42	0.74	0.49	9.04
16	-	+	+	3450	1736	1.2	1.8	0.09	0.26	0.44	8.06
16	2.2	-	-	3415	1745	1.2	1.8	0.58	0.73	0.51	8.2
16	2.2	+	-	3420	1742	1.2	1.8	0.44	0.7	0.48	7.35
16	2.2	+	+	3440	1740	1.24	1.85	0.81	0.53	0.45	8.14

As can be seen from the data in Table 36, the change in the absorption intensity of OH-groups is completely regular. In films not subjected to heat treatment, the relative intensity of this band is higher than in the original HPAC. This can be explained by the increase in the total number of hydroxyl groups due to the methyl hydroxyls of DMM, as well as by the increase in absorption in this region due to the N-H valence vibrations.

After thermal treatment of the films, especially in the presence of a catalyst, the relative integral band intensity decreases to a value lower than that of the original HPAC. This confirms the chemical interaction of DMM with the hydroxyl groups of HPAC.

We can speak with great caution about the participation of primary hydroxyl groups of cellulose in this process, since this can be judged only indirectly by changes in the intensity of the absorption band at 1430 cm^{-1}, which is associated with CH_2 -causes at the sixth (C_6) carbon atom [102].

The crosslinked acetylcellulose films obtained by the method described above have X-ray diffraction patterns not different from those of the original acetylcellulose. Figure 75 shows diffractograms of cross-linked HPAC film (7% DMM of HPAC weight) and for comparison with it the diffractogram of the original HPAC.

As can be seen from the figure, both diffractograms characterize the amorphous state of the product. The cross-linked HPAC was saponified with an aqueous 6% NaOH solution, washed with water, and boiled in water for 6x3600s. Thus, a cross-linked, regenerated cellulose film was obtained, which, unlike conventional cross-linked hydrate cellulose films, contained a cross-linking reagent distributed homogeneously.

Figure 76 shows a diffractogram of such a film and, for comparison, a diffractogram of an uncrosslinked hydrate cellulose film obtained from

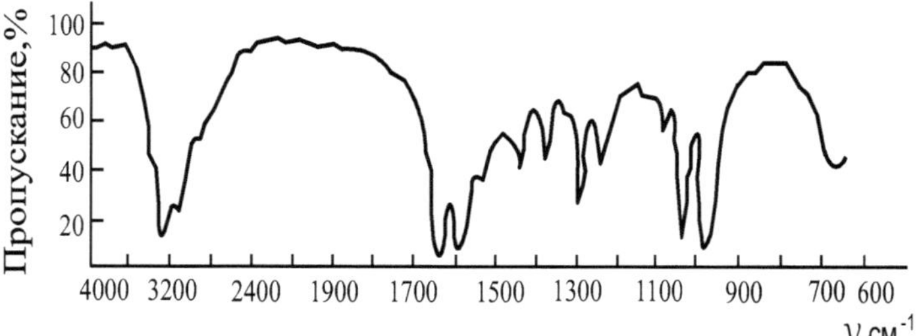

Fig. 74. IR spectrum of crystalline dimethylolurea.

Fig.75. Diffractogram of the original acetylcellulose film (1) and dimethylurea crosslinked acetylcellulose film (2).

HPAC under the same conditions. Comparison of these diffractograms indicates that in cross-linked films in the above way (7-8%. DMM from the weight of HPAC, i.e. 1 mole of DMM per 8-10 moles of anhydroglucose) the amorphous structure is not completely fixed, as a partial transition to the structure of cellulose II is observed. This suggests that transverse bridges in such cross-linked cellulose are distributed sparsely enough not to interfere with structural transformations of the substance.

The mechanical strength of such cross-linked hydrate cellulose films was high and reached, on average, 127-147 MPa in tensile strength.The elongation at break was about 20%, i.e., the cross-linked films had considerable elasticity. They did not differ much in this respect from hydrate cellulose films obtained from HPAC but not cross-linked. Usually "cross-linking" of hydrate cellulose leads to a significant drop in strength and brittleness of the sample.

As studies have shown, such cross-linked films have good chemical resistance to the action of some reagents. To determine the chemical stability of cross-linked hydrate cellulose films (8% DMM from the weight of the initial HPAC) were immersed in a solution of 40% KOH and incubated in it at 323^0 K for different times. After that, the strength of the film was tested after a certain time and swollen in a 40% KOH solution.

For comparison, non-cross-linked HPAC films were subjected to the same tests. Figure 77 shows the data on the change in strength and elongation of the film with the time of exposure to 40% KOH. As it can be seen from this figure, as a result of swelling and destruction in alkali, the uncrosslinked HPAC film almost completely loses its strength during 2-3600 s, while the crosslinked HPAC film has a strength of about 68 MPa during this period of time. The most rapid drop in strength of this film occurs during the first 6-3600 s (strength to the end of this time period was 39.2 MPa), after which the drop in strength occurs relatively slowly (after 24-3600 s 29.4 MPa). After 90-3600 s of hot alkali treatment, the film still retained a swollen strength of 4.9 MPa.

The change in elongation at break with time is shown in the same figure. As can be seen from the figure, the elongation during the first 2-3600s of exposure to 40% KOH solution increases. This process characterizes the swelling of the film. Then the elongation of the cross-linked film decreases very slowly with time. After 24-3600 s the elongation was 30%, instead of 34.6% after 2-3600 s. After 90-3600 s the elongation was another 16-18%. These data characterize the slow process of cellulose degradation when exposed to hot alkali.

Thus, the data obtained indicate a significant persistence of cross-linked acetylcellulose films in the medium of 40% KOH. It should also be noted that cross-linked hydratecellulose films obtained from HPAC have another interesting property. They have very little decrease in tensile strength when swelling in water with a significant increase in elasticity. Thus, the decrease in tensile strength when swelling such a film in water for 24-3600 s was no more than 20%, while the elongation at break was about 30%, that is higher than in the initial state (20%).

At the same time, the strength of the amorphous hydrate cellulose film obtained from HPAC decreased by more than 50%, and the elasticity of such a water-swollen film also decreased slightly (to 25-27%). Thus, a valuable property of cross-linked films is their high strength and elasticity in the swollen state.

HPAC can be rendered insoluble not only by cross-linking, as described above, but also by gentle saponification of acetyl groups, turning it into hydrate cellulose. And this process can be carried out without any destruction of the main chain.

As you know, currently a lot of attention of researchers turned to finding harmless cellulose fibers to replace viscose, their current production is technologically quite complicated and harmful from an environmental point of view.

In this connection, obtaining of hydratecellulose fibers from aqueous solutions of cellulose derivatives is a topical problem in the chemical fiber industry. For this purpose experiments on obtaining hydratecellulose fibers on the basis of HPAC were carried out. Before carrying out such a process, it was interesting to find out the influence of the degree of cellulose acetate substitution on mechanical properties of the molded fibers. Such investigations are carried out for the first time, since exactly HPAC, due to its specific properties, allows to carry out such investigations on fibers molded from solutions in the same solvent-DMF.

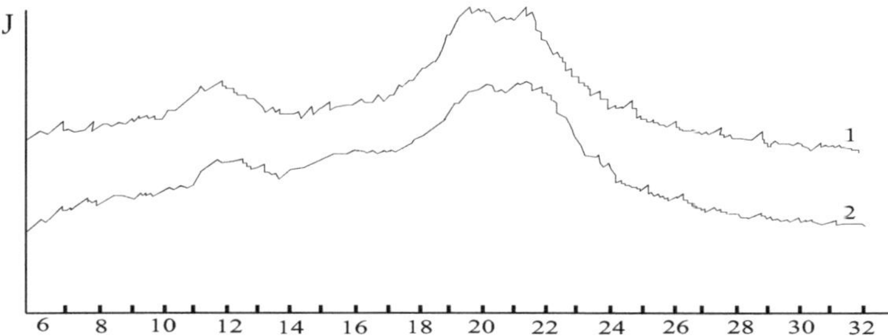

Fig.76. Diffractogram of films of saponified initial HPAC (1) and cross-linked (2).

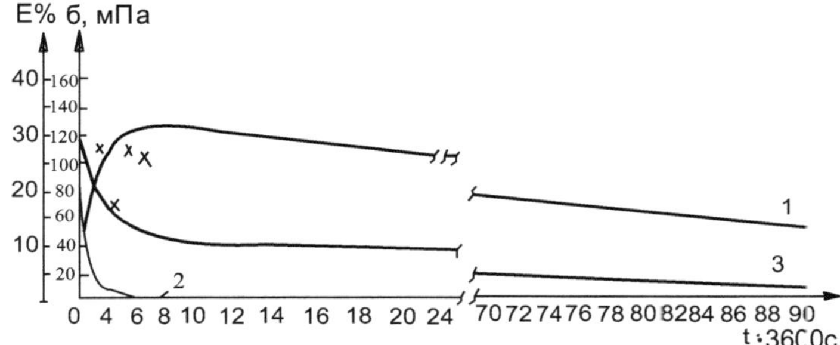

Fig.77. Variation of the breaking strength (3) and elongation of the cross-linked film (1) and the breaking strength of the original HPAC film (2) in 40% K0H at 323°K.

Cellulose acetates were synthesized with different degrees of substitution (γ=50÷250) and in the degree of polymerization (SP=210÷280). 12% dimethylformamide spinning solutions of cellulose acetates with different content of acetyl groups were prepared and fibers were molded from them under the same conditions (Vf=100%; B_Π -50%, T- 298°K).

The results are shown in Table 37.

As can be seen from Table 37, the fibers with the highest strength are with a degree of substitution of 0.97 (γ =97), although SP of these fibers is 210 against 280 of the fibers obtained from secondary acetylcellulose (γ =253). It seems to be connected with the fact that with the degree of substitution ~1(γ=100) acetylcellulose macromolecules are

179

in the most favorable position for mutual orientation in the process of plasticizing drawing.

Table 37.

Dependence of physical and mechanical properties of fibers on the degree of substitution of acetylcellulose.

The degree of substitution,	Content of bound acetic acid,%	Strength, cN/tex	Elongation, %
253	55,5	10,3	22
225	53,0	11,2	20
187	46,8	12,6	20
143	38,6	11,6	18
97	29,1	17,6	14
55	18,0	10,4	18

At this degree of substitution, apparently, the chains are still in the straightened state and due to intermolecular hydrogen bonds a certain orientation with strengthening of the fibers can be achieved. Further reduction of the degree of substitution leads to deterioration of solubility of HPAC in DMF and folding of macromolecules, which does not allow for sufficient orientation and leads to a decrease in fiber strength.

Some technical applications of water-soluble acetylcellulose and products of its modification.

3.2.3 Application of water-soluble acetylcellulose for binding hard-to-press drugs.

The study and supplementation of the set of excipients used in the manufacture of tablets still remains an urgent task of pharmaceutics. In recent decades, a number of works have been devoted to the use of cellulose and its derivatives as binding agents [103-105].

The secondary acetylcellulose recommended at this time by HFCs, as well as some other esters, are insoluble in water, in connection with which their application is significantly limited [106.,p.667]. Therefore, the study of the possibility of using indifferent HPAC is of certain practical interest. Often in the production of drugs characterized by poor particle adhesion and high elasticity, significant difficulties are felt in pressing. In tablets, cracks, delamination, and the upper surface (cap) fly off [107]. Usually in such cases the situation is resolved by increasing the amount of binding agents, the concentration of which often reaches more than 10%, which is not indifferent to the quality of the tablets. Binding ability of the used excipients not only depends on the viscosity (molecular weight), but also on the linearity of the structure, the presence of active functional groups, etc. The presence of numerous hydroxyl groups in HPAC, allows its use as a binding agent in obtaining tablets. Besides, as experiments on determination of LD-50 index showed, HPAC is practically indifferent product. The possibility of its use in tablet technology has been tested by the example of acetylsalicylic acid, amidopyrine, barbital, terpigidrate and phenacetin, as drugs that are difficult to press and have attracted the attention of many researchers for this reason [108,109]. Assessment of the suitability of HPAC as a binder was carried out by a comparative study of its mechanical strength with known natural and synthetic polymers: starch, Na-CMC, PVP, as well as known protein binder - gelatin. On the basis of these data the technological modes of making tablets of the above preparations are considered.

In obtaining the tablets, the most effective was the use of

The data presented in Table 38 demonstrate this. As the data in Table 38 show, satisfactory results for amidopyrine tablets were obtained only when a 10% HPAC solution was used. As for the application of different concentrations of PVP, starch paste, 3-5% gelatin solutions and HPAC, they do not lead to positive results. Apparently these binders in the indicated concentration are not able to reduce the elasticity of amidopyrine particles. Therefore, the tablets show cracks, peeling of the upper part. When comparing the process of pressing and quality of tablets obtained, it was found that amidopyrine tablets with gelatin are characterized by easily crumbling edges of the surface width and a slight specific smell. In the case of BPAC, tablets have a shiny, smooth surface and firm edges, which is an important indicator of the quality of the finished product. As for the ejection force, it depends on the concentration of the binding agents. Moreover, PVP has a significant anti-stick property, so the ejection forces are significantly lower compared to other binders. When pressing phenacetin tablets satisfactory results were obtained with a 5% solution of HPAC, while other binders in the same concentrations do not give a positive effect. Experiments have shown that obtaining satisfactory tablets from the above drugs requires the use of different concentrations of HPAC. Thus, while a high concentration of HPAC solution (10%) is necessary for amidopyrine, a 5% concentration is sufficient for barbital and phenacetin, a 1% solution for acetylsalicylic acid and terpinghydrate tablets, and even a 0.5% HPAC solution is sufficient for glucose.

Disintegrability and strength of tablets depend on the pressure of pressing and on the way of starch introduction to the finished granules, otherwise deterioration of disintegrability and strength of tablets is noted. As shown in the table, amidopyrine tablets should be pressed between 40.0-80.0 mn/m^2 , and other tablets can be obtained in pressure ranges of 40.0-160.0 mn/m^2 .

Table 38.

Comparative evaluation of the effectiveness of binding agents on the example of amidopyrine tablets.

№ п/п n/a	Auxiliary substances and their concentration,%	Applied forces, mn/m^2 pressing	ejection	Quality of the tablets appearance	durability	disiation
1	A) HPAC solutions 3-5	40,0	1,1	the edges are crumbling, Surface The "hat" is delaminating and flying	20	fail.
		80,0	1,7			
		120,0	2,2			
		160,0	2,8			
2	B) HPAC solution 10	40,0	1,0	smooth, colorless, without cracks or splitting	30	300
		80,0	1,4		40	720
		120,0	1,8		50	fail.
		160,0	2,2		60	-
3	PVP solutions 3-5-10	40,0	1,0	They crumble easily, unraveling And the "hat" flies		
		80,0	1,4			
		120,0	1,8			
		160,0	2,1			
		40,0	1,0			

		80,0	1,4	They		
	A) Gelatin lutions 3-5	120,0	1,8	crumble		
		160,0	2,1	easily, the "hat" flies		
	B) Gelatin-solutions, starch-7	40,0	2,8	The edges are flimsy, slightly scraped.	15	120
		80,0	3,0		25	720
		120,0	4,5		53	faili g
		160,0	5,0			
4	coahmal te solutions, 3-5-10	40,0	-	The edges are brittle, crumblin g		
		80,0	-			
		120,0	-			
		160,0	-			
5	Naa CMC solutions 3-5-10	40,0	1,0	the edges are not strong enough, slightly scraped	15	120
		80,0	1,7		25	720
		120,0	1,8			
		160,0	2,2			

Based on the studies, the following ratios of auxiliary substances involving HPACs can be recommended, which are shown in Table 39. Preparation technique consists of granulation of crushed drugs and BPAC, drying, rubbing the dry mass through a sieve with a 1 cm diameter hole, powdering with the appropriate amount of excipients and tabletting.

Thus, the use of HPAC in a relatively small concentration has a positive effect on the disintegration of tablets, which will significantly reduce the amount of loosening agents (Table 40).

2. Application of water-soluble acetylcellulose as a hydrophilic base for medicinal ointments.

Most water-soluble cellulose ethers have increased hydrophilicity, surface activity and a number of other valuable properties. Due to these properties, they are widely introduced as a base for ointments and suppositories, as thickeners for suspensions, as well as for making the drug durant. In this regard, at the present time, it seems appropriate to expand and replenish hydrophilic bases for ointments based on cellulose derivatives with new representatives, having a difference from the known ones. To this end, an attempt has been made to use HPAC as a hydrophilic base for ointments [113].

It is also relevant to note here that HPAC differs considerably in its chemical composition from other cellulose ethers that have been used for this purpose. HPAC has a much lower degree of substitution as compared to Na~KMC and MC and hence more free hydroxyl groups where there is more opportunity.

Table 39.

Ratio of drugs and excipients.

№/п	Preparations	Auxiliary substances, %			Method of administration
		VRAC	starch	calcium stearate	
1	Amidopyrine	10	1	up to 1	to pellets
2	Acetylsalicylic acid	1	3	up to 1	- "-
3	Barbital	5	3	up to 1	- "-
4	Glucose	0,5-1,0	-	up to 1	- "-
5	Terpinhydrate	1	2	up to 1	- "-
6	Phenacetin	5	5	up to 1	- "-

Table 40.

Technological characteristics of tablets using HPAC.

Preparations	Auxiliary substances,%				
	Conz Vrats; %,ml / Amount of HPAC solution, ml	starch	calcium stearate	by factory technologist.	at the recommendation. techn.
Amidopyrine	5-10/20	3-7	0,5-1,0	10-20	5-10

Glucose, acetylsalicylic acid	0,5-1,0/10	2-3	0,5-1,0	10-20	2-4
Terpinhydrate	20	-	-	-	-
Cyqualolol	5/30	5	1,0	200*	100
Acetylsalicylic acid with phenacetin, acetylsalicylic acid with phenacetin and codeine, amidopyrine with caffeine, amidopyrine with caffeine and phenobarbitol.	1/10-20	3-8	1,0	200	6-8

[x] the total amount of noparticles to form various intermolecular bonds with the other ingredients composing the ointment. Aqueous solutions of HPAC at a concentration of 5-7% have the form of an elastic gel, which are easily diluted with water. The solutions are physiologically indifferent, of neutral reaction, odorless and tasteless. stable for 25-30 days. Numerous experiments in different variants have allowed to recommend the following HPAC prescriptions as a base 6%.

I

Paraffin	40,0
petroleum jelly	40,0
HPAC in solution (6%)	20,0

II

Paraffin	40,0
castor oil	40,0
HPAC in solution (6%)	20,0

III

Vaseline	80,0
HPAC in solution (6%)	20,0

Ointments were prepared using sulfur, zinc oxide, furacilin, salicylic acid, potassium iodide, and novocaine [113].

Experiments have shown that all the above ointments can be prepared by combining equal amounts of paraffin and castor or vaseline oils, as well as vaseline with an aqueous solution of BPAC. When ointments with salicylic acid were prepared with the recommended bases I and II. a pink staining was observed one day after preparation owing to their incompatibility.

Therefore, it is advisable to prepare salicylic acid ointment on the basis of III composition. This ointment did not change during storage for 4 months. The prepared suspension ointments were studied for homogeneity according to the State Pharmacopoeia X edition and in all cases no visible particles were found. The effect of HPAC on the yield of drug substances from ointments was also studied.

Ointments were prepared with 5% iodine solution by incorporating 1 ml of iodine solution in potassium iodide into 9 g of the corresponding base. Diffusion rate observations were performed in 1,3,6,12 and 24x3600c over 24 hours. The results show that the iodine solution in potassium iodide has a high degree of diffusion from the BPAC base in composition with petroleum jelly, paraffin and petroleum jelly. Weaker from emulsion base with emulsifier T-2 and weakest from vaseline.

3. Application of water-soluble acetylcellulose for crease-free finishing of cotton fabrics.

Until now, HPAC and its derivatives have not been used in industry or any other areas of the national economy. Although other water-soluble cellulose ethers, especially carboxymethyl cellulose, have found very wide application. Therefore, expanding the number of available and environmentally friendly water-soluble cellulose derivatives is an urgent task for technology.

In this connection, having studied the properties of new water-soluble cellulose ethers obtained on the basis of available raw materials and existing basic production, it was necessary to study the possible fields of their application in the national economy. Extensive studies carried out in different directions allowed to confirm the position that water-soluble acetates and various derivatives based on them are a significant and important area of chemistry and technology of cellulose, having both theoretical and practical importance.

It is known that in the textile industry cellulose ethers are used as thickening and finishing agents. HPAC and preparations based on it may have significant advantages over the known ones. One such advantage is the ease of transition from the soluble to the insoluble state. This transition can be made by cross-linking with bifunctional reagents or by further saponification to hydrate cellulose.

The use of water-soluble acetylcellulose as a nonwrinkle appratus for both cotton fabrics and synthetic fibers can give them other useful properties that are impossible or difficult to obtain with other finishing agents. In particular, use of BPAC for impregnation of synthetic fibers with subsequent saponification or cross-linking would allow to create so-called bicomponent fibers, the surface layer of which would have high hydrophilicity, the inner layer of fiber "core" would impart high strength and elasticity. Such biocomponent fibers, in addition to their high hygienic properties, would also have reduced electrifiability. On the other hand, the apposition of cotton fabrics with water-soluble appers, which can be derived from soluble cellulose ether, could lead to a dramatic improvement in consumer properties. In general, the use of HPACs in such a large-tonnage industry would dramatically improve product quality. Further experiments have been directed to the use of HPACs for these purposes [114-115].

Currently, the textile industry uses about 60,000 tons of food starch a year for sizing, finishing fabrics and other needs.

In this regard, there is a need to find full-fledged substitutes for food products. In addition, conventional starch and glue applications are completely washed out in the first washings, so they cannot have a significant effect on increasing the fabric's wearability, and the finishing effect also disappears with washings. For this reason, it is also important to obtain an application that not only temporarily refines cotton fabrics, but also increases their resistance to abrasion, the main cause of fabric wear over time.

This section describes experiments on application of HPAC for textile appliqué. The application was carried out by impregnating the fabric with 2% acetylcellulose solution, pressed on rollers and dried at 373-378°K. After that, in the dried to a constant weight of the applied material, the amount of applied adhesive was determined.

The amount of apprette applied to the fabric was about 8-9% to the original material. The application on the fabric was further tested for its resistance to washing by the standard method [116]. In this case, the remaining applicator was almost completely washed off during the first wash (the amount of remaining applicator was 0.15% of the original material).

To increase the amount of remaining appetite, the HPACs were saponified in various ways.

For saponification with soda ($NaHCO_3$), an estimated amount of $NaHCO_3$ was injected into a 2% aqueous solution of HPAC while stirring. Afterwards, the fabric to be coated was immersed in this solution. After wringing on the rollers, the fabric was brought into the thermostat and incubated at 343°K for 3x3600s. At this temperature, the soda decomposes and the alkali released saponifies the acetyl groups of HPAC and thus regenerated cellulose remains on the fiber, which is not removed during further washing. As the results of experiments have shown, the appetite is resistant to multiple washes and is completely removed only after 25-30 washes. After each wash, the material had a good look, elegant griffin and gloss. The strength of the fabric increases by 12-14%.

The results of the experiment are shown in Fig. 78.

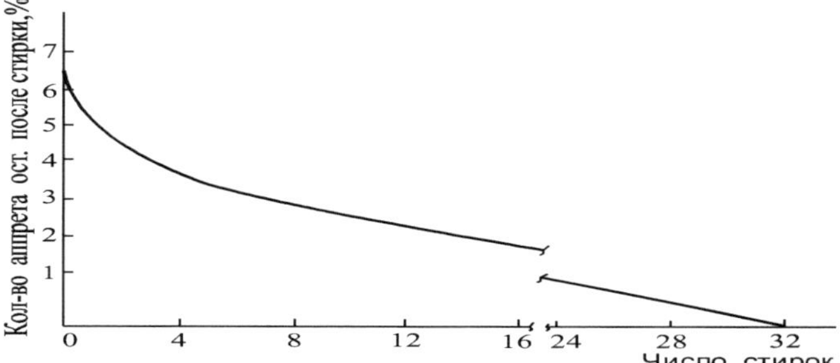

Fig.78. Changes in the amount of apprette after washing.

Thus, water-soluble acetyl cellulose can be successfully used in the textile industry for fabric finishing. Currently, much attention is paid to giving cellulosic materials the properties of crease-free wet, since cellulosic fabrics with such a property do not require ironing. However, the known methods of obtaining crease-free wet fabrics are technologically complicated. To simplify it some researchers (117, 118) proposed to soak the cotton fabric in strongly acidic solution (pH=1,8-2,0), dry it at 365-383 °K to humidity of 6-12% and keep this fabric in the roller from 14 to 30x3600s. Under such conditions of heat treatment there was a partial cross-linking of the wet swollen cellulose material and it acquired the properties of unkinkability in the wet state.

Others /119/ treated cotton fabrics pre-swollen in alkaline solutions with a crosslinking reagent. Under such processing conditions, the cellulose from form I changes into form II. Cross-linking of the cellulose macromolecules in this swollen state (form of cellulose II) stabilizes the resulting structure II and the fabric becomes unkinkable in the wet state. However, these methods are currently difficult to carry out with existing equipment and are obviously of theoretical importance.

In this regard, it is of great interest to find new processing methods or new excipients that without complicating the existing technology would make it possible to obtain cotton fabrics that have the property of not crease in the wet and dry state.

For this purpose, water-soluble acetylcellulose introduced into the impregnation solution (crosslinking reagent) was used as an excipient. In the presence of a large number of hydroxyl groups (γ=245), HPAC is able to enter into a cross-linking reaction with cellulose via dimethyl ethylene urea (DMEM). If HPAC and DMEM are dissolved in a common solvent (water), the cross-linking reaction between HPAC itself and the tissue via DMEM cannot be excluded. Therefore, the assumed cross-linking reaction can be written down as follows:

$$1.\ Öåëë—O—CH_2—N\underset{\underset{O}{\overset{\|}{C}}}{\overset{CH_2—CH_2}{\diagup\diagdown}}N—CH_2—O—ÂÐÀÖ—O—CH_2—N\underset{\underset{O}{\overset{\|}{C}}}{\overset{CH_2—CH_2}{\diagup\diagdown}}N—CH_2—O—Öåëë$$

$$2.\ Öåëë—O—CH_2—N\underset{\underset{O}{\overset{\|}{C}}}{\overset{CH_2—CH_2}{\diagup\diagdown}}N—CH_2—O—Öåëë$$

$$3.\ ÂÐÀÖ—O—CH_2—N\underset{\underset{O}{\overset{\|}{C}}}{\overset{CH_2—CH_2}{\diagup\diagdown}}N—CH_2—O—ÂÐÀÖ$$

The flow of the reaction according to the first scheme is more preferable compared to the other two reactions. The fabric samples were treated with an impregnation solution containing 180 g/dm^3 DMEM, 10 g/dm^3 AM preparation, 5 g/dm^3 polyvinyl acetate emulsion, 14 g/dm^3 urea, 8 g/dm^3 Mg(H$_2$PO)$_{42}$ -4H O$_2$ and 5 g/dm^3 BPAC. Studies have shown that the cotton fabric acquires the property of not creasing in the dry and wet states. The change in the opening angle of a fabric sample (deg.) subjected to the treatment and measured by the staff of the CNIIHBI is shown below:

№	Basics	Ducks	Amount
	In dry condition		
1	129,8	132,0	261.8
2	120,6	120,1	240.7
3	123,0	128,6	251.6
4.	56,0	68,0	124.0

sourc e			
	Wet		
1	129,3	136,1	265,4
2	125,9	126,5	252,4
3	141,3	150,0	291,3
4. sourc e	73,6	71,9	145,5

The addition of HPAC in an insignificant amount of reagent dramatically increases the effect of kneelability of the fabric in the wet state, which can only be explained by the course of the reaction according to the first scheme. If the reaction had proceeded according to the second scheme, we would not have obtained such a high effect of creaseability in the wet state, there would have been an effect of non-kinkability only in the dry state (Fig. 79).

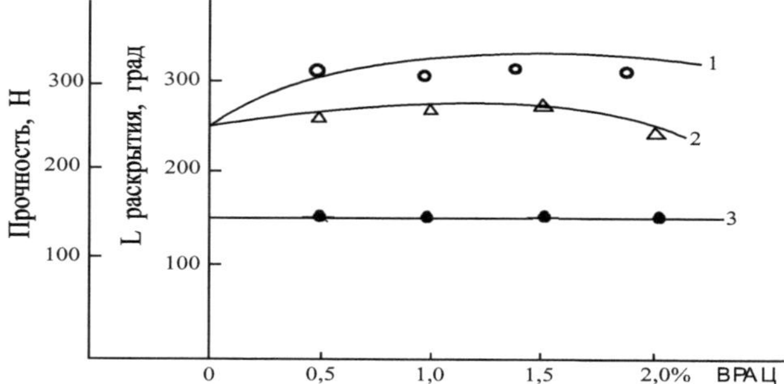

Fig.79. Influence of HPAC concentration on the change of opening angle in dry (1), wet (2) and tensile strength (3) of glass fabric.

To confirm this, a number of studies were carried out where tri- and secondary acetylcellulose with different amounts of free hydroxyl groups capable of entering the cross-linking reaction and glass fabrics were used as model substances. The model substances taken were treated according to the above formulation with the addition of different amounts of HPAC under conditions similar to those required for the treatment of cotton fabrics. In the case of glass fabrics and triacetyl cellulose, there is partial surface resin deposition in the form of "stocking" (up to 2%), and this results in a slight increase in the opening angle. But the strength of these fabrics does not change, which indicates the absence of cross-linking reaction (Fig.79).With secondary acetylcellulose, where there is a significant amount of free hydroxyl groups, the opening angle after treatment slightly increases, the strength decreases compared to triacetylcellulose and glass fabric (Fig.80). In this case there is

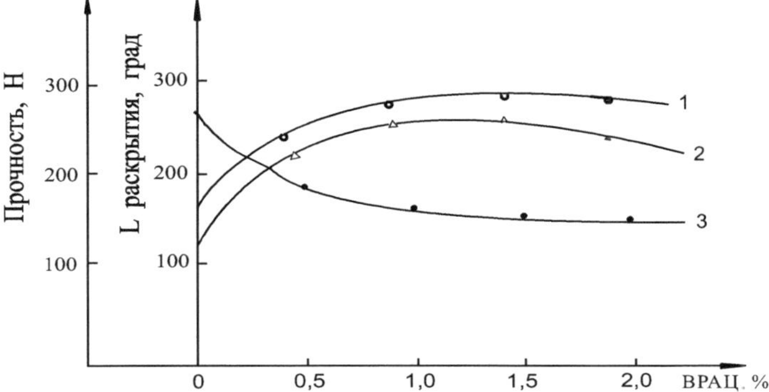

Fig.80. Effect of HPAC concentration on changes in opening angle in dry (1), wet (2) and tensile strength (3) of cotton fabric.

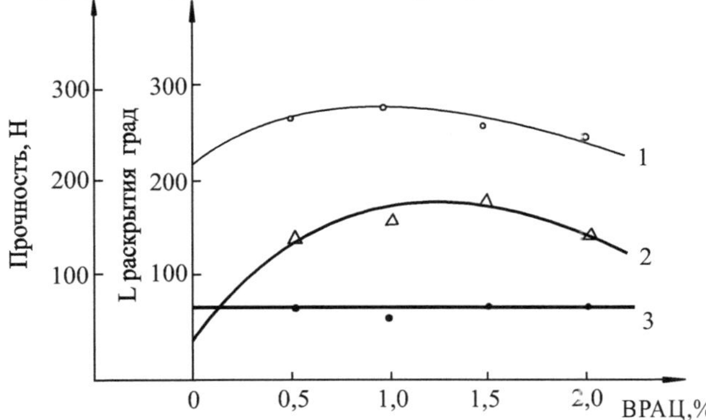

Fig.81. Effect of HPAC concentration on the change of opening angle in dry (1) wet (2) states and tensile strength (3) of acetate fabric.

partial formation of transverse bonds between the hydroxyl groups of secondary acetylcellulose and BPAC by DMEM, leading to a change in the physical and mechanical properties. The tensile strength decreases from 70 to 60 N. The opening angles in the dry state increase from 215 to

191

272°. A sharp increase in the opening angle is observed in the wet state (42-164°). After treatment of cotton fabrics with impregnating solution, the effect of non-kinking in dry and wet state increases sharply, and the tear strength decreases sharply (Fig.80).

In this case, apparently, a cross-linking reaction occurs in contrast to the above-mentioned model samples, which leads to profound physical and chemical changes.

Thus, under the described processing conditions, the reaction predominantly proceeds according to the first scheme, although the course of the reaction according to the second and third schemes is not excluded, but their proportion is significantly small compared with the first scheme.

In addition, a chemical method was used to confirm the chemical interaction of BPAC with the tissue via DMEM, i.e., the amount of bound acetic acid in the treated samples was determined and found to be consistent with the acetic acid content of BPAC. Consequently, HPAC is involved in the cross-linking process. The fabric retains its unkinkability properties in both dry and wet conditions when washed repeatedly. The results obtained are shown in Table 41.

Thus, by introducing a small amount of HPAC into the impregnating solution, it is possible to dramatically improve the kneadability properties of cotton fabrics in both dry and wet conditions.

Table 41.

Changing the properties of the non-wrinkability of the fabric after repeated washings.

Type of treatment	Weight gain,%		Total opening angle,deg			
	To Laun dry	after Laund ry	dry		wet	
			To Laundr y	after Laund ry	at Laund ry	after Laun dry
raw	-	-	134	-	143	-
treated with a known compound	7,9	6,5	243	220	168	160
treated with a preposition	10,7	10,4	270	265	267	269

**Use of water-soluble acetylcellulose to impart wet kneadability
and abrasion resistance to acetate fibers.**

It is known that acetate fibers, having many advantages over other chemical fibers (harmless production, cheapness, etc.) have a number of drawbacks that prevent their use in various industries. These disadvantages include: high electrifiability, poor resistance to abrasion, unkinkability, etc. Elimination of these disadvantages would open up the possibility of even wider use of acetate fibers in the textile industry.

To eliminate these drawbacks, an impregnating solution [120,121], usually used for modifying cotton fabrics, was also used, but with the introduction of small amounts of HPAC. The impregnating solution had the following composition, g/dm^3 :

Dimethyldioxyethylen e urea	100-160
VRAC	5-30
Catalyst	8-12
Softener	3-50
Acetone	10-20
Water	to dm^3

Fabric made of acetate threads was treated at room temperature with various compositions, the formulation of which is given in the table, squeezed to a residual moisture content of 80-100%, dried at 343-353°K and heat-treated at 443-453°K. After that the physical and chemical properties of the modified fabrics were determined. The results obtained depending on the composition of impregnating solution (Table 42) are shown in Table 43.

Table 42.

Compositions of impregnation solutions.

Components	Concentration of components in solution, g/dm^3				
	1	2	3	4	5
Dimethylenedioxyethylene urea	160	100	100	100	100
VRAC	5	30	5	5	5
Magnesium chloride	5	8	8	12	6
Quaternary salts of the condensation product of stearic acid and alkylamines					

Acetone	3	3	3	3	5
Water	10	10	20	10	10
to dm³					

As can be seen from Table 43, the introduction of a small amount of HPAC, as in the case of cotton fabrics, strongly affects the kneadability properties of fabrics made of acetate fibers in the dry as well as in the wet state. In addition, and very importantly, resistance is greatly improved.

Table 43.

Changes in physical and mechanical properties of fabrics made of acetate fibers when modifying with different compositions of impregnating solution.

Composition	Weight gain, %	Total opening angle, deg.		Abrasion, cycles
		Dry state	Wet condition.	
1	12,6	265	160	284
2	11,6	267	139	275
3	93,0	290	167	310
4	8,4	258	142	365
5	7,9	265	148	270
Source fabric		180	90	94

of fabrics to abrasion (3.0-3.5 times). It gives an opportunity to use such fabrics as lining materials instead of viscose ones used so far. One more important property of such fabrics is that after modification by proposed composition hydroscopicity of fabrics strongly increases, as can be seen from the following table 44.

Table 44.

Hygroscopicity of fabrics made of acetate fibers modified with impregnation solution at different relative humidity.

Relativity humidity, %	Water vapor sorption, %	
	original fabric	coated fabric
16	1,3	2,6
55	3,5	7,2
67	4,7	8,9
76	6,2	10,6

| 90 | 10,6 | 16,4 |

Changes in the sorption properties of modified acetate fabrics with the addition of BPAC after five and ten washings were determined. The results of the study (Fig. 82) show that the sorption properties of the fabrics modified in this way are practically preserved after multiple washes.

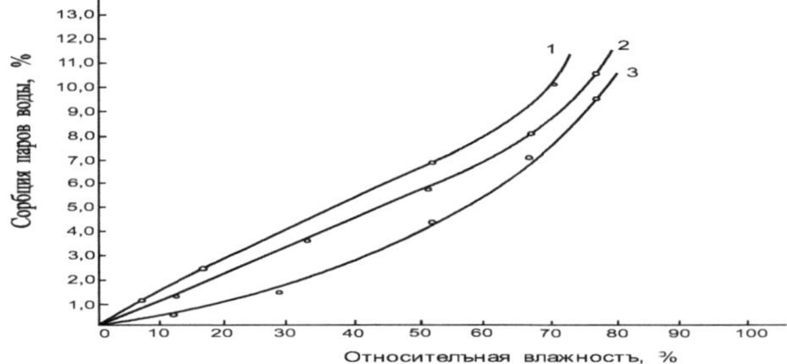

Fig.82. Isotherms of water vapor sorption of acetate fabric after repeated washings.

1. treatment with carbamol CEM+VRAC (30 g/l).
2. after 5 washes.
3. After 10 washes.

4.Application of water-soluble acetylcellulose to impart antistatic properties to acetate fibers and films.

As already noted, acetate fibers, due to their hydrophobicity, are highly electrifying, which can lead to undesirable aftereffects in production, as well as unpleasant wearing phenomena. In this regard, there is currently a search to find ways to eliminate this disadvantage in acetate fibers.

The possibility of removal of electrostatic charges, accumulated during processing of acetate fibers, by impregnation of acetate fibers with HPAC solution was shown [122-123]. For this purpose, 2,3,4 and 5% aqueous solutions of BPAC were prepared and acetate fibers were passed through this solution and dried afterwards. However, the formed layer of HPAC on the fiber in the form of a stocking after drying was completely removed during the wet treatment. In order to fix the HPAC on the fiber, it was lightly cross-linked with formaldehyde (4-8% concentration) in the presence of catalyst-chloric acid (2-4%). The results obtained are shown in Table 45. As can be seen from the table, treatment of acetate fibers with a 2% solution of HPAC .

Table 45.

Properties of acetate fibers treated with HPAC solution.

concentration VratC in p-R, %	gain after treatment, %	gain after 15 times washing,%	durability, sn/tex	lengthening,%	specific electrical resistance after 15 times washing.
Source	0	0	11,0	21,8	$6\text{-}10^{12}$
2	1,0	0,6	10,8	21,4	$6\text{-}10^{7}$
3	4,5	1,9	10,0	18,4	$4\text{-}10^{7}$
4	8,4	3,0	8,2	11,4	$4\text{-}10^{6}$
5	13,0	3,4	4,8	5,0	$4\text{-}10^{6}$

subsequent cross-linking greatly reduces (by 5 orders of magnitude) their electrifiability with little change in other indicators.

In order to obtain flexible and durable films with high electrical conductivity were dissolved in a common solvent (dimethylformamide or dioxane: ethyl alcohol in the ratios 90:10, 80:20 and 75:25) HPAC and copolymer of p-vinylbenzoic acid with maleic anhydride. Films cast from the solution were dried for 2-3600s at room temperature and treated for a day with 0.01 N NaOH solution, washed with water and dried at 80°C. The films obtained had high electrical conductivity (specific conductivity 0.2 ohm^{-1}) [123]. Such films can be used in the capacitor industry.

5.Application of water-soluble acetylcellulose as a dispersant for suspension polymerization of vinyl fluoride.

Since, as it was found earlier, HPAC combines well with such hydrophobic substances as fats, waxes, paraffins, etc., it was natural to assume that it would be a good protective colloid in processes of producing stable emulsions or suspensions of particularly hydrophobic substances, for example, in the process of suspension polymerization of fluorinated vinyl monomers.

For this purpose, experiments were performed on suspension polymerization of vinyl fluoride using Y-radiation as an initiator of the polymerization reaction in the presence of HPAC [124]. In addition to HPAC, methyl and carboxymethylcellulose were also used as a dispersant for comparison.

196

Polymerization was initiated by Co^{60} beams at initial concentration of dispersant in aqueous phase 0.01-2.00 wt. % and dose rate 20-150 rad/s. The formed polyvinylfluoride (PVF) in contrast to PVF obtained by radiative polymerization of WF in bulk is soluble in organic solvents, in particular, in dimethylformamide. The molecular weight (MM) of PVF samples was determined viscometrically and calculated using the well-known equation [125].

Figure 83 shows the dependence of the characteristic viscosity /η/. PVF obtained in the presence of HPAC, on its yield at different dose rates. As can be seen from the figure, the same pattern is observed for all dose rates, i.e., at the same dose rate, as the polymer yield increases, its /η/ and the MM of the PVF samples increase, and as the dose rate increases, the /η/ and MM of the PVF samples drop rather sharply. The decrease in /η/ and MM as the dose rate increases is due to an increase in the yield of free radicals, which results in an intensification of the growing chain breakage process, and the increase in their values with increasing PVF yield is due to the polymerization process under heterophase conditions [126].

The nature of the polymeric dispersant and its concentration in the aqueous phase have a significant influence on the properties of the formed PVF samples. Table 46 summarizes the obtained data. As can be seen from the above data, the values of /η/ and MM of the formed PVP in the case of HPAC and methylcellulose depend little on their content in the aqueous phase. A similar dependence was observed in the radical-suspension polymerization of styrene in the presence of polyvinyl alcohol [127]. However, in the case of carboxymethyl cellulose, a rather sharp decrease in /η/ and MM of PVP as its content in the aqueous phase is observed. It can be assumed that the observed difference is due to more intense chain transfer to carboxymethylcellulose macromolecules than to other dispersants.

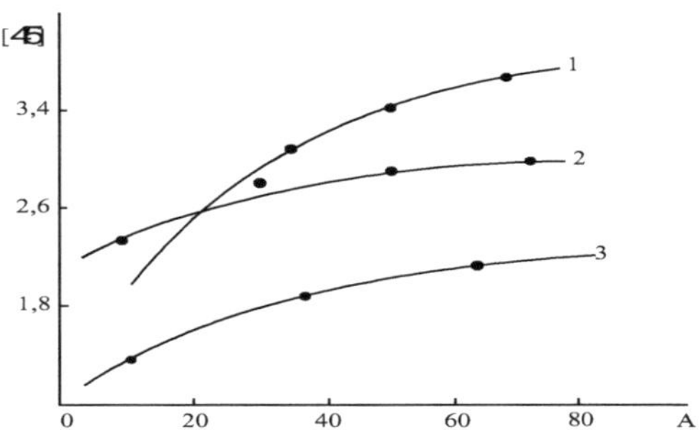

Fig. 83. Dependence of /η/ PVP obtained in the presence of HPAC on its yield A (wt%) at different dose rates: 1-20, 2-70, 3-150.

Table 46.

Dependence of characteristic viscosity and molecular weight of polyvinylfluoride on the nature of dispersant and its concentration in the aqueous phase (irradiation duration 3-3600s).

Dispersant	Concentrated dispersant in the aqueous phase, mass %	/η/	MM
Carboxymethylcellulo se	0,1	2,9	657000
	0,5	2,7	602000
	1,0	2,2	466000
Methylcellulose	0,1	2,8	630000
	0,5	2,7	600000
VRAC	0,1	3,15	730000
	0,5	3,0	687000
	1,0	3,0	687000

The monomer:aqueous phase ratio is one of the main hydrodynamic factors in suspension polymerization [127]. Table 47 presents data on the dependence of the characteristic viscosity and MM of PVP on the initial ratio of BF:aqueous phase.

Table 47.

Dependence of characteristic viscosity and molecular weight of polyvinylfluoride on the monomer:aqueous phase ratio (dispersant

concentration in aqueous phase -0.1 wt.%, irradiation duration 3-3600s).

Dispersant	Polymer:aqueous phase ratio	/η/	MM
Carboxymethylcellulose	1 1	2,1	439000
	1 3	2,9	657000
	15	2,3	493000
	17	1,6	312000
VRAC	11	2,5	560000
	13	2,8	630000
	15	2,4	520000
	17	2,2	480000

As follows from the data in Table 47, regardless of the type of polymeric dispersant used /η/ and MM of the formed PVF increase in the transition from a ratio of 1:1 to 1:3 reaching a maximum, and then decrease with a further increase in the phase ratio. The initial increase in /η/ and MM of PVP appears to be related to the effective removal of reaction heat, which prevents the phenomenon of local overheating to some extent, and their decrease with further increase in the monomer: aqueous phase ratio is explained by the strengthening of the chain breaking reaction.

LITERATURE.

Hiller L.A. The Reactions of Cellulose acetate with Acetil Acid and Water. J. Pol. Scl., 1953, vol. 10, no. 4, p. 385-423.

2. Malm C.J., Barkey K.T., Martin Salo, May D.C. Far-Hydrolyzed Cellulose Acetates. 1nd. Eng. Chem., 1957, vol. 49, No. 1, p. 79-83.

3. U.S. Pat, là 2129052, 1938, LMT.: Malm C, J., Barkey K.T., Martin Salo, May D.C. Far-Hydrolyzed Cellulose Acetates. inc. Eng. Chem., 1957, vol. 49, No. 1. pp.79-83.

English Patent No. 479239, 1938. Citation: Malm C.I., Berkey q.t., Martin Salo, May D.C. Far-Hydrolyzed Cellulose Acetates. Chem., 1957, vol. 49, no. 1, pp. 79 83.

5. Howlet Martin. Study of the rate of soponification of cellulose acetates. J. Text. ind., 1947, 38, 4, vol. 212.

6. Vos K.D., Burris P.0., Riley R.L., Kinetic study of the Hydrolys1s Pol of cellulose acetate in the ph Range of 2-10, J. Appi. Pol. Sci, 1966, vol. 10, N° 5, p.825-832.

7. Petropavlovsky G.A., Rakhmanberdiev G. Kinetics of deep hydrolysis of cellulose acetate. ZhT1X, 1966, vol. 39, 11, c. 237-240.

8. Petropavlovsky G.A., Rakhmanberdiev G. Structural and analytical characteristics of products of deep hydrolysis of cellulose acetates. Petroleum Chemistry, 1966, Vol. 39, 42, c. 423-427.

9. Rakhmanberdiev G., Petropavlovsky G.A. Products of deep hydrolysis of cellulose acetates. In: Chemistry and technology of cellulose derivatives. Vladimir, 1968, p. 235.

10. Rakhmanberdiev G., Petropavlovsky G.A., Vasilyeva G.G., Volkova L.N. Relationship between radiographic structure and mechanical properties of films and fibers obtained from water-soluble acetates. Theses of reports of XU scientific conference of IVS AS USSR, Leningrad, 1967, p. 41.

11. A.S. 0735596 (SSR.). Method for obtaining water-soluble acetylcellulose (Inoyatova A.Kh., Berenshtein E.I., Kaybusheva R.Kh., Aikhaygaev B.I.). - Published in B.I., 1980, #19.

Aikhodjaev B.I., Inoyatova A.H., Berenshtein B.I. Preparation and study of properties of primary low-substituted soluble cellulose acetates. Vysokomol. soed. 1982, A24, 06, p. 111-1322.

13. Torsu J.P., Lehrhart W.M. Properties and potential uses of Na-sulfoacetate of cellulose a new water-soluble cellulose derivative. J. Chem. and Engs, 1964, 16, J No. 4, p. 566.

14. Control of the production of chemical fibers. Edited by A.B. Pakshver and A.A. Konkin. - 606 s.o.

15. Calistru E., Gheorghin A. Complecsi metalici ca solveni1 pentru cellulosa. 11. Sollitile alcalene ale complexylei feicracid tartresodin (EWNN) siulilizarea ior in chimia cellulozei. "Eeluisi hirtic," 1964, 13, N 11-12,p. 400-405.

16. Edelman K., Horn E. Zur Herstellung und Anwendung der alkal1schen Eisenwe1nsanre-Natrium-Komplexlösung für die viskosimetrische DP-Bestimmung der Cellulose. Faserforsch und Textiltechknik, 1958, vol. 9, N° 1, pp. 493-500

17. Practical works on the chemistry of wood and cellulose. Edited by V.M. Nikitin. M.: Lesn. prom. 1965,-411

18. Timell T.E.. Effect of Deacetilation and Nitration on Normal Giycosidie Linkages in cellulose. J. Pol. Sci., 1963,p.c., N°2, pp. 109-116.

19. Haward R.N. Degradation of Ethyi cellulose in solution. J. Pol. Scl., 1950, v. 5, No. 5, pp. 635-636.

20. Sharples A. The Hydrolysis of cellulose. Part 1. The Fine Structure of Egyption Cotton. J. Pol. Sci., 1954, vol. 13, pp. 393-401.

21. Frenkel S.Ya. Introduction to the statistical theory of polymerization. Moscow: Nauka, 1965. - 267 c.

22. Rakhmanberdiev G., Sadykov M.M., Usmanova M.I., Mirnigmatova Sh. Determination of the degree of polymerization and the rate of destruction of secondary cellulose acetate in the process of hydrolysis.Uzb.chem.zhurnal, 1975, #1,p.50-52.

23. Rakhmanberdiev G., Sadykov M.M., Mirnigmatova Sh., Gafurov T. Study of kinetics of deep hydrolysis of secondary cellulose acetate. Uzbek Chemical Journal, 1975, No. 2, pp.43-46.

24. Usmanova M.I., Kamilov Yu.M. Determination of molecular weight of water-soluble acetylcellulose by osmometric and viscometric methods. Chemistry of Wood, 1977, JE 1, pp. 47-51.

25. Usmanova M.I., Rakhmanberdiev G. Fractionation and determination of Molecular weight of water-soluble acetylcellulose. Materials of 1X annual conference of NIICHTC. Tashkent, 1974, pp. 61-62.

26. Usmanova M.I., Usmanov H.U. Determination of molecular weight of standard polystyrene samples by osmotic method. Uzb, Chem. journal, 1963, No. 3, pp. 64-69.

27. Usmanov H.U., Usmanova M.I. Study of ethyl cellulose fraction solutions by osmometry and viscosimetry methods. In the book: Structure and modification of cotton cellulose. Tashkent: Fan, 1966, p. 73-81.

28. Cleverdon Mrs. D. Loker, Smith P.G. Uncertaintles 1n Measurement of Osmotic Pressure of Polymer Solutions. J.Pol. Sci., 1954, vol. 13, pp. 393-401.

29. Immergut B., Ranby B.G., Mark E.P. Recent work on melecular weight No. 11, pp. 2483-2490. of cellulose. ind. Eng. Chem. 1953, vol. 45,

30. Rafikov S.R., Pavlova S.A., Tverdokhlebova I.I. Methods for determining the molecular weights and polydispersity of high-molecular compounds. Moscow: Nauka, 1963. - 335 c.

31. O'Connor R.T., Dyfre E.F., McCal B.R. infrared spectrophotometric procedure for the analysis of cellulose and modiled cellulose. Analytical Chemistry, 1957, vol. 29, N°7, pp. 998-1005.

32. Higgins H.G. The use of potassium chloride disks in the infrared examination of fibrons cellulose and other solid materials, Australian Journ. of Chem., 1957, vol. 10, 14, pp.481 483.

33. Gerbayx R. Surla structure chimique des acetates de cellulose. 11. Etude de la band d'absorption des founctions hydroxyles dans l'infrarouge. Bull. Soc. Chim. Belg., 1957, 66, pp. 382-390.

34. Rakhmanberdiev. Determination of the position of groups in water-soluble acetylcellulose. In: Chemistry and technology of cellulose and fiber. Tashkent: Fan, 1973, pp. 190-191.

35. Hearon W.M., Hiatt C.D., Forduce C.R. Carbamates of Cellulose and cellulose acetate. J. Amer. Chem. Soc., 1943, vol. 65, no. 5, pp. 829-833.

36. Gutterman L., Wieland 1. Practical works in organic chemistry. L., 1930, - 327 p.

37. Usmanov K.U., Usmanov T.I., Suleimanova R.T. Study of the distribution of substituents in 2.3. Di-O-acetate of cellulose. Chemistry of wood, 1982, #3, pp. 7-10.

38. Usmanov T.I., Suleimanova R.T., Karimova U.G. Compositional heterogeneity and NMR spectroscopy of partially substituted cellulose acetate. Abstracts of the International Symposium on Macromolecular Chemistry. Moscow: Nauka, vol. 6, p. 118.

39. Ranby B.C., Nae R.W.Crystallisation of Cellulose and Cellulose Derivatives from Dilate Solution. 1. Growth of Single Crystals. J. Pol. Sci., 1961, vol.51, 1° 155, pp. 337-347.

40. Conrad C.M., Creeby J.I. Thermal X-ray Diffraction Study of Highly Acetylated Cotton Cellulose. J. Pol. Sci., 1962, vol. 58, no. 166, pp. 781-790.

41. Brown L., Holiday Po, Trofter 1.P. The absorption spectra of some cellulose derivatives in the 3 M region. J. Chem. Soc., 1951, vol. 1, pp. 1532-1539.

42. Gerbaux R. Sur la structüre chimiqual des cellulose. J. Bull. Soc. Chem., Belg., 1956, vol.65, pp. 270-290.

43. Zueva R.V., Zhbankov R.G., Ivanova N.V., Kozlov P.V., Podgorodetsky E.K. Cellulose acetates obtained under homogeneous conditions of acetylation and partial saponification. Cellulose and its Derivatives, 1963, pp.118-123.

44. Mironov D.P., Vasiliev B.V., Domkin V.S., Zatsepin A.G., Sokolova M.V., Malinina L.P., Pogosov Yu.L. Synthesis and properties

of cellulose acetates with different distribution of substituents. Cell. Chem. and Techn., 1977, 11, pp. 411-420.

45. Hyrtybise F.C. The analytical and structural aspects of the infrared spectroscopy of cellulose acetate. Tappl., 1962, vol. 45, no. 6, pp. 460-465.

46. Sokolov A.D. Hydrogen bonding. Uspekhi Phys. Nauk, 1955, vol. 57, issue 2, p.205-276.

47. Petropavlovsky G.A., Rakhmanberdiev G. Products of deep hydrolysis of cellulose propionates. Zhh, 1967, vol. 40, № 8, c. 1793-1797.

48. Terasaki 1. Preparation of cellulose propionate (1) (11). J. of the Society of Textile and Cellulose industries, Japan (Sen-1 Gakkelshl), 1962, vol. 18, 94, p. 331-340.

49. Hauptmann, Greore J., Remane H. Organic Chemistry. Moscow: Chemistry, 1979, - 832 p.

50. Petropavlovskii G.L., Krunchak M.M., Vasilyeva G.G. On solubility of highly substituted methyl and ethyl cellulose in water.Cell. Chem. Techn. 1972, 6, № 2, c. 135-144.

51. Mark H.F.. Polymers in Metrial Science. J. Pol. Scl., 1965, p.c., N 9, pp. 1-33.

52. Miller M., Haskell V.C. The Mechanism of Film Formation from viscose. J. Appl. Pol. Sci., 1961, vol. 5, No. 18, pp.627-634.

53. Price C.R., Haskell V.C. The Dry Casting of Viscose Film. 1. Appl. Pol. Scl., 1961, vol. 5, No. 18, pp. 635-640.

54. Nikonovich G.V., Burkhanova N.D., Usmanov H.U. On genesis of over-molecular structures of different types of hydratecellulose fibers. Cellulose Chem. Technol., 225-268. 1971, Vol. 5, No. 3, pp.

55. Trebe, Claret, Matson, Eosht, Purz, Tejrek, Paul. Study of the structure formation process of regenerated cellulose fibers. Chemistry and Technology of Polymers. 1965, c. 99-III.

56. Drish N. Improvement of fibrillar structure and properties of hydrate cellulose fibers. D.I. Mendeleev's ZhVHO, 1966, vol. 9, je 6, p. 647-653.

57. Vasilyeva G.G., Petropavlovsky G.A. Study of reactions of partial hydrolytic elimination of ether groups and subsequent crosslinking of cellulose acetobutyrate by Dimethyloethylene Urea. Jh, 1976, vol. 49, Jk 3, pp. 606-610.

58. Rakhmanberdiev G., Isadjanov B. Study of stability of water-soluble acetylcellulose solutions to the action of precipitates Tashkent: Fan, 1977, p. 117- tel. C6. "Acetate fibers", 121.

59. Rogovin 3.A. Chemistry and technology of artificial fibers. M.: 1952. -560c.

60. Geller B.E., Pakiver A.B. on interaction of chlorinated polyvinyl compounds with solvents. T., 1952, vol. 25, 11, pp.1196-1200.

61. Khamraev A.L. On the stability of alkaline solutions of carboxymethylcellulose to the action of precipitators. In book: Chemistry and chemical technology of high-molecular compounds". Tashkent: Fan, 1967, pp.68-74.

62. Nachinkin O.I. On the structure of polyvinyl alcohol fibers produced in sedimentation baths of different compositions. Chemical fibers, 1965, J1, p. 39-41.

63. Rakhmanberdiev G., Isadjanov B.U. Properties of water-soluble acetylcellulose solutions. Chemistry of wood, 1978, 1, pp.93-95.

64. Tager A.A. Physico-chemistry of polymers. Moscow: Nauka, 1963. -536c.

65. Rakhmanberdiev G., Isadjanov B. Properties of diluted solutions of water-soluble acetylcellulose. Materials of XIIMaterials of X-th annual conference of NIICHTs, Tashkent, 1976, p.60.

66. Papkov SP Physical and chemical bases of the production of artificial and synthetic fibers. Moscow: Chemistry, 1972.-Z12p.

67. E. E. Karelsky, V. I. Kozlova, L. P. Perepelkin. Some Rheological Properties of Solutions of Cellulose Simple Ethers. Chemistry and Technology of Cellulose Derivatives. Vladimir, 1971, pp.282-286.

68. Miroshnichenko I.B., Rakhmanberdiev G., Isadjanov B.U. Study of rheological properties of water-soluble acetylcellulose. Zh I X, 1978, vol. 51, No. 3, pp.713-716.

69. Gleston S., Leitzler K., Eiring G. Theory of absolute reaction rates. MOSCOW: IL, 1948. -491c.

70. Rakhmanberdiev G., Petropavlovsky G.A., Gafurov T.G., Usmanov H.U. New water-soluble fiber based on acetylcellulose. Preprints of the International Symposium on Chemical Fibers. Kalinin, 1974, Section 4. pp.150-154.

71. Rakhmanberdiev G. Imsaludnor B.U. Obtaining rhodorosolvated 1978, 4, p. 42. fibers based on cellulose acetate. Hal. Fibers,

72. Usmanov K.U., Turabayev U., Rakhmanberdiyev G., Rozahunor R.H., Khidoyatov A.A. Reactive cotton cellose and acetate fibers on its basis. Abstracts of the congress on general and applied sciences, section "Problems of forestry", Baku, 1981, v. 6, p. 190-191.

73. Kargin V.A., Slonimsky G.A. Short sketches on physical chemistry of polymers. Moscow: Khorya, 1967. - 23 c.

74. Mikitishin O.I., Tinning A.N. On the strength of oriented polomers. Physico-chemical Mechanics of Polylers. 1968, vol. 4, No. 3, pp. 256-258.

75. Peremkulova H.T., Isadhanov B.U., Nasretdinova Sh.Sh., Rakhmanberdiev G. The study of surface structure of fibers on the basis of rodorosoluble acetylcellolozi. Materials of X11-2 year1rconference of Research Institute of Htz, Tashkent, 1976, p. 6.

76. Shoshina V.I., Nikonovich G.V., Khakimova G.F., Isalkhanov B., Rakhmanberdiev G. Physico-chemical studies of fibers based on water-soluble acetylcellulose molded from dolethylformamide solutions. Theses of the brief reports of the International Symposium on Macromolecular Mass.: Nauka, 1978, vol. 6, p. 114.

77. Nikonovich G.V., Usmanov H.U. Supramolecular structure of hydratzelose fibers. Tashkent: Fan, 1974, 307 p.

78. Nikonovch G.V., Burkhanova N.D., Usmanov Kh. Cellulose Chea. Technol. 1967, 1.1, 11° 6, c. 689-709.

79. Nilkonovich G.V., Kerimova 3.G., Burchanova N.D. and Usmanov Ch.U. The supermolokular structure of viscose fibers and i'ts dependence of spiningconditlons. J.Pol.Sci., 1973, N 42, pp. 1625-1637.

80. Usmanov H.U., Nikonovich G.V. Electron microscopy of cellulose. Tashkent: Fan, 1962. - 79 c.

81. Farahova F.n., Nikonodagch G.V., Usmanov H.U. Some problems of genesis of structure of fibers on the basis of cellulose ethers. Theses of reports of Mendeleev congress. Almaata, 1975, pp. 223-224.

82. Nikonovich G.V., Farahova F.H., Burkhanov N.D., Leont'eva S.L., Usmanov H.U. Some problems of genesis of structure of acetate fibers. Abstracts of International Symposium on Chemical Fibers. Kalinin, 1977, vol. 1, c. 189-190.

83. Usmanov K.U., Razykov K.K. Light and electron microscopy of structural transformations of cotton. Tashkent: Fan, 1974, 277 p.

84. Shoshina V.I., Nikonovich I.V., Khakimov U.B., Usmanov Kh. Cellulose Chem. Technol. 1974, 1.8, 163, c. 215-246.

85. Layus L.A., Kuvshinsky E.B. Isothermal heating as a method of studying oriented solid amorphous polymers. Vysokomol. sol. 1964, vol. 6, c. 52-58.

86. Bessonov M.I., Rudakov A.P. Study of stresses arising during heating of fibers from polyvinyl alcohol. Chemical Fibers, 1964, 52, pp. 30-35.

87. Naimark N.I., Fomenko B.A. Thermomechanical properties of triacetate fiber. Vysokomol. soed. 1966, vol. 8, no. 12, pp. 2082-2086.

88. Petropavlovsky G.A., Vasilyeva T.G., Rakhmanberdiev G. Interaction of dimethylolurea and water soluble acetylcellulose. ZhPH, 1969, vol. 42, 12, pp. 2813-2819.

89. Lee VS, Anarmetova D, Gafurov T, Shakirov KA A. A new catalyst for non-abrasive finishing of cellulose materials. Uzbek Chemical Journal, Dep. in VINITI, 2/4-1974, № 788.

90. Anarmetova D., Li V.S., Gafurov T.T. Modern types of closureSocial finishing of cotton fabrics of "wash-wear" type. Uzbek Chemical Journal, Dep. in VINITI, 2/4 - 1974, 1 7870.

91. Anarmetova D., Lee VS, Khudaiberganova Z. Increase in elasticity of cellulose without loss of mechanical strength. Abstracts of the International Symposium on Macromolecular Chemistry, Moscow: Nauka, 1978, vol. 6, p. 59.

92. Gafurov T.G., Melikuziev Sh., Rakhmanberdiev G., Tashpulatov U.T., Usmanov H.U. Modification of properties of different cellulose fibers by dimethylthiourea. Uzbek Chemical Journal, 1971, 4, pp. 67-71.

93. Melikuziev Sh., Rakhmanberdiev G., Nigmankhojaeva M., Gafurov T.B., Tashpulatov U.T. Modifications of properties of hydratecellulose fibers of DMZM. Chem. Fibers, 1972, 4, p. 26.

94. Petropavlovskii G.A., Yakovenko L.K., Vasilyeva G.G. Effect of chemical cross-linking on the properties of methylcellulose films. Cell. Chem.Technol., 1976, Vol. 10, no. 3, pp. 325-373.

95. Petropavlovsky G.A., Yakovenko L.K., Vasilyeva G.G. Study of interaction of dimethylurea with water-soluble methylcellulose. 1976, т. 10, № 3, c. 323-333.

96. Vasieleva G.G., Petropavlovskiy G.A. E1genschafteb orëntierter chemisch varnetzter Filme aus wasserlöslichen Cellulosacetat. Faserforchung und Text1itechnol., 1974,25, N° 9, s. 482-486.

97. A.S.318583 (SSSR). Method for obtaining modified acetylcellulose and products based on it. (Petropavlovskii G.A., Vasil'eva G.I., Rakhmanberdiev G.). - Publ. in B.I., 1971, 20 32.

98.Petrov G.S. Artificial resins and plastics. M.: 1937. - 607 c.

99. Gonzales E.J., Benerito R.R.. Effect of the Catalyst Upon the Physical and Chemical Propertles of Cotton Cellulose Fimshed with Derivatives of Ethyleneurea. Text. Res. J., 1965, 35, 2, pp. 168-178.

100.Nikomoto K. Infrared spectra of inorganic and coordination compounds. Moscow: Mir, 1966. - 351 c.

101. ihor Lysyj and E.Zarembo. Rapid quantitative determination of sulfur in organic compounds. J. Anal. Chem., 1958, vol. 30, 1930, p. 428-430.

102. Simonian I.V. infrared spectra of reactant Amer. Dyest. Rep., 1964, v.53, no 7, pp. 15-24. monomers.

103. Zomutreev B.A., El-Banna H.M. A new method of making tablets of amidopyrine. Chem Pharm. Journal, 1970, 6, p. 46.

104. Kulesh K.F., Bugrim A.A., Konev F.A. Replacement of extemporaneous dosage forms with tablets. Pharmacy, 1965, 1, p, 19-25.

105. Mahkamov SM, Karakozova SA About the porosity of tablets and methods of its determination. Materials of the anniversary republican scientific conference of pharmacists dedicated to the 50th anniversary of the USSR. Tashkent, 1972, p. 85.

106. Borisenko Y.B., Borzunov E.E., Dehborenko V.M. Oxypropyl methylcellulose as a binder in vibrotniitvi tablets. Pharm. zhurn. (Ukr.), 1969, 2, p. 51.

107. Rubtsova VP, Shevchenko SM, Pogorelsky ЭI, Gluzman MX, Pogovitskaya SA Application of microcrystalline cellulose in the manufacture of tablets and soft gelatin forms of modern problems of pharmaceutical science and practice. Abstracts of reports of the Congress of pharmacists. Kiev, 1972, p.182.

108. State Pharmacopoeia. USSR. X ed., 1968, p.667.

109. Makhkamov S.M. Materials of the All-Union Scientific Conference on the improvement of the production of medicines and drugs. Tashkent, 1969, p. 106.

110. Kalyadin V.G. Investigation of radiation emulsion and suspension polymerization of vinyl fluoride. Dissertation ... Candidate of Chemical Sciences, Tashkent, 1982. - 166 c.

111. A.S.427710 (SSR.). Binding agent (Makhkamov S.M., Rakhmanberdiev G., Karakozova S.A., Gafurov T.G., Usmanov H.U.), Published in B.I., 1974, JP 18.

112. Mukhamedjanov M., Rakhmanberdiev G., Makhkamov S.M. Water-soluble acetylcellulose - a new binding substance for obtaining

tablets. Materials of the 1st All-Union Conference on Composite Materials. Tashkent, 1980, part I, p. 87.

113. Makhkamov SM, Rahmanberdiev B, Mukhamedjanov M. Application of water-soluble acetylcellulose in the manufacture of tablets. Theses of reports of All-Union conference "State and prospects for the development, production and use of excipients for the manufacture of medicines. Kharkov, 1982, part I, p. 66-67.

114. Rakhmanberdiev G., Mansurkhanova I., Nazirov Z.N., Isajanov B.U., Mirnigmatova Sh.M. Water-soluble acetylcellulose hydrophilic base for ointments. Medical Journal of Uzbekistan, 1978, № 12, p. 32-33.

115. A.S.443950 (SSSR). Composition for nonwrinkle finishing of fabrics of cellulose fibers. (Anarmetova D., Rakhmanberdiev I., Gafurov T.G., Usmanov H.U., Efimov A.N.). - Published in V.I., 1974, №35.

116. Anarmetova D., Rakhmanberdiev G., Aristanbekov R. Use of water-soluble acetylcellulose in modification of cellulose materials. In: Acetate fibers, Tashkent: Fan, 1977, pp. 34-38.

117. Petropavlovskii G.A., Vasilieva G.G. Low-substituted acarboxymethylcellulose and its properties as an adhesive for textile products. Zh I X, 1957, vol. 30, № 12, c. 1832-1837.

118. Anarmetova D., S-Khojaeva N.N., Aristanbekov R., Karimova N., Sardiev K.S., Leontyeva S.A. Improvement of some physical and mechanical properties of acetate fibers by spirited method. son of the International Symposium on Macromolecular Chemistry. Moscow: Nauka, 1978, v. 2, p. 42.

119. Aristanbekov R., Anarmetova D., Nikonovich T.V., Usmanov H.U. The influence of the nature of the catalyst in non wrinkled and shrinkage-free finishing of fabrics from cellulose fibers. E.I. Textile Industry. 1983, 85, c. 1-25.

120. Bogatyreva LM, Chumakova VA, Obydennikova NF New finishing preparations and textile aids. Proc. Ivan Research Institute of cotton industry, 1976, JB 35, pp. 145-152.

121. A.s. 773170 (ccCP). "Composition for modification of textile materials". (Anarmetova D., Aristanbekov R., Rakhmanberdiev G., Nikonovich G.V.) - Publ. in B.I., 1980, 1E 39.

122. Aristanbekov R.S., Khodjaeva N., Anarmetova D., Rakhmanberdiev G., Nikonovich G.V., Usmanov H.U. Improvement of the form stability of acetate knitted fabrics and a method of its evaluation.

Izv. of Higher Education Institutions. Technology of light industry, 1982, 14, pp.40-44.

123. A.S. IN 447944 (SSR.). Composition for finishing of acetate fibers (Dustmukhamedov H., Rakhmanberdiev G., Gafurov G.G., Usmanov H.U.). "Not to be published in the open press", 1972.

124. A.S. IN 245975 (SSSR.). Method of obtaining modified films (Rakhmanberdiev G., Zaitsev B.A., Rozhanovskaya G.I., Sazonova K.M., Petropavlovsky G.A., Straikhman G.A.) -published in B.I., 1969, № 20.

125. Kalyadin V.G., Sirlibaev T.S., Rakhmanberdiev G., Suspensipolymerization of vinyl fluoride in the presence of some water-soluble cellulose esters. ZhPH, 1983, Vol. 46, 52, c. 462-465.

126. Wallach W.L., Kabayma M.A. Polivlnylfluoride solution characteristics. J.Pol.Sci, 1966, vol.4, 10, pp. 2667-2674.

127. Maede N. Polymerization radiochemique du styrens en suspension Aguense. Z. Chim.phys, chimblo1,1962,vol.59,494, pp. 336-338.

CHAPTER IV. PREPARATION AND INVESTIGATION OF THE PROPERTIES OF WATER-SOLUBLE MIXED ESTERS BASED ON ACETYLCELLULOSE

In order to expand our understanding of the causes of water solubility of cellulose ethers and to confirm the previously stated assumption about such causes made when discussing the experimental data obtained on acetyl cellulose, mixed ethers based on acetyl cellulose were first synthesized and their solubilities in water were studied. In fact, if some amounts of other ether radicals, such as o-methyl groups, are introduced into cellulose in addition to acetyl groups, we can assume (although this is not obvious at first sight) that the methyl group will make the acetyl group more accessible to the water molecule than the hydroxyl group.

Thus it is known that methylcellulose is a water-soluble cellulose ester: at room temperature at $\gamma=180\text{-}220$, at low temperature (273°K) at $\gamma=300$. In other words, methoxyl radical is capable of hydration, and the change in the crystal lattice of cellulose due to the introduction of methoxyl groups (significant reduction in the strength of the crystal lattice bond due to the violation of the hydrogen bonding system) also contributes to solubility.

Based on the above, we can assume that certain amounts of methyl groups in the cellulose acetate macromolecule will increase the maximum degree of substitution by acetyl groups at which the ester will dissolve in water. It is also likely that other radicals will act in this direction, changing the interaction between the cellulose chains in the direction of its reduction while maintaining the hydrophilic balance of the macromolecule.

The following sections present experimental data that confirm the assumption and give, as it seems to us, more complete and generalized ideas about the mechanism of cellulose ethers solubility in water.

Based on these considerations, the synthesis and studies of the properties of a number of acetylcellulose-based mixed esters, such as acetylcarboxymethyl-, acetylmethyl-, acetatmalenate-, acetatephthalate- and acetate-phosphate cellulose, were first carried out.

The introduced radicals differ greatly in their chemical structure, volume, and nature, which can also provide rich information on the influence of the nature of the substituent on the hydrolysis process and on the properties of the resulting products when obtaining water-soluble mixed cellulose ethers.

4.1.Acetylcarboxymethylcellulose.

Low-mixed Na-carboxymethylcellulose (γ=5) is produced according to the method described in [1] by treating cotton cellulose with an aqueous alkaline solution of monochloroacetic acid.

For acetylation carboxymethylcellulose (H-CMC) was taken as free acid (H-CMC) and as Na-salt (Na-CMC). Acetylation and hydrolysis of the resulting syrup were carried out similarly as in obtaining HPAC, at 323°K ($\pm$0,1) [2].

The results of the hydrolysis process of acetylcarboxymethylcellulose obtained by acetylation of H-CMCP are shown in Table 48.

Table 48.

The process of hydrolysis of acetylcarboxymethylcellulose (H-CMC).

Hydrolysis time τ-3600c	Amount of added water, ml	Sensitivity of the solution, ml		Content of bound acetic acid, %
		H O$_2$	IEC	
0	-	60	-	60,16
14	17,6	-	-	-
24	-	120	-	51,72
48	27,2	-	-	-
60	50	340	108	42,82
84	100	70	70	33,98
108	80		35	-
120			15	28,00

The process of hydrolysis of acetylcarbocystmethylcellulose obtained by acetylation of Na-CMC is shown in Table 49.

Table 49.

Hydrolysis process of acetylcarbocystmethylcellulose (Na-CMC).

Hydrolysis time τx3600s	Amount of added water, ml	Sensitivity of the solution, ml		Content of bound acetic acid, %
		H$_2$ O	IEC	
0	-	60	-	60,10
16	20	-	-	-
24		-	-	-
53	32	200	-	48,00

77	100	380	94	45,00
97	100	-	65	40,00
120	100	-	45	35,00
144	50	-	35	33,20
168	-	-	23	-
180	-	-	17	30,08

Acetylation of carboxymethyl cellulose proceeds very slowly, with difficulty and requires much more time than acetylation of cellulose itself, although slight loosening of cellulose structure favors various esterification processes. This contradictory fact can be explained by the fact that the treatment of cellulose with monochloroacetic acid, on the one hand, loosens the structure and breaks the hydrogen bonds, on the other hand, introduces a radical that has a polar group-COOH, apparently, contributing to the formation of new hydrogen bonds between the chains of cellulose. The formed new hydrogen bonds worsen the solubility of the resulting ester in the esterifying medium. This structure significantly affects the hydrolysis process.

When comparing the results of the tables it can be noticed that hydrolysis time of acetylcarboxymethylcellulose obtained from Na-CMC is longer than that from H-CMC. This seems to be due to the fact that in the case of Na-CMC a certain amount (20%) of sulfuric acid catalyst in the acetylation process is consumed for conversion of Na-CMC to H-CMC and thus slows down the hydrolysis process.

Acetylcarboxymethylcellulose contained about 0.7% carboxymethyl groups (γ=5). When replacing hydroxyls with these radicals in the indicated amounts, water-soluble products with a higher content of acetyl groups can be obtained. A water-soluble product is obtained with an acetyl group content of 28-30% (instead of 18% for acetyl cellulose). This fact can be easily explained by the additional influence of ionogenic groups as a result of their electrostatic repulsion.

The resulting product is well soluble in water, forming highly viscous 6.8 and 12% solutions, also dissolves in slightly diluted with water organic liquids, in particular, in 70-80% alcohol.

4.2 Acetylmethylcellulose.

Methylcellulose (γ=50) was obtained by 0-alkylation of alkaline cellulose with dimethyl sulfate in sulfur ether medium according to the procedure described in [3].

Acetylation of methyl cellulose was performed in the same way as it was done with carboxymethyl cellulose [4,5]. After obtaining a viscous, transparent syrup of highly substituted acetylmethylcellulose, it was hydrolyzed under the same conditions as acetylcarboxymethylcellulose. The results of acetyl methylcellulose hydrolysis process proceeding at 323°K are given in Table 50.

Compared to the process of acetylation of low-substituted carboxymethylcellulose the process of acetylation of methylcellulose proceeds very easily and quickly with formation of viscous and transparent syrup of acetylmethylcellulose. In the process of obtaining methylcellulose with increasing the degree of esterification up to $Y=50$ the reactivity of the obtained ester increases due to loosening the structure by the introduction of methoxyl groups.

Table 50.

The process of hydrolysis of acetylmethylcellulose.

Hydrolysis time τ-3600c	Amount of added water, ml	Sensitivity of the solution, ml		Content of bound acetic acid, %
		$H O_2$	IEC	
0	-	60	-	46,6
14	20,95	90	-	40,8
24	-	120	-	-
36	-	175	-	35,6
46	32,0	225	165	-
65	50,0	-	76	29,74
75	50,0	-	45	-
85	50,0	-	25	-
88	-	-	18	24,4

Solubility of acetylmethylcellulose in water is achieved at hydrolysis time of 80x3600 s. If we compare with the hydrolysis time of the original triacetylcellulose and acetylcarboxymethylcellulose, we see that hydrolysis proceeds much faster and at a higher concentration of acetic acid in the hydrolyzed mixture. Solubility in water is also achieved at higher values on the acetyl radical compared to BPAC, that is, at 25% of the bound acetic acid the product is completely soluble in water.

With the introduction of methoxy groups (-OCH$_3$), which themselves are capable of forming hydrates according to the scheme (as noted on page 28).

$$-\!\!-O-\!\!-CH_3 + H_2O \longrightarrow -\!\!-O-\!\!-CH_3 \quad (\overset{H\quad OH}{\diagdown\,\diagup})$$

increases the heterogeneity of the structure and improves the solubility in water.

4.3. Cellulose acetate maleinate.

A number of works, mainly patents [4,5], are devoted to the description of the synthesis and properties of mixed ethers based on cellulose acetates with dicarboxylic acids. The methods described in the literature for the preparation of mixed cellulose esters with dicarboxylic acids, mainly aimed at obtaining products with the highest possible degree of substitution by dicarboxylic acid. For example, Aikhodzhaev and his collaborators quite thoroughly studied methods of obtaining highly substituted cellulose acetate maleate (AMC) [6-8]. Fibers and films /9,10/ with improved physical and mechanical properties in comparison with fibers and films obtained from triacetylcellulose were obtained. However, apart from the works [11,12], there are no works devoted to obtaining and researching the properties of water-soluble products based on these cellulose ethers in the world practice. Water-soluble cellulose acetate maleinate (VAMC) was obtained by saponification in homogeneous medium of AMC. AMC was obtained by the method described in [6].

Secondary industrially produced acetylcellulose with the following characteristics was used to obtain AMC: bound acetic acid content of 54.0% and degree of polymerization of 280.

The reaction of esterification of free hydroxyl groups of acetylcellulose with maleic anhydride proceeds according to the following scheme:

$$[C_6H_7O_2(OCOCH_3)_x(OH)_{3-x}] + O\underset{C-CH}{\overset{C-CH}{\diagup\ \diagdown}}\ \longrightarrow$$

$$\longrightarrow [C_6H_7O_2(OCOCH_3)_x(OCOCH=CH-COOH)_y(OH)_{3-x-y}]$$

Subsequently, the obtained AMC was dissolved in 60% acetic acid and after complete dissolution was hydrolyzed at constant temperature and under continuous stirring in the presence of the catalyst $HClO_4$ at different amounts by weight of AMC.

During the hydrolysis process, as in the case of acetate cellulose hydrolysis, acetic and maleic acids are detached and the solubility of the ether in the hydrolyzing medium changes. In order to maintain the solubility of the ether and to carry out the hydrolysis, water was added to the hydrolyzing medium at intervals and the hydrolysis reaction was monitored by taking separate samples of the solution to check its solubility in water.

After the hydrolysis reaction, the solution was diluted with 60% acetic acid and precipitated in acetone. It was washed several times with a precipitant to neutral reaction and dried. The content of bound acetic acid and maleate groups was determined in the obtained product [7,13]. VAMC contained bound acetic acid 26-28% (γ=75-85) and maleate groups 8%(γ=15).

Table 51.

Kinetics of deep hydrolysis of cellulose acetate maleate.

Temperature, ^{0} K	Hydrolysis time, τ-3600c	CH_3 C OOH, (%)	CH_3 C OOH, (γ)	HOOC- CH=CH- COOH (%)	HOOC- CH=CH- COOH (γ)	COCOCH3	COCOSH=CH- COOH
						FROM^{-1} - 3600	
	0	51,3	222	10,85	26		
313	6	40,5	152	10,25	22	0,048935	0,031608
	12	32,7	114	9,02	17	0,053271	0,031884
	20	27,5	93	7,65	14	0,048810	0,031685
	3	40,8	183	10,55	24	0,092318	0,053646
323	6	38,8	144	9,61	20	0,091703	0,052828
	10	26,8	88	8,12	15	0,092836	0,053831
	2	40,6	152	10,35	23	0,158619	0,106314

333	4	35,6	128	9,28	19	0,147125	0,103223
	6	30,7	105	8,3	16	0,150270	0,101426
	8	28,0	94	7,23	14	0,143786	0,102780

The kinetics of the hydrolysis of AMC at different temperatures is given in Table 51. The effect of temperature on the reaction rate is shown in Fig. 64 as a dependence of ln K on the inverse of the absolute temperature 1/T.

The activation energies for the hydrolysis of acetyl and maleate groups were found to be 47.32 kJ/mol and 61.45 kJ/mol, respectively.

As can be seen from Table 54. the difference in the saponification rate of acetyl and maleate groups, as well as different values of activation energy can be explained by the fact that in the presence of acid catalysts during hydrolysis the unsaturated double bond in maleate groups, having electron-accepting properties, draws electrons from carboxylic carbon and oxygen atoms, which makes it difficult to attack the ester bond with the proton [8].

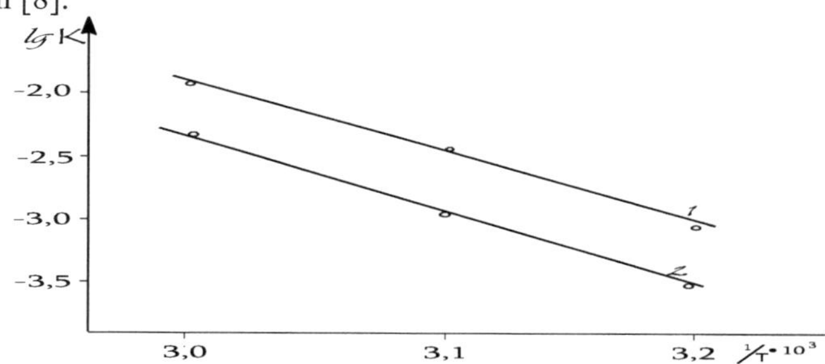

Fig.84. Dependence of the logarithm of the rate coefficient of hydrolysis of cellulose acetate maleate on the inverse value of the absolute temperature:1 - acetyl groups; 2 - maleate groups.

The rate constant of degradation of AMC and the change in the degree of polymerization during hydrolysis at 313^0 K was also determined (Table 52).

Table 52.

The rate constant of degradation of AMC at 313^0 K.

216

Hydrolysis time τ-3600c	$|\eta|$	SP	SPn= SPv/2	K, C^{-1} - 3600	Kcr, C^{-1} - 3600
0	-	280	140	-	-
5	1,06	254	127	$2,60 \cdot 10^{-5}$	$2,64 \cdot 10^{-5}$
12	0,96	230	115	$2,49 \cdot 10^{-5}$	
20	0,84	200	100	$2,85 \cdot 10^{-5}$	

IR spectroscopy was used to determine the chemical composition of VAMC.

In the IR spectrum of maleic anhydride (Fig. 85.c) there are absorption bands belonging to the C=0 groups at 1860 and 1790 cm^{-1} and the absorption band of the CH group at the double bond is at 2060 and 3130 cm^{-1} , the valence vibrations of the C=0 bond in the 1630-1580 cm region^{-1} .

In the region of 1280-1070 cm^{-1} there are absorption bands corresponding to C-O-C bonds.

In the infrared spectrum (Fig. 85.b) of AMC the intensity of the absorption band of hydroxyl groups decreases. New absorption bands appear at 1650 and 1590 cm^{-1} .

As is known in the IR spectrum of cellulose and its derivatives in the region of 1650-1620 cm^{-1} there is an absorption voice of sorbed water. In this region there are also absorption bands of C=O bonds. To find out what the absorption band at 1650 cm^{-1} refers to, a differential IR spectrum was taken, i.e. the spectrum of acetyl cellulose with the same content of acetyl groups in cellulose acetatmaleinate was calculated from the WAMC spectrum. In the differential spectrum (Fig. 85.d) there is an absorption band at 1640 cm^{-1} corresponding to the C=C bond, in addition to the absorption band at 3030 cm^{-1} relating to CH oscillations at the double bond. In the region of 1815-1840 cm^{-1} there is an absorption band corresponding to the C=0 groups of maleic anhydride.

During hydrolysis of AMC for 3,6,10-3600 s and up to the water-soluble state, an increase in the absorption band intensity of hydroxyl groups bound by hydrogen bonding is observed in the IR spectrum (Fig. 86) of the water-soluble product. The absorption band intensity of the C=0 group decreases with increasing hydrolysis time. The intensity of

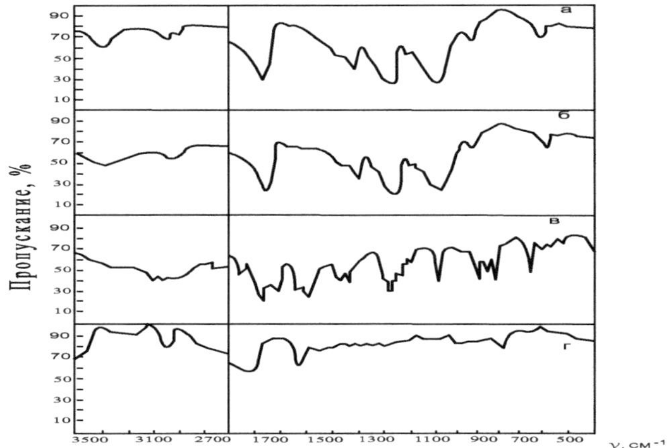

Fig. 85 IR spectra:

a/ DAC (original)

b/ cellulose acetate maleinate

c/ maleic anhydride

d/ differential spectrum of WAMC - DAC (initial).

absorption at 1650 cm^{-1} . During hydrolysis of AMC a part of acetyl groups is saponified and hydroxyl groups are formed instead of acetyl groups. Therefore, in the IR spectrum of the saponified product the absorption band intensity of hydroxyl groups increases and the absorption band of acetyl groups decreases.

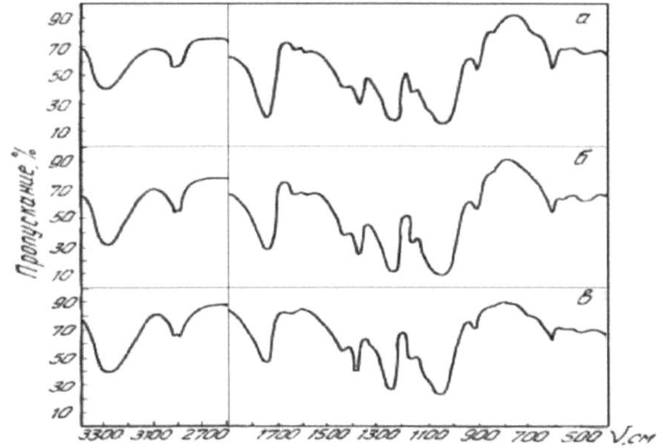

Fig.86.Changes in the intensity of the analytical absorption bands of cellulose acetomaleinate with the time of hydrolysis at 323^0 K.

a. after 3-3600 s hydrolysis

б. -"- 6 -"-

в. -"- 10 -"-

From the above it can be assumed that the decrease in the absorption band intensity of hydroxyl groups and the appearance of new absorption bands indicate a chemical interaction of AC with maleic anhydride due to the breakage of the C-O bond with preservation of the double bond.

X-ray pictures of the structure of AMC samples as hydrolysis proceeded and acetyl and maleate groups were taken.

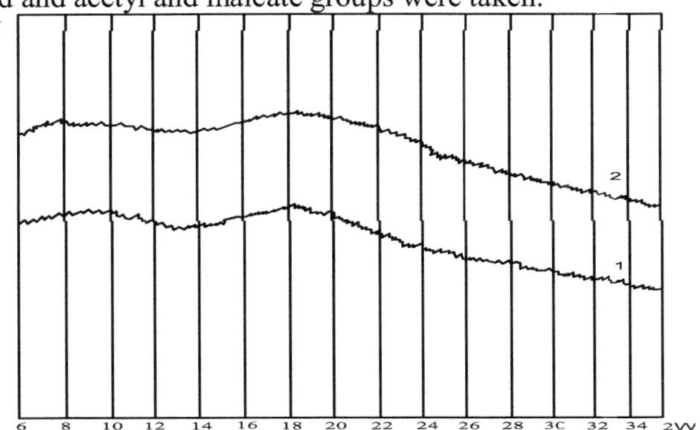

It was shown that the initial AMC is characterized by an amorphous structure (Fig. 87, curve 1). All intermediate X-ray diffraction patterns and X-ray diffraction patterns of the final product containing 24% acetyl and 8% maleate groups and soluble in water (Fig. 87, curve 2) are also X-ray patterns of amorphous substance.

Study of the properties, diluted and concentrated solutions of water-soluble cellulose acetate maleinate .

As noted above, water-soluble cellulose acetate maleinate (VAMC), unlike BPAC, has double bonds and carboxyl groups in its composition besides acetyl and hydroxyl groups. The presence of double bonds and carboxyl groups in VABC allows a number of polymer-logical transformations.

In particular, it can be used to obtain a range of new types of physiologically active polymers by attaching them to the double bonds of halogens (to make them self-sterile) and to the carboxyl groups of some drugs (to make them biologically active).

Its diluted and concentrated solutions were studied before forming fibers based on WAMC [14]. Data on the viscosity of diluted solutions in water and DMF at a temperature range of $298-343^0$ K are given in Table 53.

Table 53.

Changes in the characteristic viscosity and the Huggins constant of VAMC solutions as a function of temperature.

Solvent	Temperature, °K	Characteristic viscosity $|\eta|$	Huggins Constant К'
Water	298	1,92	0,51
	303	1,80	0,48
	313	1,65	0,46
	323	1,53	0,42
	333	1,38	0,38
	343	1,05	0,27
DMF	298	1,78	0,48
	303	1,70	0,42
	313	1,50	0,38

	323	1,39	0,30
	333	1,28	0,27
	343	1,05	0,24

*VAMC contained, 26-28% (γ =85-95) of bound acetic acid and 8% of maleate groups (γ=15).

As can be seen from the table, dimethylformamide solutions have comparatively lower viscosity than aqueous solutions. Figure 88 shows the change in the size of the molecular ball of VAMC depending on the nature of the solvent and the temperature.

It can be seen from Fig. 89 that ΔE grows with increasing polymer concentration almost linearly for both solutions. The size of VAMC microparticles was also investigated by the turbidity spectrum.

The results are shown in Table 3.29.

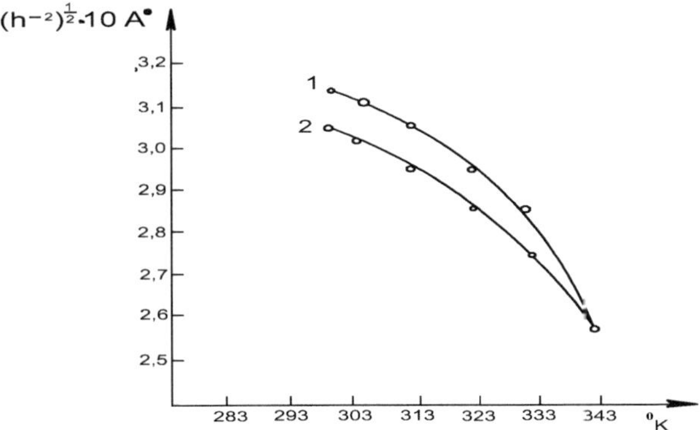

Fig. 88. Dependence of the RMS radius of the molecular ball of VAMC on the temperature in aqueous solution (1) and in DMF solution (2).

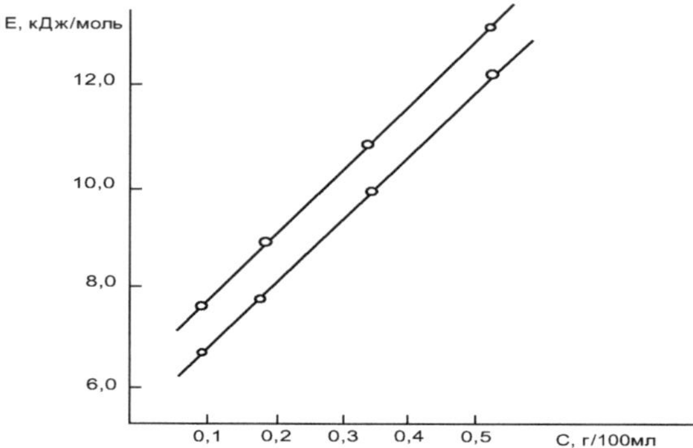

Fig.89. Dependence of apparent activation energy of viscous flow of VAMC solution in water (1) and in DMF (2) on concentration.

Table 54.

Changes in the size of VAMC microparticles as a function of concentration.

Concentration solution %	Aqueous solutions		DMF solutions	
	Radius,mg,in nm	Number of Mgc per $1см^3$r-value, $W-10^7$	Radius mgc, in nm	Number of Mgc per $1см^3$r-value, $W-10^7$
0,5	425	3,03	1,03	0,001
1	550	2,25	180	0,207
3	700	1,08	740	0,260
5	610	4,27	840	1,950
10	-	-	930	1,090
12	850	2,55	-	-

As can be seen from Table 54, the size of VAMC microgel particles in both solvents increases with increasing polymer concentration.

As in the case of HPAC, concentrated solutions of VAMC were also investigated in a similar way [15]. Based on the data obtained, the values of the apparent activation energy of the viscous flow were calculated ΔE

and, as a qualitative characteristic, the "viscous volume" V^* or the average dimensions of the kinetic units.

Keeping VAMC aqueous solutions obtained by dissolving the polymer at room temperature for 20-3600 s leads to a significant increase in viscosity (about 5 times) - from 6.0 to 30 Pa s. When such a solution is heated to 313^0 K and shear stresses are applied to it, the spatial network is destroyed; when shear stresses are removed, the solution structure is slowly restored, but the original viscosity values are no longer reached (Fig. 90. curves 1-4).

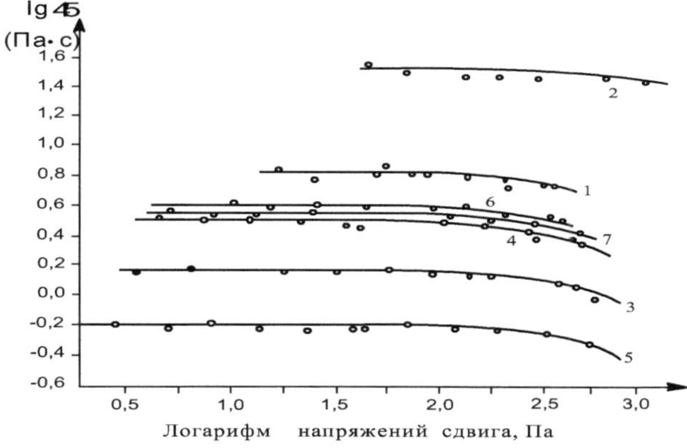

Fig. 90. Flow curves (dependence of the logarithm of effective viscosity on the logarithm of shear stresses) of aqueous 10% VAMC solutions prepared at room temperature as a function of their storage time (1-12-3600s; 2,3,5-20-3600s; 7-144-3600s) and measurement temperature (1,2,7-298°K; 3-313°K;5-333°K). Curves 4 and 6 correspond to the solution measured again at 298°K at 1 h after measuring it at 313°K (3) and 333°K (5), respectively.

The viscosity of solutions of the same drug obtained by dissolution in a water bath at 333-353°C is lower. Apparently, at room temperature there is no complete dissolution and the gel-forming centers remain in the solution, which cause the formation of a strong spatial grid during storage. When the polymer concentration in solution is up to 10%, the viscosity of aqueous solutions is higher than that of dimethylformamide solutions, but the viscosity of aqueous solutions containing 16% VAMC at 293^0 K is lower than that of dimethylformamide solutions (Fig. 91).

However, the viscous flow activation energy for aqueous solutions of WAMC is higher than that for dimethylformamide solutions at all polymer concentrations studied (Table 55). The apparent activation energy of the viscous flow of WAMC is greater than that of BPAC, and this difference can be explained by the presence in WAMC of a small number of maleate groups that are larger than the acetyl groups.

Table 55.

Values of the apparent activation energy of the viscous flow of HPAC and VAMC in different solvents.

Rasvoritel	ΔE_σ kJ/mol at polymer concentration				
	5,0%	8,0%	10,0%	12,0%	16%
VRAC					
DMFA	-	26,8±0,3	-	31,8±0,3	32,0±0,3
DMSO	-	33,4±0,30	-	36,4±0,4	38,9±0,4
Water	-	33,4±0,30	-	38,0±0,4	42,2±0,4
WAMC					
DMFA	-	-	27,8±0,08	-	35,4±0,33
Water	34,0±0,32	35,0±0,029	37±0,37	-	45,4±0,67

Figure 91 shows that the destruction of the spatial mesh in both aqueous and dimethylformamide solutions of VAMC occurs approximately at the same values τ. As the solution concentration increases, the activation energy of the viscous flow increases, indicating that the intermolecular interactions in the solution intensify. In dimethylformamide solutions of VAMC, the "viscous volume," which is a measure of mobility and size of structural elements, increases as the polymer concentration increases (Table 56).

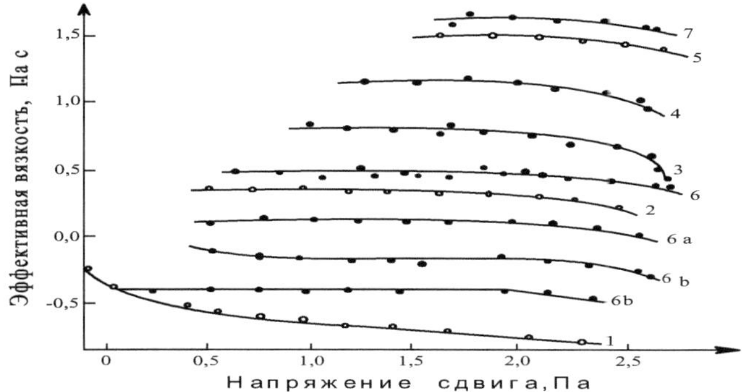

Figure 91. Flow curves of aqueous (1-5) and dimethylformamide (6,7) solutions of VAMC as a function of temperature (1-7-293⁰ K 6-a-313⁰ K; 6b-331⁰ K; 6c-353⁰ K) and polymer concentration (1-4.9%; 2-8%, 3-9.85%; 4-12%; 6a,b,c, 10%: 5.7-16%).

As can be seen from Table 56 in aqueous solutions as the concentration increases, the sizes of kinetic units and their mobility first decrease and then increase. This phenomenon can be related to the behavior of macromolecules in solution in the following way: at low VAMC concentration in water, the forces of interaction between hydroxyl and carboxyl groups inside macromolecules and with water molecules play a major role, macromolecules are swollen balls. Increasing the polymer concentration up to 12% apparently results in redistribution of polymer-polymer and polymer-solvent bonds; the probability of formation of strong hydrogen bonds between carboxyl groups of neighboring macromolecules increases, which leads to a more or less uniform distribution of macromolecules. From a certain value of polymer concentration the forces of polymer-polymer interaction dominate and larger supramolecular formations are formed in the solution. Apparently, its structure-location of hydroxyl, acetyl and maleate groups in a certain sequence-plays a significant role in such behavior of VAMC in aqueous solutions.

Table 56.

Variation of the "viscous volume" of VAMC as a function of concentration, shear stress and temperature.

τPa	Concentration, %	*V 10^{16} (see³) at temperature			
		293°K	313°K	333°K	353°K
DMF					

	10	130	142	151	154
10	12	132	144	155	163
	16	137	145	157	166
	10	12,9	13,9	14,9	15,6
100	12	13,3	14,3	15,3	16,0
	16	14,1	15,1	16,1	17,1
O n e w o r k					
	5	212	228	244	-
	8	183	196	215	221
10	10	181	196	210	221
	12	177	189	200	213
	16	202	212	225	244
	5	21,7	23,2	24,6	-
	8	18,6	19,8	21,3	22,4
100	10	18,3	19,7	20,9	22,4
	12	17,9	19,0	20,3	21,5
	16	21,2	22,5	23,7	25,6

In the studied solutions of VAMC, increasing the temperature leads to an increase in the average size of the kinetic units. It is likely that increasing the temperature leads to the phenomenon of "pushing" of VAMC macromolecules due to the presence of mutually repulsive side acetyl groups in its chain.

As a result, kinetic units are preserved in the solution, which are loosened without destroying, increasing in volume, i.e. in this case the structure is not disturbed during heating, which weakens the effect of thermal motion /15/.

When shear stresses are applied, structural formations are destroyed; a tenfold increase in shear stress leads, respectively, to a tenfold decrease in "viscous volume".

Forming of fibers based on cellulose acetate maleate.

After a detailed study of the properties of dilute and concentrated solutions of VAMC, fibers based on it were molded [16,17].

Since water-soluble fiber based on VAMC was obtained for the first time, it is of interest to study the process of fiber formation and the influence of different molding parameters on this process.

WAMC with the following characteristics was used for the experiment:

bound acetic acid content -26% (γ=85)
Maleinite group content -8% (γ=15)
degree of polymerization (SP) - 190

At fiber molding, as noted earlier, the main role is played by the process of polymer deposition in the bath, and it is of great importance when choosing the composition of the deposition bath and the optimal conditions of fiber formation. Therefore, the stability of aqueous and dimethylformamide solutions of VAMC to the action of organic and inorganic compounds was studied. Processing of obtained data on coagulation ability allows to conclude that many organic precipitators, mainly alcohols and some solutions of salts and mineral acids have the highest precipitation ability. These series of studies have established that, as in the case of HPAC, it is preferable to use acetone and isopropyl alcohol as precipitating agents, which have sufficient coagulation ability and provide good fiber properties. Further the influence of concentration of spinning solution, temperature and composition of sedimentation bath, as well as nozzle and plasticizing drawings on the properties of obtained fibers has been studied.

The most stable forming process proceeds at forming from 10-12% aqueous and 15-16% dimethylformamide solutions of VAMC. Decreasing the concentration of the spinning solution leads to unstable molding with partial breaks and "adhesions", to obtaining fibers with unsatisfactory characteristics, and increasing the concentration of the solution increases its viscosity and the stability of molding decreases again.

The temperature of the precipitation bath varied in a relatively narrow range (288-303°K) due to strong evaporation of acetone and isopropid alcohol. Also, as the temperature of the deposition bath increases, the conditions of fiber deposition become more stringent, which leads to a simultaneous decrease in both fiber strength and elongation. Studies of the effect of the composition of the deposition bath on the molding process showed that molding in organic baths and baths with 5% water content (Table 57) proceeds satisfactorily.

Increase of water content (10%) leads to deterioration of coagulation ability of the bath and partial dissolution of fibers, and thus, the strength of forming fibers decreases. Forming of fibers in the sedimentation bath containing isopropyl alcohol differs from the previous ones by more uniform course of diffusion processes and sticking of fibers is excluded.

Table 57.

Changes in the mechanical properties of fibers depending on the composition of the precipitation bath.

Composition of the precipitation bath	Tensile strength, cN/tex		Elongation, %	
	Aqueous solution	DMF r	Aqueous solution	DMF r
Isopropyl alcohol	6.8	8.2	18.6	20.0
Acetone	4.3	5.4	14.0	16.4
Isopropyl alcohol + water				
95:5	5.6	6.2	25.4	23.0
90:10	4.0	4.8	24.2	20.5

The results of the study of the effect of die drawing are shown in Table 58.

Table 58.

Influence of die drawing on mechanical properties of fibers based on VAMC.

Mechanical fiber gauges	Filial extraction, %					
	0	50	70	100	150	200
For fibers derived from from aqueous solution						
Vulnerability, cN/tex	4,2	4,0	5,3	8,0	7,7	7,0
Elongation,%	24,3	23,0	22,4	22,6	20,4	20,4
For fibers derived from from the DMF r.						
Strength, cN/tex	5,5	5,5	7,0	8,5	8,2	6,0
Elongation,%	25,0	23,6	22,4	20,2	19,5	21,4

As can be seen from the table, the strength of the fiber after nozzle drawing slightly increases. At wet forming, as it was noted earlier at forming of fibers based on HPAC, weakly oriented fiber structure appears as a result of nozzle drawing. Therefore, the necessary stage of obtaining the fiber with higher mechanical properties is plasticizing drawing [16].

The study of plasticizing drawing showed the known pattern of inverse relationship of strength and elongation (Table 59).

Table 59.

Influence of plasticizing drawing on mechanical properties of VAMC fibers.

Mechanical performance of fibers	Plasticizing extract. %			
	0	50	75	100
For fibers derived from aqueous solution				
Strength, cN/tex	6,2	7,6	8,2	5,7
Elongation,%	24,3	22,2	20,3	22,6
For fibers obtained from DMF solution				
Strength, cN/tex	5,5	8,2	8,8	7,6
Elongation,%	25,0	20,0	18,7	22,0

Improvement of mechanical properties of the fiber during plasticizing drawing is accompanied by some compaction of the structure.

Based on the data obtained, the optimal molding conditions can be considered: concentration of the spinning solution:

10-12% water

DMF 15-16%

Precipitation bath temperature 293-295°K

Filter extraction 70-100%

Plastification extraction 50-75%

Such molding conditions provide a fiber with a strength of 9-10 cN/tex with an elongation of 18-20%.

The electron-microscopic examinations of the surface structure of the obtained fibers on the basis of VAMC were also carried out. Depending on conditions of their forming, characteristic electron-microscopic pictures of surface structure of fibers were observed. At zero value of both spinning and plasticizing drawings the surface structure of the fibers is presented more heterogeneous. As the degree of drawing increases, the optical structure evens out. At relatively high degrees of drawing (about 200% of drawing and 100% of plasticizing), as a rule, optically homogeneous fiber structure is revealed.

The obtained water-soluble fibers based on cellulose acetate maleinate do not have sufficient strength at this stage of the study to use them as suture materials, but they can be used as resorbable tamponade materials. In addition, the presence of reactive groups (double bonds and carboxylic groups) in their composition makes it possible to perform a number of polymer-logical reactions in order to obtain physiologically active fibers. For example, water-soluble fibers containing various amounts (from 5 to 12%) of chemically bound iodine, tubazide, tetracycline, sarcolysin, lagohylus, and other drugs were obtained.

4.4.Cellulose Acetate Phthalate.

Synthesis of highly substituted cellulose acetate phthalates (PAC) is devoted to a large number of works [17-20]. AFCs are obtained by the action of phthalic anhydride on partially saponified cellulose acetate in the presence of organic bases [21]. A method of obtaining AFC [22] in acetic acid medium by phthalation of cellulose acetates is described. The process of cellulose acetate phthalization [23] was studied depending on temperature, amount of catalyst (potassium acetic acid), type of cellulose acetate, and physical and mechanical properties of AFC films of different composition were studied. Due to the solubility of AFC in a large number of organic solvents and the ability to form films, their application area is quite wide insulating materials, washable lacquer coatings, special chromatographic paper, capsules, coatings for medical products, etc. However, the method of obtaining water-soluble AFCs is also not described in the world literature.

Therefore, obtaining and studying the properties of water-soluble AFC may be of interest since it will provide rich information on the influence of substituents (in this case phthalate groups) on the hydrolysis process of acetylcellulose and, in addition, it is possible to obtain products (films and fibers) on its basis that have interesting properties.

Water-soluble cellulose acetate phthalate (WAPC) was obtained by hydrolysis of highly substituted APC [12]. AFC, in turn, was obtained by the method described in [24] by interaction of secondary acetylcellulose containing 54.3% sv. acetic acid (γ=240) with flalic anhydride in glacial acetic acid medium, in the presence of anhydrous Na acetate catalyst.

The reaction proceeds according to the following scheme:

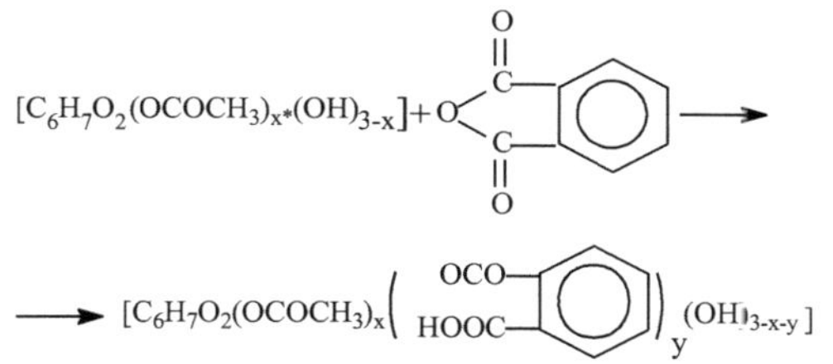

$$[C_6H_7O_2(OCOCH_3)_{x*}(OH)_{3-x}] + O\!\!\begin{array}{c}\underset{\parallel}{C}\overset{O}{} \\ \underset{\parallel}{C} \\ O\end{array}\!\!\bigcirc \longrightarrow$$

$$\longrightarrow [C_6H_7O_2(OCOCH_3)_x\left(\begin{array}{c} OCO\!\!-\!\!\bigcirc \\ HOOC\!\!-\!\!\bigcirc \end{array}\right)_y (OH)_{3-x-y}]$$

After completion of the reaction, the AFC was precipitated in distilled water, washed to a neutral reaction, dried, and the content of bound acetic acid and phthalate groups was determined, which were 53.2% and 9.7%, respectively.

Subsequently, the obtained highly substituted AFC was dissolved in 60% acetic acid and after complete dissolution underwent hydrolysis in the same manner as in the case of hydrolysis of AMC. After the hydrolysis reaction, the solution was diluted with 60% acetic acid and precipitated in acetone. The content of bound acetic acid and phthalate groups was determined in the obtained product. WAFC contained 24.2-25.0 % bound acetic acid and 9.1-9.2 % phthalate groups. Table 60 shows the kinetics of the hydrolysis of AFC at 313^0 K.

Table 60.

Kinetics of AFC hydrolysis.

T^0 K	Hydrolysis time τ-3600c	CH content COOH$_{CB}$, %	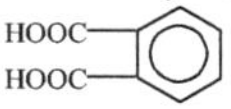 Content, % HOOC– HOOC–	K CH$_3$ COOH	K 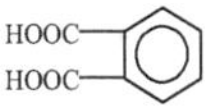 HOOC– HOOC–
	0	53,2	9,7		
	16	42,7	9,5		
313	35	32,8	9,3	0,022400	0,008583
	60	24,2	9,2		

IR spectra of AFC were taken. In the IR spectrum (Fig.92.a) of phthalic anhydride, the absorption band of the C=O group is at 1854, 1875 and 1700 cm^{-1} , and the C-O-C ether groups at 1265 cm^{-1} .

In addition, there are absorption bands of the benzene ring at 1590, 1500 cm^{-1} and strain vibrations at 750 and 720 cm /41/.$^{-1}$

In the infrared spectrum of AFC (Fig.93.b) the intensity of the absorption band of hydroxyl groups decreases as compared to DAC and it shifts towards low frequencies, the absorption bands of 1854 and 1775 cm⁻¹ disappear, and the absorption band of 1700 cm⁻¹ shifts towards high frequencies to 1730 cm⁻¹ . New absorption bands appear at 1600, 1490,790, 750 cm⁻¹ .

To clarify the disappearance of the absorption bands in the 1854 and 1775 cm region⁻¹ and the shift of the absorption band in the 1700 cm region⁻¹ the IR spectrum of the mechanical mixture of DAC with phthalic anhydride was taken. In the IR spectrum (Fig. 92.b) of the mechanical mixture of DAC with phthalic anhydride, absorption bands appear at 1854, 1775 cm⁻¹ , and the absorption band at 720 cm⁻¹ does not shift.

The absence of the 1854 and 1775 cm absorption bands⁻¹ in the infrared spectrum of AFC is apparently due to chemical interaction of the components with the breakage of a complex ether bridge. An attempt was made to quantify the content of the phthalate antidrylate in the final

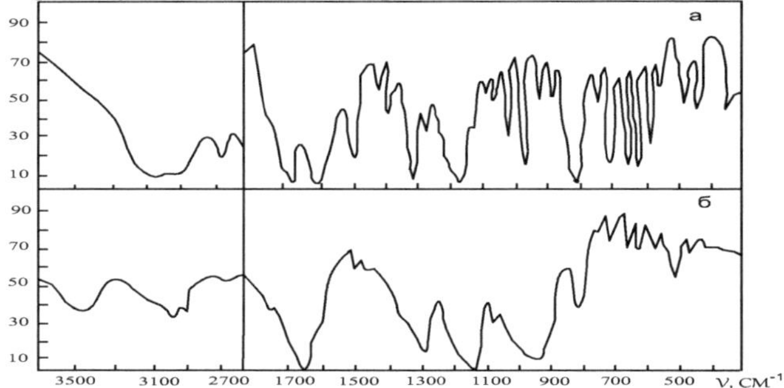

Fig. 92. IR spectra of phthalic anhydride (a) and the mechanical mixture of phthalic anhydride with DAC (b).

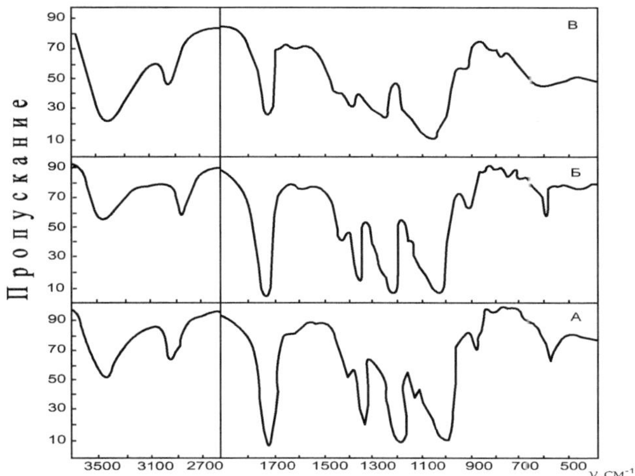

Fig.93. IR spectra of DAC (a), AFC (b) and WAFC (c).

hydrolysis product of WAPC. For this purpose, mechanical mixtures were prepared with the content of phthalic anhydride from 3 to 20% relative to WAPC. The absorption bands at 750 and 607 cm^{-1} were used for determination. The ratio of the optical densities of these absorption bands to the phthalic anhydride content gives a straight-line dependence (Fig. 94).

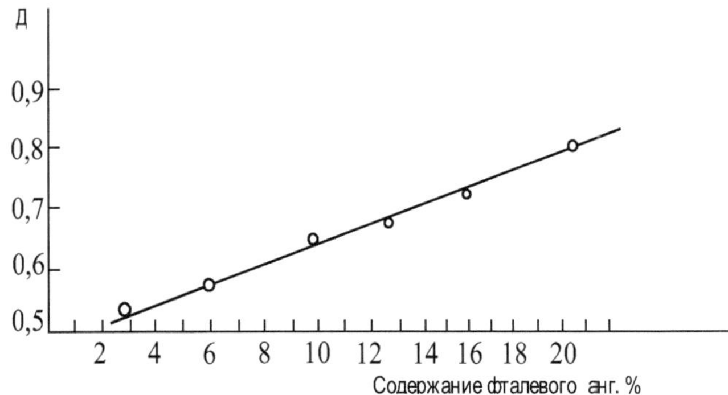

Fig.94.Change of the ratio of optical densities depending on the content of phthalic anhydride.

As the analysis shows, the content of phthalic anhydride, unlike that of AMC, almost does not change during hydrolysis. Thus, if in the highly substituted AFC insoluble in water the content of phthalic anhydride was 9.7%, then in the final product soluble in water after hydrolysis for 60x3600 s the content of phthalic anhydride was 9.2%. The obtained results agree with the results described in [25], where the resistance of phthalate groups to acid hydrolysis (saponification) is explained by the fact that the benzene ring of bound phthalic acid, possessing

electron acceptor properties, draws electrons away from the carbonyl carbon atom and the carbonyl oxygen atom, with the result that the proton is unable to attack the complex ether bond.

Thus, the chemical composition of cellulose acetate phthalates can also be quantified by infrared spectroscopy. In order to determine the effect of the third radical on the hydrolysis rate of acetyl groups, we compared the values of the rate constants of hydrolysis of acetyl groups in acetyl cellulose, acetatmaleinate cellulose, and acetate phthalate cellulose.

The rate constant of hydrolysis (K) of acetyl groups in acetyl cellulose is 0.023 at 313^0 K, in acetate maleate cellulose 0.048 and in acetate phthalate cellulose 0.022, that is, they differ in magnitude from each other.

Thus, in cellulose acetate maleate the acetyl groups are detached at a rate 2 times greater than in the original acetylcellulose. This can be explained by the effect of the "neighboring group" replacing the hydroxyl in the pyranose cycle of the cellulose ether.

The difference in the rate constant of hydrolysis of acetyl groups in acetyl cellulose compared with the rate constant of hydrolysis of acetyl groups in acetate maleate cellulose can be explained, apparently, by the fact that maleate groups loosen the structure due to mutual repulsion of carboxyl groups and thereby provide an easy proton attack complex ether bond of acetyl groups.

During hydrolysis of acetyl groups in cellulose acetate phthalate, the rate constant of hydrolysis turned out to be equal to the rate constant of hydrolysis of acetyl groups in acetyl cellulose. Obviously, in this case, the steric tension in the cellulose acetate phthalate chain due to the large volume of the substituent prevents the realization of the neighbor effect [26].

4.5.Cellulose acetate phosphate.

Cellulose acetate phosphates are of interest in that along with solubility in water, products based on them (fibers and films) have high antimicrobial activity and fire resistance. There is one patent in the world literature on cellulose acetate phosphates [27].

Water-soluble acetate phosphate cellulose (WAPoC) was obtained by etherification in a solution of secondary acetylcellulose $POCl_3$ in the presence of pyridine at 293-295°K. Hydrolysis was carried out in the presence of -NH OH.$_4$

The mechanism of the esterification reaction of acetylcellulose with trichloride phosphate and further hydrolysis, apparently, can be represented as follows:

1. Etherification:

$$C_2H_7O_2\!\!\begin{array}{l}-OCOCH_3\\-OCOCH_3\\-OH\end{array}+POCl \xrightarrow{\text{ï ëðèäèí}} C_2H_7O_2\!\!\begin{array}{l}-OCOCH_3\\-OCOCH_3\\-O-\overset{O}{\underset{Cl}{\overset{\|}{P}}}-Cl\end{array}+HClC_6H_5N$$

2. Hydrolysis:

$$C_2H_7O_2\!\!\begin{array}{l}-OCOCH_3\\-OCOCH_3\\-O-\overset{O}{\underset{Cl}{\overset{\|}{P}}}-Cl\end{array}+NH_4OH \xrightarrow{\text{ï ëðèäèí}} C_2H_7O_2\!\!\begin{array}{l}-OH\\-OCOCH\\-O-P-ONH_4\\ \ \ \ \ ONH_4\end{array}+HClC_6H_5N$$

The content of acetyl groups and phosphorus was determined in the final product after thorough washing with alcohol. Table 61 shows the kinetics of hydrolysis of cellulose acetate phosphate at different temperatures.

Table 61 shows that cellulose acetate phosphate is soluble in water with a bound acetic acid content of 18.0% and phosphorus 3.0%. This ester was further used to obtain films and fibers having both antimicrobial and flame retardant properties.

Table 61.

Kinetics of hydrolysis of cellulose acetate phosphate.

Temperature, ^{0} K	Hydrolysis time, τ-3600c	Content of bound acetic acid, %	Phosphorus content, %
293	0	54,4	
	3	42,8	
	8	39,0	
	15	33,3	

	21	30,5	
	25	28,1	
	28	24,3	
	30	20,3	
	32	17,8	3.08 (plant in water)
	3	45,0	
	8	40,5	
308	12	34,1	
	17	19,0	2.55 (plant in water)
	3	43,4	
313	8	32,1	
	14	18,3	2.87 (plant in water)

In order to obtain water-soluble fibers with their own antimicrobial properties (no sterilization is required), fibers based on cellulose acetate phosphate were molded for the first time. The resulting fiber had a tensile strength of 6-7 cN/tex, elongation 12-14%. The content of phosphorus in the fiber was 3.6%, the content of acetate groups 18-19%, it had antimicrobial and oestrogenic properties at the same time. The antimicrobial properties of the obtained fibers were tested on such microbes as s.tyghi, S.Parb., E.Coli 913, Sh. Feexneria 20, Sh. Baydi II., St. achernes 1104, achernes 1039 and their antimicrobial activity against these microbes was established.

LITERATURE

1. Petropavlovsky G.A., Vasilyeva G.G. Low-exchange carboxymethylcellulose and its properties as an adhesive for textile products. 111, 1957, т. 30, c. 832-837.

2. Rakhmanberdiev G., Asilova G.Y., Sultanova M. Preparation of water-soluble acetate-mixed cellulose ethers. Co.: Acetate fibers. Tashkent: Fan, 1977, v. 6, p. 47-51.

3. Rudneva T.I., Antonovsky c.d. Guide to practical works on the chemistry of wood and cellulose. Ed. by Nikitin, N I:-90.

4. PAT. 2856399 (USA). Method of preparing dicarboxyicacidester of the lower fatty acid ester and the ethers of cellulose containing free and esterified hydroxyl groups./ Mench J., Filkerson, Lapham - Published in Official Gazette of Mat. U.S., 1958, vol. 735, no. 2, p. 470.

5. Pat. 2852507 (USA). Method of preparing powdered salffree cellulose acetate-dicarboxyiates. / Carlton L., Emerson J. -Published in Official Gazette of Mat. U.S., 1957, vol. 734, 13, c. 760.

6. Kayumova H.N., Kryazhev V.N., Aikhodjaev B.I., Pogosov Y.L. Study of conditions for synthesis of aceto-maleinates of cellulose in acetic acid medium. Uz. Chem. journal, 1971, 5, p. 67-70.

7. Kayumova H.N., Kryazhev V.N., Aikhodzhaev B.I. Polymer-logical transformations of cellulose acetomaleinates. Uzb. Chem. journal, 1972, № 4, p. 55 57.

8. Kayumova H.N., Kryazhev V.N., Aikhodzhaev B.I., Pogosov Y.L. Study of conditions for synthesis of cellulose acetomaleinates. Uzbek Chemical Journal, 1971, 4, pp. 61-66.

9. Svistunova R.P., Aikhodjaev B.I., Pogosov Y L. Synthesis and properties of acetomegacrylates of cellulose. Plastics, 165, 6, pp. 57-59.

10. Rozmahunov R.U. Preparation and study of the properies of fibers and films from acetometacrylates of cellulose: Author's dissertation of candidate of chemical sciences, Tashkent, 1969. - 2 c.

11. A.S. 659574 (SSSR.). Method for the preparation of acetomalein ether in N.I., L.: 1951. - 90 c. cellulose. (Rakhmanberdiev G., Mukhamadaliev D., Suleimanova M.A., Usmanov H.U.). opubl. in B.I., 19776, % 16.

12. Miroshnichenko I.B., Rakhmanberdiev G., Mukhamadaliev D., Mukhamejanov M. Properties of concentrated solutions of cellulose acetomaleinate. Chemistry of Wood, 1981, 166, pp. 68-71.

13. 244, Eliseeva L.M., Berenshtein E.I., Aikhodzhaev B.I., Pogosov Y.l. va m. M-A., Usmanov H.U.). -published in B.I., 19776, % 16.

14. Mukhamedjanov M., Mukhamadaliev D., Rakhmanberdiev G. Properties of water-soluble acetate mixed cellulose ether solutions. Materials of the Hu-annual conference of NIICHTZ, Tashkent, 1980, p. 25.

15. Miroshnichenko I.B., Rakhmanberdiev G., Mukhamedaliev D., Mukhamedjanov M. Properties of concentrated solutions of cellulose acetomaleinate. Chemistry of Wood, 1981, #6, pp.68-71.

16. Mukhamedjanov M., Sultanova M., Rakhmanberdiev G. A new water-soluble fiber based on aceto-mixed cellulose ether. Proceedings of

the XXU-annual conference of Research Institute of Irrigation and Chemical Technology, Tashkent, 1980, p. 4.

17. Peremkulova H.T., Rakhmanberdiev G., Nasritdinova Sh., Razinov K.H. Effect of formation conditions on the surface structure of water-soluble fibers based on acetomaleinat cellulose. Materials of X11th Annual Conference of Research Institute of Irrigation and Chemical Technology, Tashkent,1977, p. 37-39.

16. Mikitishin O.I., Tininny A.N. On the strength of oriented polomers. Physico-chemical Mechanics of Polylers. 1968, Vol. 4, No. 3, pp. 256-258.

17. Malm G.I. and Charles R.F. Cellulose esters of Dibasic Organic Acids. J. ind. Eng. Chem., 1940, vol. 32, N 3, p. 405-408,

18. Rosenthal L.V., Burdygina T.I. Lacquer layers on the basis of colloidal dispersion of soot in acetobale in cellulose ethers. zh1X, 1959, vol. 32, issue 9, p. 2080-2084.

19. Ushakov S.A., Klimova O.M., Lubetsky S.G. Synthesis and study of properties of phthal esters of oxyethyl cellulose. JLX, 1956, vol. 29, vol. 3, pp. 438-447.

20. Sakurada Fujiwara. Alkyd resins modified with unsaturated fatty acids. J. Chem. Soc., Japan, 1955, 58, no. 11, pp. 934-987. R.Chem. 1957, 30 15, 52899.

21. Malm Golo, Mench Jów. Brazelton, Pulkerson, H1att G.D. Cellulose Derivatives, Preparation of Phthalic acid esters of cellulose. ind. Eng. Chem., 1957, vol. 49, № 1, p.84-88.

22. A.S.213794 (SSSR.). Method for obtaining low-substituted acetylcellulose (Ivanov V.V., Kubayenko T.M., Pavlova G.S., Tsokolaev R.B., Chernov L.M.). - Published in B.I., 1968, 1% 11, p. 20.

23. Kryazhev V.N., Pogosov Y.L., Smirnov B.P., Naimark N.I. Study of acetylcellulose phthalation and properties of cellulose acetophthalates. In: Chemistry and Technology of Cellulose Derivatives. Vladimir, 1971, pp. 102-109.

24. Kryazhev V.N., Pogosov Yu.L. On peculiarities of saponification of acetophthalic cellulose esters. Vysokomol. soed. 1970, X11-B, 9, p. 683-686.

25. Zhbankov R.G. Infrared spectra of carbohydrates structure. Minsk: Nauka-Tehnike, 1972. - 456 c.

26. Litmanovich A.D., Plate N.A., Aghasandyan V.A., Noa O.V., Yun E., Kryshtof V.I., Lukyanova N.A., Lemoshchenko N.V., Kreshetov

V.V. Kinetics of hydrolysis of polymethacrylic acid esters. VMS, v. (A) ХУШ, В5, p. 1112-1122.

27. Pat. I0638 (Japan). Method for the preparation of water-soluble cellulose phosphate esters. Maksima Shoji, Ziroi Hideshi, Hiro 992. Kanari. Published in RZHim, 1969, No. 10.

CHAPTER V. THERMODYNAMIC AND CO-CHEMICAL PROPERTIES OF WATER-SOLUBLE CELLULOSE ETHER SOLUTIONS

5.1. Thermodynamics of interaction of water-soluble cellulose ethers with water.

Application of the laws of thermodynamics to solutions, including solutions of polymers, which include solutions of cellulose ethers, is extremely important. This gives us information on the thermodynamic affinity of the solvent to the polymer, allows us to understand the causes of this affinity, i.e. the role of entropy interaction energy, estimates the thermodynamic stability of the binary polymer-solvent system and its changes with temperature, predicts the system decay upon cooling and heating, and finally allows us to connect the thermodynamic parameters of dissolution with the basic elements of the structure of most dissolving polymers-flexibility of chains, packing density, physical state, molecular weight, etc.

Researchers dealing with the thermodynamics of polymer solutions are always interested in questions about whether substances are able to mutually mix or mutually dissolve. In terms of thermodynamics, this means: Will the process go spontaneously in the direction of mixing of the components or not? Is there an affinity between them.

Processes occurring at constant pressure and temperature (and dissolution usually occurs under these conditions) proceed spontaneously in the direction of decreasing Gibbs free energy (G), i.e. the final state of the system should be characterized by lower values of this parameter than the initial one. Consequently, the solution must have a lower free energy than the sum of the free energies of the components:

$$\text{G Solution} < \Sigma\, G^0{}_i\, .n_i$$

The difference between these quantities is called the free energy of displacement

$$\Delta G^m = G \text{ solution} - G\Sigma^0{}_i\, .n_i$$

and in the case of spontaneous dissolution it should be negative

($G\Delta^m < 0$). The measure of the directionality of the process is also the change in the partial molar free energy of each i-component at constant pressure P, temperature T, and constant number can all other components n_i

$$\mu_i = G_i = (G\partial/\partial\ n)\, n_{iP,T,i-1}$$

The chemical potential of the components in solution μ_i should be less than its chemical potential before dissolution μ^0_I and, therefore, the difference of these values should be negative ($\Delta\mu_i < 0$).

$$\mu_i < \mu^0_i$$

The process proceeds in the direction of coupling of the components only if there is affinity between them. Therefore, the affinity and directionality criteria are the same ($G\Delta^m < 0$; $G\Delta^m < 0$): the greater the absolute value of these differences, the greater the thermodynamic affinities between the components, the more complete is the process of their interaction.

Solvents are divided into good and bad according to their thermodynamic affinity. Good solvents are those solvents whose interaction with polymers is accompanied by large changes in chemical potentials or the free energy of the entire system. In good solvents, contacts between polymer chain links and solvent molecules are energetically more favorable than contacts between polymer chain links and solvent molecules. Thermodynamically bad solvents are considered to be solvents, the interaction of which with high molecular weight compounds leads to small changes in chemical potentials of the components and the free energy of the system. In bad solvents, interaction between links of macromolecules is preferable. In the i-solvent, long-range order interactions take place: attraction between individual chain links, their repulsion upon approaching and interaction of polymer chain links with solvent molecules compensate each other. As a result, the state of the polymer-solvent system is characterized by equilibrium between the energy of Brownian motion of the units and rotation energy barriers around the valence bonds of the chain, and a macromolecule of sufficient length forms a freely coiled ball whose size is called unperturbed. Consequently, in order to evaluate the quality of the solvent one should know the values $G\Delta_i$ and $\Delta\mu_i$. This can be done by determining the saturated vapor pressure of the solvent over the solution, by measuring the osmotic pressure of the solution or swelling pressure, and by determining the so-called second virial coefficient A_2 , which is an important parameter evaluating the quality of the solvent and is related to the osmotic pressure of the solution and the value $\Delta\mu_i$.

At: $A=0$ - perfect solvent, at $A_2 > 0$ - good solvent, $A_2 < 0$ - bad solvent [1].

Improvement of cellulose solubility in water, when its hydroxyl groups are partially replaced by other radicals, is due to the destruction of strong hydrogen bonds blocking the interaction of cellulose with water. The solubility of cellulose derivatives depends not only on the amount but also on the properties of the substituent radical. Therefore, it was expedient to study the sorption properties and thermodynamics of interaction of water-soluble mixed cellulose ethers with water [2].

The thermodynamic affinity of the polymer with the solvent was varied by changing the hydrophilic-hydrophobic balance of cellulose macromolecules. To study the effect of polar and nonpolar substituents on the rheological properties, the following aceto-mixed cellulose ethers with different degrees of substitution were investigated: Water-soluble acetylcellulose (WRAC)$\gamma_{ац}$ =0.55; water-soluble phthalatyl cellulose acetate (WAPC)$\gamma_{ац}$ =0.72;$\gamma_{фт}$ =0.08, content$\gamma_{он}$ =2.2 water soluble acetomaleinate cellulose (VAMC)$\gamma_{ац}$ =0.71,$\gamma_{мал}$ =0.06,$\gamma_{он}$ =2.2; water soluble aminoacetate cellulose (VAAC) with$\gamma_{ац}$ =0.65,$\gamma_{амин}$ =0.51, g$_{OH}$ =1.84 The degree of polymerization of all tested samples is 200-220.

Water vapor sorption was studied using spiral McBen scales with a sensitivity of -1.3-10^{-3} m/kg at 293^0 K and a residual pressure of 10^{-3} - 10^{-4} Pa. When the amount of sorbed water reached equilibrium values, the equilibrium vapor pressure was measured and sorption isotherms were plotted. The heat of mixing was determined by an upgraded DAK-I-IA calorimeter. Samples were dried to constant weight and placed in the working cell of the microcalorimeter before the experiments. After establishing thermal equilibrium, the samples were brought into contact with water vapor. The thermal effect was recorded on the integrator display and recorded in the form of curves on the diagram.

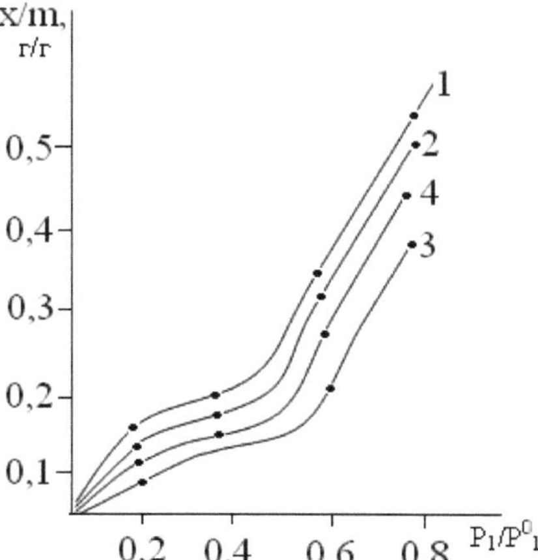

Fig. 95. Isotherms of water vapor sorption by water-soluble cellulose ethers at 25⁰ C: 1 - VAMC, 2 - VAPC, 3 - VAPC, 4 - VRAC.

Fig. 95 shows isotherms of water vapor sorption for all studied samples. S-shaped form of curves with convex initial segment is caused by simultaneously occurring processes: physical adsorption in polymer pores and its swelling. It follows from Fig. 5.1.1 that mixed cellulose ethers have a high sorption capacity in water, and the value of equilibrium sorption depends on the nature of the substituent radical and decreases in the series VAMC> VAPC> VAPC> VRAC.

The sorption isotherms were used to calculate the thermodynamic affinity of water to aceto-mixed cellulose ethers. The difference of chemical potentials of 1g of water in the swollen polymer phase was calculated $\Delta\mu^0_1$. The calculation was carried out according to the equation

$$\Delta\mu_1 = \frac{RT}{M_1} \ln \frac{P_1}{P_1^0}$$

where M_1 is the molecular weight of the sorbent.

Then by the Gibss-Duegem equation values $\Delta\mu_2$ were determined and by the equation g $=\Delta^m\omega_1\Delta\mu_1\omega_2\Delta\mu_2$ +the average Gibbs energy of polymer-water mixing was calculated. Dependence gΔ^m on polymer volume fraction in polymer-water system is shown in Fig.5.1.2.

243

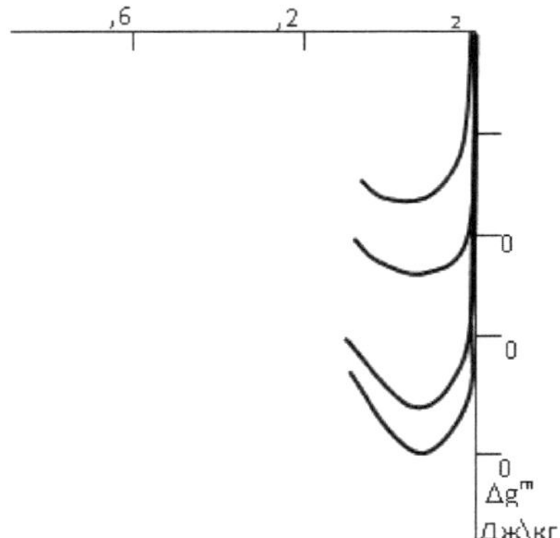

Figure 96. Concentration dependence of the mean free energy of displacement GΔ^m of water with: 1-VAMTS; 2-VAMTS; 3-VAMTS; 4-VATTS.

For all studied systems gΔ^m< 0, i.e. the process of water sorption proceeds spontaneously. However, for cellulose aminoacetate the absolute value of Gibbs energy as compared to the initial water-soluble cellulose acetate decreases, i.e. thermodynamic affinity to water
or its hydrophilicity deteriorates. Introduction of phthalyl and maleic groups into the acetyl cellulose macromolecule significantly increases the affinity of cellulose acetate for water. Apparently, because of the relatively larger volume of phthalyl and maleic groups, the supramolecular structure of acetylcellulose loosens to a greater extent than in the case of substitution of hydroxyl groups for amine groups.

It is known that the interaction of cellulose with water is accompanied by an exothermic heat effect. As studies have shown, the equilibrium effect of interaction of water with mixed cellulose ethers is also exothermic. However, for cellulose amino acetate, a separation of the overall process of interaction with water into endo- and exo-effects was observed, proceeding with a subsequent enhancement of the exo-effect (Table 95). Apparently, nitrogen and hydrogen atoms in the macromolecule contribute to the formation of additional hydrogen bonds in aminoacetyl cellulose.

The endothermic stage, as a rule, is accompanied by hydrogen bond breaking in cellulose during water sorption; the exothermic stage is caused by solvatation of formed free OH-groups. A change in the signs of the heat effect in the interaction of HAAC with water is apparently due to the sequence: solvation - breaking of hydrogen bonds - solvation of free polar groups.

The difference between the systems under study is due to the thermodynamic parameters of the interaction, which carry the main information about the mechanism of interaction and the causes of affinity between the components of the system (Table 62).

Table 62.

Thermodynamic mixing parameters for acetate-mixed cellulose ether-water systems at 293 K.

Sample	Δ	Δ	ΔH	T SΔ
	kJ/kg polymer			
VAMet.C	30,8	3,2	-	28,8
VRAC (ex.).	32,6	4,1	-	28,5
VAAC	33,1	4,4	1,2	29,7
WAF$_0$ C	37,2	8,5	-	28,7
N-VKMC	43,7	8,6	-	35,1
WAMC	45,4	9,7	-	35,7
WAFC	45,7	10,2	-	35,5
Na- AKMC	49,1	24,4	-	34,3

As can be seen from Table 62, the ratio of the values HΔ^m and T SΔ^m is such that the Gibbs energy of mixing of the studied cellulose ethers with water is negative in the whole range of compositions and the type of the given concentration dependence indicates thermodynamic stability of solutions (Fig. 96).

The improvement in the solubility of water for HPAC with the introduction of phthalyl and maleic groups is indicated by an increase in the negative energy component of the Gibbs potential and a decrease in the Flory-Huggins h interaction parameter$_i$ (Fig. 97).

Direct application of the Flory-Huggins equation to the cellulose ethers under study shows that the value of h_i is not constant but depends on the moisture content, especially during the initial sorption period. Typically, the value of h_i is 0.5 for the solubility limit of the liquid. Water is the solvent for the cellulose ethers under study, and the values of h_i should lie below 0.5, which in reality is not the case. These changes are related to the deviation of the structure of the real solution from that accepted in the original model.

245

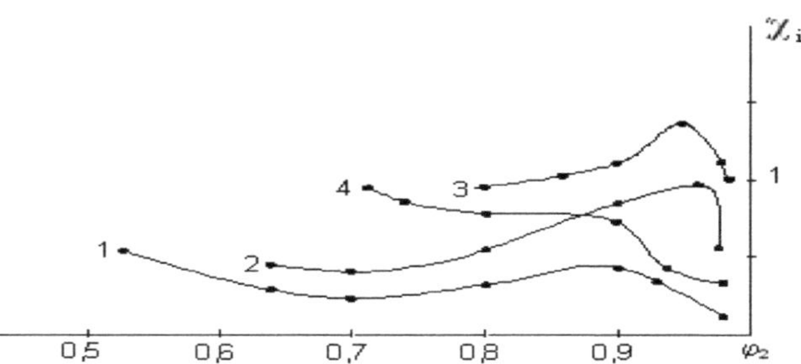

Fig. 97. Dependence of the Flory-Huggins interaction parameter h_i on the polymer volume fraction: 1-VAMC; 2-VAFC; 3-VAFC; 4-VRAC.

The compatibility of the systems WAMC-water and WAPC-water is determined by a change in the enthalpic mixing factor and its contribution to the free energy of formation of the systems becomes predominant. In this case high values ΔH^m related to the energy interaction between sorbate molecules and active groups of the sorbent determine the preferential localization of sorbate molecules in space. For the WAAC-water system, there is a gain in the free energy of mixing the solvent with cellulose ether, which is determined mainly by the entropy of mixing. Positive values and increasing entropy indicate a preferential disordering of the polymer and solvent structure.

Thus, thermodynamic analysis has shown the dependence of the affinity of the water-soluble acetate-cellulose-water system on the new group introduced into it and the change in the competing interaction of homogeneous and heterogeneous molecules. The sorption of water into acetomaleinate and acetatophthalate cellulose is mainly due to the presence of thermodynamic affinity leading to hydrophilic hydration accompanied by a significant thermal effect. The aminoacetyl cellulose-water system is characterized by a weakening of the energy interaction and an increase in entropic effects and, consequently, by a relatively lower affinity of the ester to water.

The obtained data on the effect of thermodynamic affinity of cellulose derivatives with the solvent on the properties of its diluted and concentrated solutions in aqueous media make an additional contribution to the current understanding of the physico-chemical properties of solutions of semi-rigid-chain polymers.

Thermodynamics of cluster formation in cellulose and its esters depending on the degree of substitution and the chemical nature of substituents.

The problem associated with the study of the structure and sorption properties of natural polymers and, first of all, cellulose, is of topical importance. The relevance is determined, first, by the wide use of these polymers in the chemical, pulp and paper, textile industries, as well as in medicine and in microbiological production. Second, cellulose and other natural polymers are reproduced annually, making their reserves inexhaustible. At the same time, cellulose belongs to the rigid-chain polymers, in which the intermolecular bonding is very strong internally. The supramolecular structure of cellulose is extremely complex, it is determined by the ratio of crystalline and amorphous areas, each of which has specific heterogeneity and reactivity. All this dictates the use of the most modern methods of analysis when studying the structural and sorption properties of cellulose and its derivatives.

Numerous sorption studies of cotton fibers, cellulose and its derivatives testify to the non-monotonic character of sorption of water vapor and other solvents [2,3]. Consideration of such self-organized processes in conditions of strong non-equilibrium, leading to hierarchical structures, is possible only on the basis of modern non-equilibrium thermodynamics and synergetic approaches [4]. Despite the abundance of literature devoted to the study of solvation and sorption, at present there is no single theory which would give a complete picture of water sorption by polymers in different regions of relative vapor elasticity. Combining the conventional solution theory with the cluster theory allows the results of water sorption studies in polar polymers to be interpreted. The total amount of water in the polymer can be represented as the sum of uniformly distributed and clustered. It is possible to establish the beginning of clustering and to estimate the values formed by the Zimm-Landberg relation [5]. According to the theory, the value G_{11}/V_1 characterizes the intensity of association of sorbate molecules, the beginning of which corresponds to $G_{22}/V_1 > 1$, which indicates the preferential interaction of sorbed molecules among themselves.

Table 5.1.2 shows the results of theoretical calculations of thermodynamic parameters, characteristics of capillary-porous structure and cluster formation function according to the experimental data of sorption and pycnometric studies [6,7].

Table 5.1.1. shows that the most developed surface has the initial cotton fiber and AC withγ =1.6. Thermodynamic calculations showed that the main contribution to the interaction energy in the case of cotton fiber and cotton cellulose is contributed by the entropic component of Gibbs energy, while in the case of MCC, flax, and wood cellulose, the enthalpic contribution predominates. Intensive clustering is observed for all samples of cellulose and cotton fibers. Water sorption by cellulose changes radically when hydroxyl groups are partially replaced. Hydrophilicity and crystal structure play an essential role.

Low-substituted acetylcellulose derivatives have high sorption capacity as compared to the high-substituted ones, and MCC has the lowest sorption capacity. The highest hydrophilicity is observed for acetylcellulose withγ =1.6 andγ =0.55 for which the values of thermodynamic parameters of polymer-solvent interaction are most negative and the value of h_i is low, i.e. close to ideal.

The results obtained showed that in the case of low-substituted acetylcelluloses, clustering is observed only in the region of high water vapor activity, which agrees with the literature data on clustering in the region of low relative pressures, which seems to be associated with the formation of hydrogen bonds between macromolecules of polymer and water molecules and the possibility of forming mobile fragmented clusters, whose presence reduces aggregation of formed clusters.

Works devoted to direct determination of molecular characteristics of cellulose ethers are few, of which the systematic studies of Kamide and H. Akbarov with collaborators [8-10] should be particularly noted. The difficulty lies in the fact that the propensity of cellulose acetates to form microgel particles, even in dilute solutions, leads to a sharp increase in the molecular weight and size of macromolecules. The presence of microgel in the solutions distorts the Zim diagram and the double extrapolation method proves to be unsuitable for finding the molecular parameters of cellulose acetates. Studies have shown that the molecular mass of cellulose acetate free of microgel is $(0.68 - 0.75) \cdot 10^5$, whereas the molecular mass of microgel particles is $(1.29-3.24) \cdot 10^6$, and the RMS inertia radii of molecularly dispersed polymer vary from 180-290 Å for VAC, TAC and VRP in different thermodynamic quality solvents [11]. The mass fraction of the microgel varies in processes of 0.26-0.35, with rms radii of inertia of 125-160 nm (1250-1600 Å). With little difference in the molecular weights of the "true" polymer solution, the differently substituted cellulose acetates have approximately equal mass fractions of microgel particles, and as the molecular weight increases, the propensity to aggregate increases dramatically [12].

Table 63.

Values of thermodynamic parameters, capillary-porous structure characteristics and clustering function of cellulose and its acetates samples.

№	Sample	$R\,A_{\text{ср},0}$	W_o, see $/g^3$	$S_{уд}$, m $/g^2$	Δ_i J/g	$ч_i$	$\Delta G /V_{111}$
1	Cotton fiber	26,44	0,24	178,9	24,30	1,29	7,72
2	Cotton pulp	21,86	0,13	116,0	17,00	1,45	10,90
3	Linen pulp	25,73	0,17	132,2	11,90	1,55	6,95
4	Wood pulp	30,70	0,22	145,9	21,30	1,86	7,20
5	MCC	17,20	0,12	143,7	16,20	1,67	10,59
6	VRACγ =0.55	32,98	0,18	111,6	24,8	1,20	-3,20
7	ACγ = 0.90	57,24	0,32	111,8	11,0	0,80	-3,00
8	ACγ = 1.2	32,56	0,12	76,8	18,2	1,00	-2,50
9	ACγ = 1.6	15,48	0,26	335,9	37,0	1,10	-2,10
10	ACγ = 2.1	39,08	0,18	76,8	18,2	1,20	7,50
11	WAMETZ	34,6	0,19	114,3	30,8	1,5	-2,8
12	VAAC	36,2	0,21	117,1	33,1	1,3	-3,8
14	WAFoC	36,7	0,23	115,3	35,5	0,9	-3,8
15	WAFC	34,4	0,26	120,2	45,7	0,9	-3,9
16	N-VACMC	38,1	0,22	118,4	43,7	1,0	-2,8
17	WAMC	33,2	0,24	119,5	45,4	0,9	-3,6
18	Na-VACMC	34,8	0,27	124,6	49,1	0,8	-4,7

Conformational properties of macromolecules of aceto-mixed cellulose ethers in solution.

The dependence of the viscosity of polymer solutions on their molecular weight was first shown by Staudinger, then by Mark and Howwink. The logarithmic dependence of the characteristic viscosity on the molecular weight does not always have a linear form, which is

249

determined both by the ratio of the contour length and the Kuhn segment and by the quality of the solvent. It is known that for flexible-chain polymers in a good solvent, the exponent, α in the Mark-Kuhn-Hauvinck equation increases with increasing molecular weight, which is determined by volume effects and approaches 0.5 in Q-solvent. For rigid-chain polymers in the region of relatively low molecular weight, the value of α can, changes in the range of 1.0-4.7. In the case of branched or globulated polymers, the size of the macromolecular tangle grows with an increase in molecular weight to a degree less than 0.5. The work [13] summarizes the literature data on the determination of Mark-Kuhn-Hauvinck constants for cellulose acetates of different degrees of substitution by light scattering of solutions. It is shown that the values of the exponent of the Mark-Kuhn-Hauvinck equation vary in the range ~0.6-0.8. It is noted that such high values of the exponent α are related not so much to volume effects and the permeability of macromolecules by solvent molecules as to the skeletal stiffness of cellulose acetate macromolecules.

The theoretical approaches developed for rigid-chain polymers were shown in [14] to be applicable to the determination of unperturbed by volume exclusion sizes of semi-rigid-chain cellulose acetates by light scattering of solutions in non-ideal solvents. Various methods for determining macromolecule sizes unperturbed by volume exclusion from extrapolation data from measurements of the molecular characteristics of polymers in thermodynamically good solvents by solution light scattering, viscometry, double ray refraction, gel permeation chromatography, membrane osmometry and sedimentation are known. In the case of flexible-chain polymers, the Stockmeyer-Fixaman, Flory-Fox methods, as well as various extraction variants proposed by Tanner-Berry, Bauman, Kamide-Kawan, and Kamide-Miyazaki are most common [13-17]. These methods are also often used to determine the unperturbed sizes of cellulose ether macromolecules. To describe the conformation of semi-rigid-chain polymers, the persistence chain model and the extrapolation methods of Kuhn, Benoit, Hirst, Hirst-Stockmeyer, and Yamakawa based on it have been developed. In papers [15-17] the results of application of these techniques to cellulose acetates of various degrees of substitution are given. The authors conclude that the literature data on the determination of unperturbed excluded volume sizes of cellulose acetates using various extrapolation techniques proposed for flexible-chain polymers and on the basis of worm-chain theory are ambiguous. Based on these considerations, it is recommended to apply to semi-rigid-chain cellulose acetates the techniques for determining the unperturbed dimensions of rigid-chain polymers, while making some corrections for the degree of thermodynamic interaction in the polymer-solvent system depending on the molecular weight of cellulose acetates.

From the above data on the study of molecular characteristics of cellulose ethers it follows that in the literature there are works devoted mainly to cellulose acetates of different degrees of substitution, while mixed cellulose ethers have not been studied practically. In this connection an urgent need arose to study the properties of their solutions and to determine the molecular characteristics depending on the chemical composition of mixed cellulose ethers and the degree of esterification. Such studies are necessary because recently water-soluble cellulose ethers have an extremely important practical importance.

It is known that the physicochemical and colloidal properties of solutions of high molecular weight compounds depend on the degree of polymerization, the degree of esterification, the nature of the solvent and the medium temperature.

This section presents the results of conformational properties of macromolecules in dilute solutions of mixed cellulose ethers, by viscometry [18]. There was water-soluble acetyl cellulose (WAPC) containing 18% bound acetic acid, water-soluble cellulose acetate phthalate (WAPC) containing 24-25% bound acetic acid and 9-10% phthalyl groups.., Water-soluble acetoammonium phosphate cellulose (WAPCo C), containing 20% of bound acetic acid and 8% of phosphate groups; water-soluble aceto carboxymethyl cellulose (WACMC) in H- and Na-forms, containing 28-30% of bound acetic acid and 0.7% of carboxymethyl groups. The degree of polymerization of the above mixed cellulose esters was 190-220.

Using the values of characteristic viscosity of aqueous solutions of aceto-mixed cellulose ethers on temperature, the viscous flow activation energy, the effect of the nature of the substituted functional group on the value of the rms distances between their ends ($'h^2 \,'^{1/2}$) and Huggins constants by the Flory formula $/\eta\ /=F'h^2 \,'^{3/2} /M$, whose values indicate the size of balls of macromolecules (Tables 64, 65 and Fig. 66) were determined.

The monomer links of the polymer chain of polyelectrolytes contain ionogenic side groups, as a result, macromolecules acquire a number of characteristic electrical, conformational and hydrodynamic properties.

251

As can be seen from Table 64, partial substitution of hydroxyl groups of HPAC with such ionogenic groups as phthalyl, carboxymethyl or phosphate ones leads to increased viscosity of the solutions of obtained aceto-mixed cellulose ethers.

Among the samples studied, the WAMC solution has a relatively low characteristic viscosity value, while the VAPhOC solution has the highest one. Since WAAFC is a salt whose macromolecules easily dissociate into phosphate anions, the latter are strongly hydrated, resulting in an increase in viscosity.

The carboxyl groups of H-VACMC slightly dissociate, hence, its macromolecules are weakly hydrated. Since Na-VACMC is easily dissociated in solution, the viscosity of the solution increases noticeably when H is replaced by Na. The viscosity of WAPMC solution is higher than that of H-WACMC. This can be explained by the fact that the acidity index of phthalic acid is higher than that of acetic acid.

Table 64.

Dependence of the characteristic viscosity of aqueous solutions of cellulose ethers [η] on temperature and their viscous flow activation energy.

Sample	$[\eta]$ at temperature, ^{0} C						E, kJ/mol
	25	30	40	50	60	70	
WAMC	1,48	1,39	1,30	1,22	1,15	1,12	4,85
VRAC	1,50	1,41	1,32	1,24	1,17	1,09	5,14
VAAC	1,57	1,48	1,42	1,34	1,28	1,21	5,24
WAFoC	1,61	1,56	1,51	1,46	1,1,46	1,35	5,38
N-VACMC	1,75	1,68	1,61	1,53	1,53	1,40	5,54
WAFC	1,85	1,75	1,68	1,61	1,73	1,48	7,82
Na-ACMC	2,13	1,95	1,87	1,80	1,95	1,68	7,12
WAAFoC	2,21	2,15	2,05	2,01	1,98	1,88	9,48

Increasing the temperature of the medium leads to a decrease in the viscosity of the studied solutions, which is associated with the destruction of the hydrate shell of macromolecules and the structure of the solvent.

The effect of the nature of the substituted group on the size of the tangle of ester macromolecules in solution was assessed by the value of the RMS distances between their ends (Fig. 98).

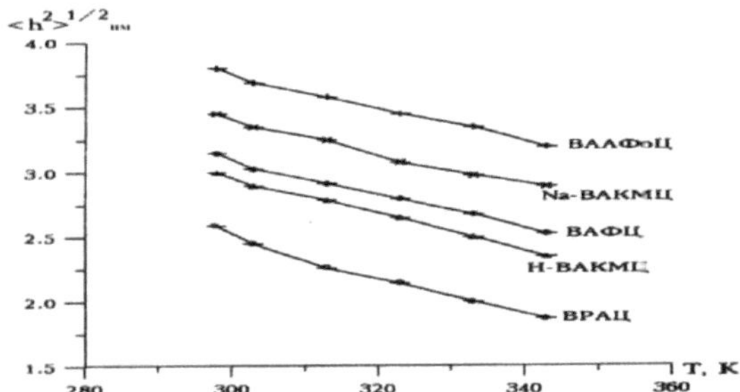

Figure 98. Dependence of the RMS size of the macromolecular ball of cellulose ethers 'h² '¹ᐟ² on the temperature in aqueous solution.

As can be seen from the figure, the size of the balls of macromolecules of mixed cellulose ethers is larger than that of HPACs and is directly proportional to the viscosity of their solutions, since the distance between the ends of the macromolecules increases as the polarity of the substituents increases.

The size of balls of cellulose ether macromolecules decreases as the temperature of solutions increases. Apparently, increasing the temperature of the medium leads to destruction of hydrate shells of functional groups of macromolecules, as a result of which their flexibility increases.

Fig. 98 also shows that the flexibility of the macromolecules of Naa VACMC and VAAFC is little dependent on temperature, since they are stronger polyelectrolytes and easily dissociate, forming carboxylic anions along the macromolecules, which give them rigidity.

To estimate the equilibrium flexibility of the macromolecules of the studied cellulose derivatives, the values of Huggins constant K were calculated. The results of calculations are given in Table 65.

It is known that the value of the Huggins constant is directly proportional to the flexibility of polymer macromolecules, i.e. the lower the K, the less flexible the macromolecule is. For example, during carboxymethylation of HPAC the K value decreases from 0.40 to 0.38, and during phthalation - to 0.34. The most rigid is a macromolecule of VAAPhoc, which has a Huggins constant at 298^0 K with a value of 0.26.

The macromolecules of WAFC and Na-VACMC occupy an intermediate position in terms of flexibility among mixed cellulose ethers.

Table 65.

253

Dependence of the Huggins constant K on temperature.

Sample	K at temperature,^{0} K					
	25	30	40	50	60	70
WAMC						
VRAC	0,42	0,43	0,44	0,46	0,47	0,48
VAAC	0,40	0,41	0,43	0,44	0,46	0,47
WAFoC	0,39	0,40	0,42	0,43	0,45	0,46
N-	0,37	0,38	0,41	0,43	0,44	0,45
VACMC	0,35	0,37	0,39	0,40	0,41	0,43
WAFC	0.34	0,34	0,36	0,36	0,37	0,37
Na-	0,30	0,30	0,32	0,32	0,33	0,33
VACMC	0,26	0,26	0,27	0,27	0,28	0,29
WAAFoC						

With increasing temperature in all cases there is a slight increase in the Huggins constant, indicating an increase in equilibrium flexibility of macromolecules, as a result of the destruction of the hydrate shell. The calculated values of K correlate well with the results calculated values 'h^2 $_{1/2}$.

To estimate the intermolecular interaction in solutions, we calculated the temperature coefficient of viscosity, from which, in turn, we can calculate the apparent activation energy of viscous flowΔ E by the equation

$$\frac{d \ln[\eta]}{d\left(\frac{1}{T}\right)} = \frac{\Delta E}{R}$$

The valueΔ E is also easily determined graphically by the dependence ln [η] = f (1/T) (Table 65).

As can be seen from the above data, with increasing temperature, the viscosity of the solutions decreases, i.e., the temperature coefficient of viscosity is positive (Table 65). The values of the viscous flow activation energy of cellulose ether solutions also depend on the nature of the substituted functional group, and the higher the polarity of the substituted functional group, the greater the viscous flow activation energy.

Thus, solution viscosity of water-soluble mixed cellulose ethers, conformation and flexibility of their macromolecules depend on the nature of substituents. According to the value of solution viscosity and macromolecule stiffness, the studied cellulose ethers can be arranged in the following series WAMC < VARAC < VAAC < N-VACMC < WAFC < Na-VACMC < VAAFoC. From this we can conclude that WAMC and VAPC are not electrolytes, and the polyelectrolyte properties of the given mixed cellulose ethers increase from left to right.

The experimental data obtained on the effect of thermodynamic affinity of cellulose and its aceto-mixed esters with the solvent on the properties of its dilute and concentrated solutions in aqueous media make an additional contribution to the current understanding of the physicochemical properties of solutions of semi-rigid-chain polymers.

In order to scientifically predict the possibility of polymer dissolution in a particular solvent and to predict the properties of concentrated polymer solutions, a comprehensive approach that takes into account the physical and chemical properties of the polymer, the solvent and their mutual influence during interaction, i.e. the mechanism of polymer dissolution, is necessary.

The presented material concerning the influence of intermolecular interactions of macromolecules with the solvent on the rheological properties and structural organization of cellulose derivatives solutions and the possibility of variation of this contribution by changing the hydrophilic-hydrophobic balance of macromolecules allows one to draw conclusions about how the technological characteristics of the solutions can be adjusted with regard to the requirements for the final products obtained.

Macromolecule conformational state and optical activity of solutions of water-soluble aceto-mixed cellulose ethers.

Optical activity, which characterizes the peculiarities of the spatial structure of molecules, has long been known.

It is also known that the peculiarities of the spatial structure of molecules largely determine the various properties of the substance and play an important role in the processes of its synthesis. The value of optical density is determined by the double optical rotation dispersion (DOV) method [19]. As applied to cellulose and its derivatives, this method is poorly studied. There are only a few works [20-25] where almost all material on determination of optical activity of cellulose and acetic acid ether with different content of acetyl groups obtained by saponification of cellulose triacetate and for some of its other esters is collected.

However, for water-soluble cellulose ethers with different functional groups, this method has not been applied. In this connection, it was of interest to study how the substitution of some functional groups in the pyranose cycle of cellulose macromolecules by other functional groups affects the optical activity of water-soluble acetoacetate-mixed cellulose ethers solutions.

To explain the reason for the dependence of the optical activity of cellulose derivatives solutions, when some functional groups are replaced by others in the macromolecule, the solution of water-soluble cellulose derivatives and for comparison the solution of d-glucose and cellobiose in different solvents were studied by DOV method together with the employees of the "High-molecular compounds" department of Saratov State University. The rotational constants of cellulose ethers were calculated from DOV data.

The optical activity of the solutions was estimated by the value of specific rotation calculated according to the formula:

$$[d]_{\lambda}^{t} = \frac{\alpha \cdot 100}{l \cdot c};$$

where: α - the observed angle of rotation of the polarization plane, deg.

λ - wavelength of incident light, nm.

l - is the path length of light in the optically active medium, decimeter.

c - concentration of the optically active substance, grams per deciliter of solution.

t - temperature of the studied solution, ^{0} C.

Table 66 shows the values of specific rotation of cellulose solutions and its various esters at different solvents and wavelengths.

The data in Table 66 show that compounds of the same functional composition and dissolved in the same solvent differ in magnitude and sign of specific rotation. The difference in the optical density of the same functional composition-glucose, cellobiose, and cellulose in the same solvent-can be explained by the three most likely variables: the spatial position of the functional groups, conformation, and reisomerization.

Table 66.

Value of specific rotation of cellulose and its esters solutions at different solvents and wavelengths.

Solvent	Samples	Wavelength, nm					
		579	546	495	435	414	400

Cadoxen	Glucose	+4	+2	+2	-2	+3	+3
	Cellobilose	-6	-8	-11	-18	-19	-21
	Cellulose	-44	-54	-66	-84	-95	-100
Tetrachloroe thane	b-pentaacetyl-d-glucose	+21	+26	+50	+50	+73	+82
	b - octaacetylcell lobiose	+10	+11	+18	+32	+43	+54
	Triacetylcellulose	-4	-6	-10	-15	-18	-25
Water	Acetylcellulose (HPAC)	0	-2	-5	-19	-21	-29
	Aminoacetate cellulose (AVAC)	+3	+4	+4	+3	+2	+2
	Acetaphthalate cellulose (VAPC)	-14	-15	-24	-39	-49	-54
	Acetocarboxymeth ylcellulose (VACMC)	-3	-5	-8	-22	-24	-32

Since it is difficult to differentially assess the contribution of each variable to the optical activity, it is reasonable to represent the position of the functional groups and the conformation of the pyranose cycles as the concept of the conformational state of the pyranose cycle. In this case, the optical activity will be determined by the conformational state of the pyranose cycle and rotational isomerism. However, for glucose the rotational isomerism is excluded, while for cellobiose and cellulose it is generally accepted to be the same-the pyranose cycles are rotated relative to each other by 180^0. Hence, we can conclude that the optical activity of the solutions in the series under consideration is most likely determined by the conformational state of the pyranose cycles.

The rotational constants A and λ_0 were calculated to prove more convincingly the role of the pyranose cycle conformation in the optical activity of the solutions of the considered systems. The constant A is responsible for the stereochemical state of the polymer, and λ_0 characterizes the contribution of functional groups to the optical rotation.

The data in Table 67 show that the rotational constant A for a number of carbohydrates of the same functional composition-glucose, cellobiose, and cellulose (dissolved in the same solvent) differs dramatically.

Table 67.

Rotational constants of cellulose ethers.

Connection	AND· 10^{-5}	λ_0, nm	Solvents
Glucose	-8	392	Cadoxen
Cellobiose	-21	321	Cadoxen
Cellulose	-113	383	Cadoxen
TAC	-20	327	Acetic acid
Glucose	+150	361	Water
VRAC	+11	343	Water
VAAC	+125	369	Water
WAFC	-33	327	Water
WACMC	+24	320	Water

For water-soluble cellulose derivatives of different functional compositions, the constant A also differs in value and sign. The constant λ_0, on the other hand, has practically similar values for all the systems studied.

The close values of λ_0 for compounds of the same and different functional compositions show that a change in the ratio of functional groups has no significant effect on the specific rotation and, conversely, the wide interval of changes in the constant A for compounds of both the same and different functional compositions may indicate a large difference in the conformation of the pyranose cycles.

It should be noted that the optical activity of the same compound in different solvents is different. Since the solvents cannot cause any changes in the compounds other than the redistribution of the electron density in the cycle and hence a change in its conformation, the difference in the optical activity of the studied systems in different solvents confirms the relationship between the optical activity and the conformation of the pyranose cycle.

Based on the data obtained, we can conclude that along with the replacement of some functional groups in the pyranose cycle of the cellulose macromolecule by others, conformational changes also occur, which determine the optical activity of their solutions.

5.2 Effect of the chemical composition of water-soluble aceto-mixed cellulose ethers on the surface tension of their aqueous solutions.

It is known that the surface properties of hydrophilic colloidal systems, including high molecular weight compounds, as well as other manifestations of adsorption activity depend on the time of formation of the surface layer of molecules of the dissolved substance.

Consequently, a reasonable conclusion about the surface activity or inactivity of cellulose derivatives can be made only by studying the surface tension of their solutions using methods that allow the complete formation of the adsorption layer at the solution-air interface.

Based on measurements of the surface tension of aqueous solutions of Na-CMC at the liquid-gas interface, a number of researchers [26,27] have concluded that Na-CMC practically does not change the surface tension of water.

In this regard, it was of interest to study the surface-active properties of aceto-mixed cellulose ethers and to establish the regularities of the influence of different substituents introduced into the structure of HPAC macromolecules on these properties, since the established regularities will affect the values of thermodynamic values when studying the colloid-chemical properties of difilic polymers. [28].

In this chapter, the influence of the chemical composition of water-soluble acetate-mixed cellulose ethers on the surface activity at the solution-saturated solvent vapor interface, at different concentrations of their aqueous solutions, was studied, taking into account the kinetics of adsorption layer formation.

In order to compare the properties of the solutions of synthesized cellulose ethers with other known cellulose ethers, the properties of aqueous solutions of methylcellulose and sodium salt of carboxymethylcellulose were studied.

The surface tension of aqueous solutions of cellulose ethers under study was determined by the "maximum bubble pressure" method on a Rebinder apparatus at 20^0 C. The instrument was calibrated to water (solvent). The surface tension was calculated by the formula:

$$\sigma = \sigma_\text{в} \cdot \frac{H}{H_\text{в}}$$

where: σ - surface tension of the solution, n/m;

$\sigma_\text{в}$ - surface tension of water, n/m;

$H, H_\text{в}$ - is the difference of liquid heights in the gauge at the moment of air bubble slip in the studied solution and inlet, respectively.

The influence of the chemical composition of the macromolecule of water-soluble cellulose ethers and the concentration of their aqueous solutions on the surface activity at the interface between the phases of aqueous solution and saturated solvent vapor was studied taking into account the kinetics of the adsorption layer formation.

The results of the measurements are shown in Table 69 and Fig. 99.

As can be seen from Table 67 for almost all investigated cellulose ethers the value of surface tension depends on the duration of bubble formation (t) in min; the value of surface tension σ , solution concentration (C), characterizing the kinetics of surface layer formation.

To a lesser extent, this dependence is manifested for simple and complex homo-mixed cellulose ethers containing only ionogenic substituents (CMC), as well as for mixed cellulose ethers containing a certain amount of non-ionogenic substituents (AMC, AKMC). To the greatest extent, the value of surface tension depends on the duration of surface

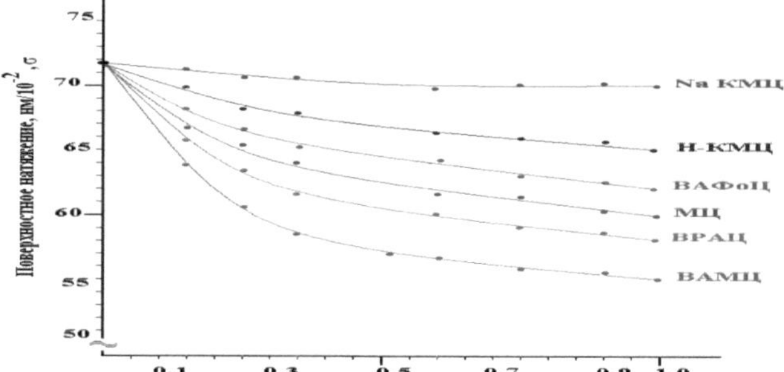

Figure 99. Dependence of surface tension (σ) of aqueous solutions of cellulose ethers on concentration.

tension depends on the duration of formation of the adsorption layer on the interface for MC and AFoC, i.e., for those esters that significantly reduce the surface tension of water.

The dependence of the surface tension of aqueous solutions of cellulose ethers on concentration obeys the following empirical formula:

D - surface tension of cellulose ether solution with concentration C.

G* is the minimum value of surface tension for a given cellulose ether.

α, β - coefficients depending on the chemical composition of the cellulose ether macromolecule.

By experimentally determining G and G* for several concentrations of cellulose ether solutions, one can graphically and computationally find

260

valuesα andβ , which allow one to establish values of the surface tension of the solution for any average values of concentration.

Table 68 shows the obtained valuesα andβ for the studied cellulose ethers.

As can be seen from Table 68, the coefficientsα and G^* are inversely related, i.e. cellulose ethers having the lowest value of G^* are characterized by the highest valueα and vice versa, while the coefficientsα andβ change in direct proportion to each other.

Table 68.

Valuesα andβ for different samples of cellulose ethers.

Equation coefficients	Cellulose ethers under study								
	AMC	MC	AC	AAC	AF_0C	AMC	AFC	Na-ACMC	Na-KMC
α	21,10	20,20	11,2	10,50	18,10	7,40	6,30	4,1	2,2
β	0,86	0,82	00,75	0,62	0,65	0,60	0,56	0,41	0,32
Γ^*	49,0	50,6	52,9	60,8	63,4	68,2	69,1	69,5	70,6

G*- *for 1% solution of cellulose ethers, at 20^0 C.*

Thus, if the coefficientα characterizes the ability of a given cellulose ether to reduce the surface tension of the solvent, then the coefficientβ determines the character of the change in surface tension with a change in concentration for each cellulose ether.

The results showed that CMC, AKMC, AMAC and AFC have no or insignificant surface activity at the interface between the phases of aqueous solution and saturated steam.

AMC, MC and AAC have the greatest surface activity. AF_0 CD and AAC occupy an intermediate position. Consequently, cellulose ethers reduce the surface tension at the interface between aqueous solution and saturated steam if they contain ether groups which do not dissociate into ions in water ($-OSH_3$; $-OSOSH_3$, $-OS_2 H_5$). Cellulose ethers containing only ionogenic groups have practically no surface activity.

Using acetylcarboxymethylcellulose as an example, we investigated the dependence of the surface tension of aqueous solution on the ratio of the degree of substitution of ionogenic and non-ionogenic groups in the anhydroglucose chain of cellulose.

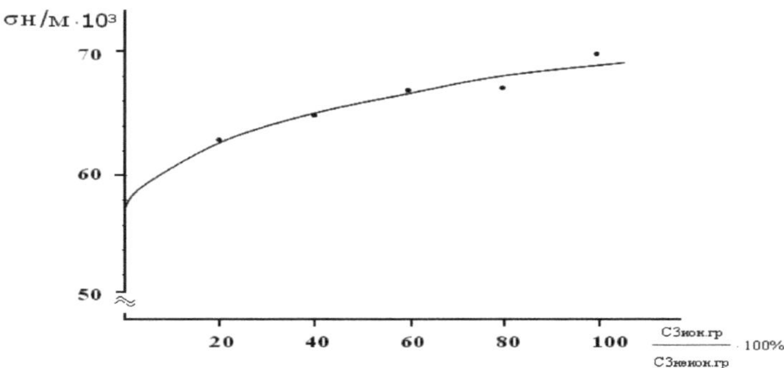

Fig. 100. Dependence of surface tension of aqueous solution of acetylcarboxymethylcellulose on the ratio of ionogenic groups in the macromolecule.

As can be seen from Fig. 100, the surface tension of the solution decreases with increasing content of nonionogenic groups, i.e. the adsorption capacity of the macromolecule increases.

For water-soluble cellulose derivatives of different functional compositions, the constant A also differs in value and sign. The constant λ_0, on the other hand, has practically similar values for all the systems studied.

The close values of λ_0 for compounds of the same and different functional compositions show that a change in the ratio of functional groups has no significant effect on the specific rotation and, conversely, the wide interval of changes in the constant A for compounds of both the same and different functional compositions may indicate a large difference in the conformation of the pyranose cycles.

It should be noted that the optical activity of the same compound in different solvents is different. Since the solvents cannot cause any changes in the compounds other than the redistribution of the electron density in the cycle and hence a change in its conformation, the difference in the optical activity of the studied systems in different solvents confirms the relationship between the optical activity and the conformation of the pyranose cycle.

Based on the data obtained, we can conclude that along with the replacement of some functional groups in the pyranose cycle of the cellulose macromolecule by others, conformational changes also occur, which determine the optical activity of their solutions.

Thus, by changing the ratio of the number of ionogenic and non-ionogenic groups in the macromolecule of aceto-mixed cellulose ethers, their solubility and the surface tension of solutions can be adjusted.

Dependence of equilibrium surface tension (γ· 10⁻³ n/m) of aqueous solutions of cellulose ethers on concentration (t - time of establishment of adsorption equilibrium, min)

C, Cong. solution.	Cellulose ethers under study															
	MC		CMC		ACMC		AMC		AFC		AMAC		AC		AAC	
	t	Γ	t	Γ	t	Γ	t	Γ	t	Γ	t	Γ	t	Γ	t	Γ
1,0	11,1	50,6	9,88	70,6	9,65	67,0	11,2	49,0	12,0	69,1	17,60	68,2	0,63	55,7	34,55	60,8
0,5	27,3	52,7	10,4	71,6	15,5	68,1	25,3	50,0	14,3	70,7	25,9	70,4	0,67	56,0	38,22	61,3
0,25	30,1	54,3	12,5	71,6	17,6	69,4	28,0	50,0	22,6	71,4	31,05	71,7	0,73	57,0	43,66	62,6
0,10	32,9	57,6	13,4	71,7	18,5	70,0	30,9	54,2	35,0	71,7	35,75	71,7	0,76	60,4	465,1	63,2
0,05	35,3	60,2	14,2	72,0	19,6	70,6	34,3	55,0	48,0	72,0	38,03	71,7	0,78	63,3	54,5	64,1
0,01	42,3	64,4	19,10	72,0	22,4	71,0	39,0	61,8	-	-	-	-	0,95	66,2	58,40	67,8
0,005	56,7	64,7	-	-	27,2	71,7	46,7	2,0	-	-	-	-	0,80	69,4	67,2	70,1

thermodynamic parameters of adsorption of cellulose ether macromolecules at the "aqueous solution-saturated solvent vapor" interface.

It was interesting to study the adsorption of water-soluble mixed cellulose ethers at the interface "aqueous solution - saturated solvent vapor" using the example of three different in chemical structure substituting radicals of water-soluble mixed cellulose ethers, such as acetyl methyl cellulose (VAMC), acetyl carboxymethyl cellulose (VACMC) and acetyl cellulose (APAC), which have surface activity [29].

The relationship between surface tension, surfactant adsorption and concentration is established by the fundamental Gibbs adsorption equation:

$$\Gamma = -\frac{C}{RT} \cdot \frac{d\sigma}{dc}$$

Where: Γ - Specific adsorption of the dissolved substance, mol/cm^2 .
C is the concentration of the dissolved substance, mol/L.

The Gibbs equation was used to calculate the specific adsorption of the studied cellulose ethers at the "aqueous solution-saturated solvent vapor" interface.

The calculation results shown in Fig. 20 indicate a low adsorption activity of water-soluble cellulose ethers at the above interface.

Steric hindrance and the presence of hydrogen bonds due to the specific structure of cellulose ethers probably make it unfavorable to change the free energy of the system downward due to adsorption of cellulose derivative macromolecules on the "aqueous solution - saturated solvent vapor" interface.

The surface activity of conventional surfactants is the greater the adsorption potential, which, in turn, increases with increasing hydrocarbon radical. For homologous surfactant series, the relationship between surface activity is established by the Duclos-Traubet rule:

$$\Gamma = \frac{G_{n+1}}{G_n} = 3,0 \div 3,5$$

While the nature of the polar group in the case of typical surfactants has little effect on the surface properties, for cellulose ethers the adsorption activity is determined precisely by the nature of the substituent group.

It is known that cellulose has no surface activity with respect to water, i.e. it is a surface-indifferent substance.

It was shown above that cellulose monoesters (e.g., MC), which contain only substituted nonionogenic groups and mixed cellulose ethers containing a certain amount of nonionogenic substituents along with nonionogenic substituents, have surface activity. However, cellulose ethers containing only non-ionogenic functional groups also differ in their ability to adsorb at the interface "aqueous solution - saturated solvent vapor", but they adsorb better than mixed cellulose ethers (Fig. 101).

As can be seen from Fig. 101, non-ionic VAMC is better adsorbed from aqueous solution than HPAC, and VAMC ranks last in adsorption capacity, since the number of non-ionic substituents is less than 50%.

Further thermodynamic parameters of adsorption of synthesized esters were calculated. It is known that for nonionic surfactants the surface pressure value $p = -\sigma_0 \sigma$ (where σ_0 and σ surface tension

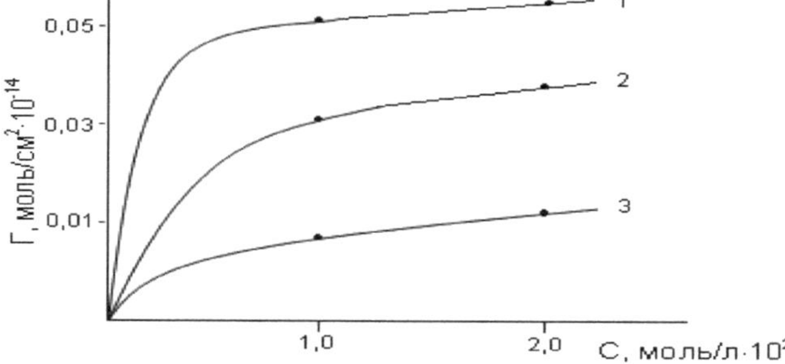

Fig. 101. Adsorption isotherms of water-soluble mixed cellulose ethers on the "aqueous solution - saturated solvent vapor" interface 1-VAMC, 2-VARAC, 3-VACMC.

solvent and solution, respectively) is related to the adsorption parameters: Henry's constant (K_r), the adsorption equilibrium constant ($K_л$), and the maximum monolayer capacity (G_∞). These parameters were calculated using the following formulas:

For diluted solutions: $\dfrac{\pi}{RT} = \Gamma = K_r \cdot C$

For higher concentrations:

$$\frac{\pi}{RT} = \Gamma_\infty \cdot \ln(1 + KC), m.\kappa.KC \rangle 1$$

$$\frac{KT}{\pi} = \frac{1}{\Gamma_\infty} - \frac{1}{\Gamma_\infty \cdot K_\Lambda \cdot C}$$

These formulas can be used to calculate the values of K_Λ and G_∞

To determine the thermodynamic parameters we used the following relations: for the Gibbs energy of adsorption

$$\Delta G_1 = -RT\ln K_r$$
$$\Delta G_2 = -RT\ln K_\Lambda$$

for surface activity p $= RT\ln K_r$

$\Delta H = RT_2^2$ (d ln K_Λ /dT)

for the adsorption energies $\quad \Delta S = (\Delta H - \Delta G)/T$

The results are shown in Table 6.2.1. and 6.2.2.

Table 68.

Dependence of surface tension of mixed cellulose ether solutions on temperature (C=1.0 g/dl).

№	Cellulose esters	WAMC	VRAC	WAFoC	N-VACMC
1.	298	55,6	63,4	62,8	65,7
2.	303	53,7	61,2	59,6	63,6
3.	318	51,5	59,5	58,4	63,8
4.	328	50,6	58,4	57,7	64,1
5.	338	50,1	57,8	57,2	64,3

The data in Table 68. indicate that the synthesized cellulose ethers are characterized by a decrease in the surface tension of solutions with increasing temperature. Surface tension of Na-VACM solution decreases with increasing temperature in the initial stages and remains practically constant. This is apparently due to an increase in the degree of dissociation of the carboxyl group with the formation of -COO$^-$ and Na$^+$ ions, which contribute to an increase in the surface tension of water.

By linear solution of the Langmuir equation based on surface tension data the values of adsorption equilibrium constants K, maximum Gibbs adsorption G∞ Henry's adsorption constant K =KG$_{r\infty}$, and surface activity p =RTlnK$_r$ were found . From the temperature dependence of the adsorption constant (K) the thermodynamic parameters of adsorption were determined.

A comparative analysis of the data in Table 68 shows that the surface activity increases with the increase of hydrophobic functional groups (-CH$_3$) in the macromolecule of mixed cellulose ethers. It should be noted that in all cases with the increase of hydrophobic groups in the macromolecule the surface activity increases with increasing temperature.

The values of Henry's and Langmuir's constants increase both with increasing hydrophobic group content in the macromolecules of cellulose ethers under study and with increasing temperature. The size of the landing sites of the molecules for all the studied samples is almost the same and increases with increasing temperature.

It is shown that the adsorption of the studied compounds at the liquid-gas interface is well described by the Langmuir equation. The values of free energy GΔ_a of adsorption were also determined using equation II of the law of thermodynamics:

$$\Delta G = H - T\ S_a\Delta_a\Delta_a$$

The results of calculations are shown in Table 6.2.2. ValuesΔ H$_a$ for all cellulose ethers are positive. However, for this class of surfactants, the change in the free energy of the system GΔ_a was negative, which indicates the thermodynamic advantage of their dissolution in water. Characteristically, the change in the entropy of adsorption for these compounds is the same and close to zero. The results indicate that the adsorption of the compounds studied at the liquid-gas interface is an enthalpically controlled process due to the presence of a number of functional groups in the macromolecules.

Table 69.

Thermodynamic parameters of adsorption of mixed cellulose ethers at the liquid-gas interface.

№	Samples	T, K	BY_r 10^3	p, nm^2 /mol	$K_{л}$	$D\infty$ 10^{20}	S_j 10^{20}	Δ kj/mol	ΔH_a kJ/mol	ΔS_a kJ/mol
1	NACMC	298	1,32	3,10	76	1,82	88	9,68	16,15	0.086
		308	1,39	3,76	78	1,94	89	10,82		0.087
		318	1,45	4,24	92	1,82	93	11,71		0.086
		328	1,46	4,52	93	1,63	97	12,15		0.086
		338	1,46	4,37	94	1,44	102	13,60		0.088
2	WAFoC	298	5,71	14,31	291	1,90	86	11,02	10,46	0.073
		308	5,76	14,62	302	1,87	89	11,77		0.072
		318	5,62	14,76	310	1,61	100	13,00		0.073
		328	6,28	16,15	386	1,46	102	13,98		0.074
		338	6,78	17,12	402	1,47	104	14,28		0.073
3	VRAC	298	7,18	15,70	217	1,72	92	12,64	8,34	0.071
		308	7,23	16,51	265	1,71	89	13,25		0.070
		318	7,52	17,73	286	1,68	94	14,42		0,071
		328	8,12	17,92	298	1,65	98	15,14		0,071
		338	8,36	18,34	333	1,61	106	15,45		0,070
4	WAMETZ	298	12,14	24,31	550	1,92	87	19,68	6,15	0,086
		308	12,35	24,75	561	1,85	91	20,83		0,087
		318	12,83	24,96	565	1,82	94	21,40		0,086

| | | 328 | 13,27 | 25,38 | 578 | 1,78 | 97 | 22,25 | | 0,086 |
| | | 338 | 13,50 | 26,12 | 583 | 1,87 | 108 | 23,60 | | 0,088 |

LITERATURE.
Bartenev G.M., Frenkel S.Ya. - T.: Chemistry, 1990. - C. 349-364.

2. Miagkova N.V., Sagdieva Z.T., Rakhmanberdiev G., Sidikov A.S. Thermodynamic properties of water-soluble mixed cellulose ethers solutions // Chemistry of natural compounds. -T., 1996. - № 1. - C. 102-106.

3. Bocek A.M., Kalyuzhnaya L.M. Features of interaction of cellulose and cellulose acetates with water when changing the system of hydrogen bonds in them and hydrophobic balance of macromolecules // Journal of Applied Chemistry. - 2002. - T. 75. - №6. - C. 1007-1011.

4. Bugnin S.V., Astapenko E.P., Belyaeva E.V., Khripunov A.K., Tsvetkov V.N. Hydrodynamic and conformational properties of cellulose acetomiristanate molecules // High-molecular sols. - 1999. -T. 41 A. - № 6. - C. 1021-1026.

5. Akbarov H.I. Molecular characteristics of molecularly dispersed polymer and nanoparticles in cellulose acetate solutions // UzB Chem. - 2004. - № 3. - C.3-7.

6. Akbarov H.I., Tillaev R.S., Umarov B.U., Sagdullaev B.U., Khalikov A., Sidikov A.S. Clustering in samples of cellulose and its acetates of different degrees of substitution // Chemistry of natural compounds. Special issue. - 2002. - C. 39-40.

7. Sidikov A.S., Rakhmanberdiev G. Thermodynamics of cluster formation in various samples of cellulose and its derivatives // Conf. "Actual problems of chemistry, physics and technology of polymers": Abstracts - T., 2009. - C. 156-158.

8. Bocek A.M., Yusupova L.D., Zabivalova N.M., Petropavlovsky G.A. Rheological properties of aqueous solutions of H-carboxymethylcellulose with additives of various nature // Journal of Applied Chemistry. - 2002. - T. 75. -№ 4. - C. 659-663.

9. Akbarov H.I., Tillaev R.S., Umarov B.U., Sagdullaev B.U. Molecular characteristics of cellulose acetates in different solvents. The science of polymers on the threshold of 21st century. -Tashkent, 1999. - C. 80-81.

10. Akbarov H.I. Constant of the Mark-Kuhn-Hauvinck equation for cellulose acetates of different degrees of substitution // Uzb Chem. - 2004. - № 4. - C. 24-27.

11. Akbarov H.I. Undisturbed dimensions of cellulose acetate macromolecules in non-ideal solvents // Dokl. of the Academy of Sciences of the Republic of Uzbekistan. - 2004. - № 3. - C. 59-62.

12. Akbarov K.I., Tillaev R.S., Musaev D.A., Umarov B.U. Capillary-porous structure and thermodynamic properties of cotton fibers of different genotypes // Uzbek Chemical Journal. -2000. - № 5. - C.22.

13. Akbarov H.I., Tillaev R.S. Sorption properties of microcrystalline cellulose // Bulletin of Tashkent State University. -1999. -№ 2. - C.23.

14. Akbarov K.I., Tillaev R.S. Thermodynamic properties, molecular packing density and capillary-porous structure of microcrystalline cellulose // Chemistry of natural compounds. Special issue. - 1999.- C. 58-61.

15. Rashidova S.Sh., Voropaeva N.L., Pahorukov Yu.V., Ruban I.N. Synergetic model of water sorption by polymeric systems // Synergetics of macromolecular hierarchical structures: International Conference. - Tashkent, 2000.

16. Wessling B. Cristical shear-the instability tea son for the creation of dissipative structure in polymer // Phys. Chem. - 1995. - V. 191. - P.119.

17. Water in polymers / Edited by Rowland S.M. - M.: The World, 1984. -555 c.

18. Rakhmanberdiev G.R., Fedyakova N.A., Myagkova N.V., Sidikov A. Optical rotatory of water-soluble cellulose derivatives // Chemistry of natural compounds. - New York. 1997.- P.734-736.

19. Fedyakova N.A. Stereochemical transformations in the processes of obtaining and processing of cellulose acetates: Ph. -Tashkent, 1991.

20. Fedyakova N.A. Optical activity and rheology of cellulose acetate solutions // Chemical fibers. - 1993. - № 5. - C. 14-16.

21. Fedyakova N.A., Rakhmanberdiev G.R. Effect of substituents on optical activity of cellulose derivatives // High-molecular sols. -1994. - T. 36 Б. - № 3. - C. 496-498.

22. Shipovskaya A.B., Kazmicheva O.F., Timofeeva G.N. Dispersion of optical rotation of chitosan solutions // Structure and dynamics of molecular systems: Collection of papers. - Ufa: Institute of Physical Mechanics and Mathematics of the Ural Scientific Center of RAS, 2002. - Vol. IX. - T. 2. - C. 293-296.

23. Shipovskaya A.B., Kazmicheva O.F., Timofeeva G.N. Phase state and optical activity of systems natural polysaccharide mesophageal

solvent // "Polymers - 2004": Proc. of the Third All-Russian. Kargir conf. - Moscow, 2004. T. 1. - C. 420.

24. Shipovskaya A.B., Kazmicheva O.F., Timofeeva G.N. Liquid crystalline state and optical activity of cellulose ethers // Izvestiya Sarat. univ. Ser. chemistry, biology, ecology. - 2005. - T. 5. - № 1. - C. 72-78.

25. Shipovskaya A.B., Kazmicheva O.F., Shmakov S.L., Shchegolev S.Yu. Anisotropy of optical activity of ordered phases of cellulose acetates // High-molecular compounds. - 2009. - T. 51 A. - №7. - C. 1109-1121.

26. Tretennik V.Y., Pechska L.N., Ovcharenko F.D., Kruglitsky N.N. Thermal stability of disperse systems. - Kyiv: Naukova Dumka, 1975. - C. 29-33.

27. Timokhin I.M., Zaikanova L.I. Chemical modification - method of increasing thermal stability of disperse systems // Scientific conference on oil and gas problems. Gubkin MINH and GP: Abstracts. - Moscow, 1977. - C. 73.

28. Sidikov A.S., Yusupkhojaeva E.N., Rakhmanberdiev G.R., Sh.S. Arslanov. Dependence of the surface activity of water-soluble cellulose ethers on their chemical composition // Kimyo va kimyovii technology. - 2007. - №1. - C. 43-45.

29. Sidikov AS, Yusupkhojaeva EN, Rahmanberdiev G.R. Thermodynamic properties of water-soluble mixed cellulose esters // New achievements in chemistry and chemical technology of plant raw materials. Proc. of All-Russia Conf. - Barnaul: Publishing House of Altai State University, 2007. - C. 164-168.

CHAPTER VI. PHYSIOLOGICALLY ACTIVE POLYMERS BASED ON WATER-SOLUBLE CELLULOSE ESTERS

In recent decades, advances in the chemistry of high-molecular-weight compounds have been increasingly used to solve problems in biochemistry, biology, and medicine.

One of the main directions of polymer chemistry in this field of research is the production of new polymers and chemical modification of already known polymers in order to obtain polymers and polymeric substances with biological activity. Back in 1960 S. N. Ushakov [1] showed that the effect of a number of drugs can be prolonged if they are administered in solution together with polymers. In this case, the higher the molecular weight of the polymer and its concentration, the longer the action of such drugs and the better their solubility, reduced toxicity and side effects of drugs. The prolongation effect and decrease in toxicity are due to the fact that the drugs are more or less bound to polymers and at lower concentrations in the body, diffusion of the drug from the site of administration becomes more difficult, and the polymer acts as a "depot" of the drug substance.

Prolongation of the action of the drug substance enables its better use, lengthens or eliminates completely the concentration fluctuations of the active substance in blood and tissues that take place during periodic drug administration. Stability of the active substance in the blood reduces the number of intakes and injections, which is of great importance for the convenience of patient care.

The easiest way to prolong the effect of drugs by simultaneous administration of them with polymers, for example, novocaine, insulin, adrenaline are administered this way [2-5]. Using mixtures of drugs with polymers, it is possible to transform the effect known for a given drug substance. For example, administration of apricot gum (0.5% solution) together with anesthetics not only prolongs anesthesia, but also enhances it [6]. In a number of cases, the polymer, when administered simultaneously with drugs, increases its solubility and reduces its toxicity /7/.

Due to the emergence of new polymers with diverse properties, chemical addition of drugs to polymers, which previously had only theoretical interest, now seems to be the most promising.

Studies carried out in this direction have established the suitability for these purposes of a number of synthetic and medicinal polymers, both of which can be used as auxiliary medicinal materials-basis for plasters, pastes, ointments, blood substitutes, prolongers, detoxifiers, drugs of direct therapeutic effect /8/. This purpose corresponds mainly to hydrophilic polymers of linear structure.

The use of high-molecular-weight compounds to obtain drug-eluting polymers is of great interest. Currently, synthetic polymers, namely polyvinylpyrrolidone (PVP), are used for this purpose along with polymers of natural origin dextran [9-10]. Polyvinylpyrrolidone (PVP), introduced into practice during the Second World War, occupies the main place among synthetic blood substitutes. A known blood plasma substitute, peristone, is a 3.3% solution of PVP with a molecular weight of 30-40 thousand, containing mineral salts. Low molecular weight polymers of PVP (M.V.=10,000-14,000) can be used as detoxifiers, which is explained by the formation of covalent bonds of PVF with some toxins and possibly viruses and subsequent removal from the kidney.

When films based on vinylpyrrolidone and methyl acrylate copolymer are implanted both intramuscularly and intraperitoneally, at the first stage an increase in the weight of the samples is observed due to swelling followed by biodegradation, leading to complete resorption of the samples after 12-14 months [11].

When using a particular polymer, accurate knowledge of all the parameters that determine the polymer's target properties is required.

For example, it has been established that the molecular weight of most polymers used for plasma substitutes should be 30-60 thousand. When the molecular weight decreases, the time of solution circulation in the bloodstream is reduced and the polymer is eliminated from the body before there is a complete recovery of blood volume and composition [11]. On the contrary, when the molecular weight increases, there is a danger of accumulation and deposition of the polymer in the tissues of various organs. When toxic substances enter the body from the outside, as well as in other cases of intoxication, it is effective to use plasma substitutes with a molecular weight of about 10.000-14.000 that can bind toxins and accelerate their elimination from the body through the kidneys.

Polymers possessing ion-exchange properties are capable of influencing the electrolyte content in various biospheres and can be used not only for diagnostic studies [12,p.307], but also for the treatment of diseases associated with electrolyte disequilibrium in the extracellular fluid. The introduction of ion-exchange polymers into the body has its own difficulties-side effects that are a consequence of the low selectivity of ion-exchangers with respect to the ions to be removed [13-16].

No less important and interesting is the application of water-soluble complexing polymers for the purpose of introducing missing trace elements into the body. It is known that trace elements (copper, cobalt, zinc, cadmium, manganese, magnesium, etc.) are part of enzymes where they are bound in complex compounds and catalyze physiological processes [17-20].

A disturbance in the ratio of various trace elements in the organism leads to a functional disorder of its systems [20,p.29]. For the introduction of microelements in the body it is advisable to use not simple but intracomplex salts, providing gradual arrival of a microelement to the place of its location in the organism /17/. In recent years, researchers are increasingly interested in the use of prolonged-acting drug compounds obtained by chemically attaching drugs to polymers. Polymeric materials with ion-exchange properties (ion-exchange resins) are used to synthesize such compounds.

In this case, salts are formed that can release a "free" drug substance during decomposition. The rate of decomposition depends on the nature of the polymer and the drug substance, as well as the type of physiological environment. Thus, it has been found that 0.1 H hydrochloric acid quickly elutes drug substances such as ephedrine from cation-exchange resins containing carboxyl groups, while the decomposition rate of drug salts and polymers containing sulfogroups in 0.1 H hydrochloric acid is much lower. Therefore, polymers containing carboxyl groups cannot be used to make oral drug compounds [21].

P. Malek obtained the so-called "antibiolymphins" by the interaction of antibiotics containing major groups, for example, streptomycin and neomycin and polymers containing acidic groups of polyacrylic and polyurethane acids (sulfates and phosphates, polysaccharides). When they were used, a high concentration of the antibiotic in the lymph was achieved [22,23].

Various polymers can be used as carrier polymers for the prolonged action of drugs. S.N. Ushakov et al[24,25] synthesized polymers of drug action based on crotonic acid, methylolcrotonamide and a number of drugs intended for the treatment of cardiovascular diseases, tuberculosis, tumor formations, etc. Thus, a blood anticoagulant [26,27], an ester of peltanic acid and PVA, was synthesized by the interaction of PVA with the lactone di-14-oxycoumarin-3 acetic acid (peltanic acid).

Antituberculosis polymers were obtained by the interaction of p-aminosalicylic acid chlorohydride and PVA /28/. By the interaction of PVA and PASC diacetate chlorohydride at the interface with subsequent saponification of acetyl groups, the content of aminosalicylic acid was increased to 30%. This drug has an effect on Mycobacterium tuberculosis and is retained in the body much more than conventional PASC. As established by I.M. Rabinovich in experiments with this drug tested on animals [29], the concentration of PASC is retained for 10-12 days.

However, due to the deposition of PVA in the liver and lymph nodes and its negative effect on the body, it was abolished as a basis for drug substances. S.N. Ushakov and E.F. Panarin [30,31] synthesized polymeric penicillin amides and hydrazides by the interaction of mixed penicillin acid anhydrides and carbonic acid monoethyl ether with polymers containing hydrazide or amino groups (copolymer of vinyl alcohol and acrylic acid hydrazide or copolymer of vinyl alcohol and vinylamine). In this case, the formation of an amide or hydrazide bond occurs at a low temperature, eliminating the possibility of drug inactivation and thermal degradation of the polymer.

The inclusion of salsolidine in the vinylpyrrolidone copolymer by amide bonding leads to the simultaneous realization of two effects: prolongation and reduction of toxicity, and by amine bonding to prolongation and increase of toxicity and activity. Despite the encouraging results of tests in medicine, the use of a number of polymers as carriers of drugs should be treated with extreme caution. The practice of work with some drugs has shown that often only years and extensive testing can give the final conclusions about a new drug. For this purpose, it is advisable to use polymers that have passed such tests; polysaccharides are among such polymers: they are part of living organisms and are well assimilated or well combined with them. A number of polysaccharides have been successfully applied in medical practice. The use of starch gave positive results in obtaining high molecular weight drugs [32] Anti-tuberculosis drugs were obtained by condensation of polysaccharides oxidized with sodium metaperiodate with thiosemicarbazide, aminobesaldehyde-thiosemicarbazide, isonicotinic acid hydrazide and p-aminosalicylic acid.

The most widespread among polysaccharides is dextran, which has been well studied and used as plasma substitutes [33]. The results obtained in the study of the mechanism of dextran-based drugs, its distribution and behavior in the human body are of great interest, because they can be used when working with other polysaccharides. A. E. Frome established the dependence of dextran excretion rate on its molecular weight [10]. An increase in molecular weight leads to a decrease in the rate of dextran excretion with urine, but does not significantly affect its content in the blood, since it begins to be deposited in the tissues of various organs.

The possibility of attachment of various drugs to dextran chemically opened up further prospects for its use. A.D.Vernik and co-workers [34] obtained preparations containing chemically bound tubazide and novocaine on the basis of dialdehyde- and dicarboxyldextran. Products of interaction of dialdehyde-dextran with tubazide were obtained according to the following scheme:

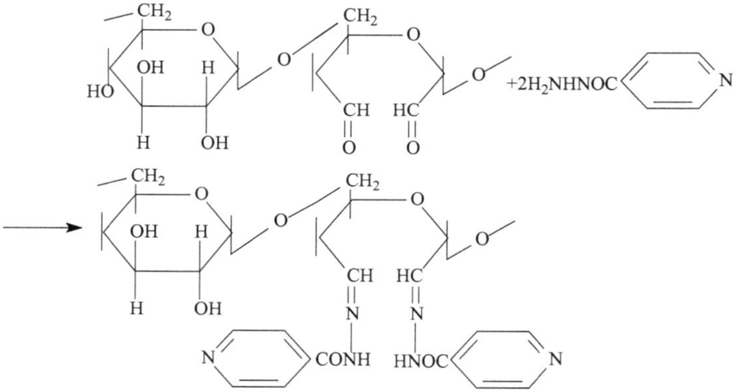

The formation of a salt compound using dicarboxyldextran and novacaine as examples can be represented as follows:

$$\text{[cellulose structure]} + H_2N\text{-}C_6H_4\text{-}COOCH_2CH_2N(C_2H_5) \longrightarrow$$

$$\longrightarrow \text{[cellulose structure]} \quad \overset{-}{C}OOH(C_2H_5) \cdot \overset{+}{N}CH_2CH_2OOC\text{-}C_6H_4\text{-}NH_2$$

Recently, cellulose and especially its hydrophilic water-soluble derivatives, in particular, its esters and ethers, have been of interest to researchers working with medical polymers. High water solubility, ease of production and physiological indeterminacy of cellulose ethers ensure their relative harmlessness.

It should be noted that cellulose esters in the production of drugs have long been used in medical practice.

Most often they are part of the protective coating of solid dosage formulations. For example, cellulose acetate phthalate [35-39], oxypropyl cellulose [40], oxypropyl methyl cellulose /41/, oxyethyl cellulose [42], acetyl cellulose [43, p. 667], acidic cellulose acetate /44/ are used to obtain gastric juice-soluble coating. Preparation of carboxymethylcellulose preparations for blood substitute solutions is also known [45]. Methylcellulose and Na-carboxymethylcellulose solutions containing levomycetin and furacilin can be used as bactericidal film-forming fluids in treatment of cuts and small skin lesions, as well as antiseptic water-soluble compositions obtained by mixing the polymer with iodine until complete homogenization [46].

The use of methylcellulose as a base for ointments is also of interest for practical dermatology, because this base can be used to produce dosage forms for external application. Ethyl cellulose, as a therapeutic agent, was used for functional gut diseases and only in the beginning of the 1970s cellulose esters, also as synthetic polymers, began to be used for synthesis of biologically active drugs, mainly with antibiotics [47].

The method of combining cellulose ethers with synthetic polymers is used more often, particularly in the manufacture of solid dosage forms, such as tablets [48]. Microcrystalline cellulose is also recommended for making pharmaceutical tablets [49].

Water-soluble cellulose esters are used to produce drug capsules. While in the 1950s capsules were obtained by immersing the heated form in water-soluble, for example, methylcellulose [50] and drying the extracted form, in recent years capsules have been obtained by extrusion with combined protective layers, such as polyvinylpyrrolidone as the main component and an outer layer of cellulose acetate [51]. One method of protecting the drug substance is the microencapsulation method, where cellulose derivatives are also used.

In the mid 50's an attempt was made to use carboxymethylcellulose with different degrees of polymerization and substitution in obtaining plasma substitutes and anticoagulants [52]. However, the attempt was unsuccessful because of the possibility of deposition of the obtained preparations in the body.

The use of cellulose itself directly for prolongation of drugs is devoted to the least number of works, meanwhile, the possibilities of cellulose in this field of application are also quite promising. The development of cellulose chemistry makes it possible to obtain a variety of preparations on its basis, and the development of FAP science, which has identified a number of mandatory requirements and regularities, makes it possible to avoid many errors and obtain preparations that are appropriate for the intended purpose.

Chemical attachment of the drug substance to the polymeric carrier can be carried out by ionic, coordinational or covalent bonds.

Since the therapeutic effect of therapeutic drugs, including polymeric ones, is determined by the structure and structure of the drug substance, its attachment to the polymer should not occur along the functional groups that provide therapeutic activity. When synthesizing cellulose-based drugs, the greatest difficulty is obtaining chemically homogeneous derivatives because the drug substance can be introduced into cellulose only by first introducing new reactive groups into it. Another problem is the study and selection of conditions of interaction of modified cellulose with drug substances. In this connection, works devoted to the study of the mechanism of such interactions will be of great interest.

Thus, during condensation with nucleophilic drugs in the 2-3-dialdehyde links, aldehyde groups react in the hydrated form, while in the 2,3-dialdehyde 6-0-carboxymethyl links the reaction proceeds with the reduction of the C_2 -C bond$_3$ of the pyranose cycle; the maximum addition of a drug substance is no more than one mole per oxidized glucopyranose cycle [53]. To prove this conclusion, the authors performed condensation of the reaction products with hydroxylamine and found that, on average, out of 25 oxidized glucopyranose rings, 17 have a 2,3-dialdehyde structure and the remaining 8-2,3-dialdehyde 6-0-carboxymethyl structure:

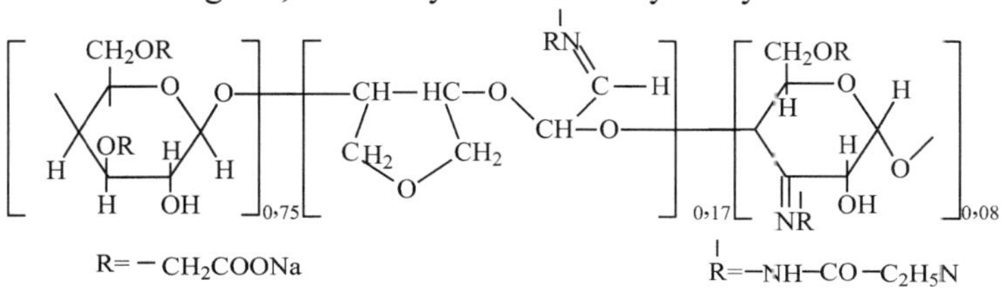

On the basis of dialdehyde carboxymethylcellulose Kh.U. Usmanov, Sh. [54-56] synthesized and studied water-soluble tuberculostatic active preparations containing tubazide, cycloserine, and streptomycin.

Analysis of the literature on FAP leads to the conclusion that the information accumulated in this field allows us to meet many practical needs of various fields of medicine already at this stage. In this case, it is advisable to use polymers and drug substances sufficiently proven in practice for their synthesis (FAPs). Cellulose and its derivatives are not amenable to enzymatic degradation in the body, as some authors state [57], because there are no enzymes like cellobiase and β-glucosidase (contained in snail digestive juice) in the human body, which could subject β-4-glucoside bonds to enzymatic degradation. But it is safe to say that any other decomposition of cellulose under the influence of hydrolytic influences produces products that are akin to and not harmful to the body as compared to synthetic polymers. In other words, the decomposition of cellulose in the body produces products that are similar to metabolic products in functionality (e.g., acetyl and "OH" groups in acetyl cellulose).

Therefore, the thesis that polysaccharides are completely non-toxic to the body has not been fully proven at this time. A polysaccharide substance is not toxic when it is introduced into the gastrointestinal tract, for the simple reason that it usually does not dissolve like cellulose. If introduced into the bloodstream, however, the insoluble cellulose can cause vessel blockage. The same is true of cellulose ether if it is easily saponified in the blood and turns into insoluble cellulose. Some even stable cellulose ethers are unacceptable for other reasons. For example, MC has pyrogenic properties, oxyethyl cellulose can coagulate proteins, etc. Therefore, with respect to the use of water-soluble cellulose esters in medicine, not all of them, but some of them have advantages and useful properties. The latter include water-soluble acetylcellulose, because it has significant positive qualities that distinguish it from other water-soluble cellulose ethers. For example, it can be obtained absolutely chemically pure compared to CMC. The presence of hydrophilic-hydrophobic functions due to hydroxyl and acetyl groups gives it permeability through the lipid membranes of body cells. It is stable at blood plasma pH. In addition, as was established in [58], acetylation and hydrolysis processes take place in the organism under the action of the COA (Coenzyme) enzyme. In this regard, water-soluble acetylcellulose is the only and suitable object where these processes can be used in the right direction in order to regulate the residence time or resorption of its products (fibers, films) in the body.

All these advantages of water-soluble acetylcellulose led to the conclusion that it is advisable to use it to obtain physiologically active polymers and resorbable fibers and films, as well as non-resorbable (sutures, dressings and tampons) materials based on oxidized cellulose and low-molecular weight drugs.

Thus, one of the main directions of use of cellulose derivatives, especially its water-soluble esters, is their use as a polymeric base for obtaining physiologically active polymers for the purpose of prolongation of drugs, in protective coatings of medicinal tablets for various purposes, as well as resorbable and non-resorbable suture, tamponage and dressing materials.

6.1.Medical fibers based on water-soluble cellulose ethers

Besides those fields of national economy, which needs can be satisfied by few types of large-tonnage artificial fibers produced on the basis of cellulose and its derivatives, there are such fields, for which fibers with specific properties are required. An example is the use of fibers based on cellulose derivatives in medicine and, in particular, in surgical practice, as tampon and suture materials.

Currently, the creation of the most rational surgical suture material is one of the unsolved problems of surgery. Among the various indicators of suture materials, resorbability in the body is always one of the most desirable qualities of sutures. Dissolving suture materials carry certain mechanical loads until the suture overgrows or the area of surgical intervention in an organ of the living body heals. After a certain period of time (usually two to four weeks) the material has to be completely resorbed and the subsequent loads have to be borne by the body tissues.

The only absorbable suture material currently used in surgery is catgut, a material of protein nature derived from the intestines of cattle. However, the raw material base is not sufficient to meet the medical need for it. In addition, catgut can cause allergic reactions in some patients. It takes 2-4 weeks to resorb in the body, and the rate of resorption is regulated by the conditions of tanning and the thickness of the thread [59,62 p.86-88]. Due to the scarcity of thread, the imperfection of its production process and the difficulty of obtaining particularly thin surgical threads in recent years, more and more attention is paid to chemical and protein fibers based on collagen (protein of covering tissues), fibroin (from alcopopsis), zine (muscle protein), etc. [60]. The absorbable suture materials on the basis of synthetic polymers are known on the basis of polyvinyl alcohol [61,60 p.35-42] and polyglycolic acid /62/. All the indicated fibers are capable of hydrolytic cleavage, but many of them are used only in the presence of acid-alkali catalysts or enzymes. The ability of protein materials to resorb in the body depends sharply on their structure, conformation of macromolecules, presence of side substituents, degree of crystallinity, molecular weight and other reasons [60].

Strong and elastic natural silk threads are widely used for stitching wounds. However, being protein substances, such threads can fix microorganisms on themselves and serve as a source of wound festering.

Non-absorbable suture materials are currently represented by a large set of filament yarns and monofilaments from polycaproamide, lavsan, polypropylene, polytetrafluoroethylene, etc. The fact that they remain in the tissues for a relatively long time is not in their favor, since a foreign body can always be the basis for late suppuration and other complications. In addition, they cause pain when they are removed.

Thus, existing suture materials of both natural and chemical origin do not exhaust the problem and new ways are needed to obtain other better types of medical fibers.

Studies of available and indifferent polymer of natural origin-cellulose and its derivatives are of great importance in this direction. In this connection, obtaining water-soluble fibers on the basis of cellulose derivatives plays a great role. The number of works devoted to obtaining water-soluble fibers based on cellulose derivatives in the world literature is very, very small. Oxyethylcellulose fibers are produced by wet molding from 4-15% oxyethylcellulose solutions in 8% aqueous solution of NaOH [63] in a sedimentation bath, containing the following composition:

N_2 SO_4 , % - 4-15

Na_2 SO_4 ,%-13-25

bathtub temperature, °K - 298-338

molding speed, m/min - 30-40

filter extraction, % - 4-80

After plasticizing drawing by 60% in the bath of close composition and drying under mild conditions the fiber had the strength of 18-23 cN/tex with the elongation of 7,5-8%. The strength characteristics of oxyethyl cellulose fibers largely depend on the molecular weight of the initial product and the drawing ratio.

It is also possible to mold oxyethylcellulose fibers in acetic acid [64], sulfatammonic [65] and other baths. Depending on forming conditions, on the composition of the deposition bath and on the multiplicity of plasticizing drawing strength of oxyethylcellulose fibers can vary in the range from 5 to 20 cN/tex at elongation of 15-25%.

Fibers based on carboxymethylcellulose are produced by molding 6-8% CMC solutions in 6-8% NaOH solution in a precipitation bath containing acid-salt solutions. Organic aqueous precipitation baths containing aliphatic alcohols have also been suggested [66-68]. Usually carboxymethyl cellulose, used for fiber molding, has the following characteristics: degree of polymerization 600-800, degree of esterification Y = 70-80. However, there are works, where in special conditions carboxymethylcellulose, water-soluble with the degree of substitution (C3) of 0.3-0.4 (Y=30-40) was obtained [69]. This sample of carboxymethyl cellulose differed from the usual ones in that it had a more uniform distribution of substituents. The fiber obtained from these samples and formed under normal conditions had the following characteristics: strength 17-18 cfu/tex and elongation 15-16% [70].

Cellulose sulfate fibers, which have a degree of substitution from 0.15 to 0.5 (Y = 15-5) and the degree of polymerization 250-350 are formed from a 10% solution of NaOH in the precipitation baths containing, as in the case of forming of carboxymethylcellulose fibers, acid-salt or water-organic solutions. However, in all cases, the use of acid-salt baths is difficult because of the difficulty of washing from salts. Therefore, it is advisable to use as a precipitation bath isopropyl alcohol 45-70% concentration and acetic acid (10%) and water at a molding temperature of 283-303^0 K, at a molding speed of 1-3 m / sec. The fiber molded under these conditions and after plasticizing drawing (10-50%) had a tensile strength of 13-14 cH/tex with elongation of 11-12% [71].

There are also reports about production of fibers based on carboxyethyl cellulose [72], methyl cellulose [73] and ethyl cellulose [74]. However, studies on obtaining of fibers resorbable in organism and having physiological activity simultaneously are almost absent in world literature, except for one work [75], where possibility of obtaining of fibers based on carboxymethylcellulose, resorbable in organism and having antimicrobial activity is proved. Meanwhile, the development of research in this direction, as mentioned above, is extremely important for medical practice.

Low mechanical properties of fibers based on cellulose derivatives and great loss of strength in the wet state obviously allow their application only in the form of resorbable dressings or absorbent cotton, i.e. in those cases when there are no any significant requirements to the level of mechanical properties.

In addition to fibers, so-called drug films (DF) are widely used in medical practice in the treatment of various eye diseases.

Significant advances have been made in the treatment of eye disease in recent times; however, despite the measures taken, a large number of people are now suffering from this disease.

Numerous antibiotics, anesthetics, sulfonamides, vitamins, corticosteroids and other medications are a powerful tool in the fight against infectious eye diseases, allowing in many cases, without surgery to cure and completely preserve the patient's vision.

Previously existed therapeutic agents (water solutions for drops and injections, ointments) did not allow to realize the potential of drugs. The main disadvantage of the formerly available eye medications is their low therapeutic efficacy: the drops provide the required therapeutic concentration only within $12 \cdot 10^2$ sec, and ointments $18 \cdot 10^2$ seconds after injection. Up to 80% of the total amount of medication is removed from the eye cavity in the first minutes after injection. Drops must be administered up to 12 times and ointments 6-8 times per day to achieve the desired therapeutic effect.

In order to increase the duration of action and convenience of use of medicines for treatment of eye diseases their various prescription forms and substances prolonging action of a preparation are offered. Back in

1911 the Russian military doctor I.I.Budzko in his work [76] performed at the Department of Ophthalmology of the Military Medical Academy recommended using in the field conditions the so-called eye tablets representing medicines in a compressed form for introduction of a bag. Eye tablets contained cane or milk sugar, sometimes starch, as a base. Despite a number of advantages over solutions, eye tablets did not find practical application, as they were poorly tolerated by patients due to the irritating effect of their base on the tissues of the eye.

Many authors in the treatment of eye diseases, in order to prolong the action of drugs, in particular antibiotics, during their local application suggested using different fats and oils [77]. In 1961 Y. F. Maychuk proposed polyvinyl alcohol (PVA) as a base to prolong the action of drugs used in ophthalmology. Later, numerous studies by different authors showed that films based on PVS containing antibiotics and other drugs create a higher concentration of them compared to aqueous solutions and ointments and retain the drug activity 2-4 times longer than aqueous solutions [78,79].

However, a significant disadvantage of PVS films is the need to remove them from the conjunctival sac after $7.2 \cdot 10^3$, $10.8 \cdot 10^3$ sec. after application. In [80] the prolongation properties of biosoluble polymers for ocular drug films (ODDs) - polyacrylamide of different molecular weight, polyvinylpyrrolidone (PVP) and copolymers of acrylamide with some monomers of acrylic and vinyl series - were evaluated.

Based on the above polymers and their compositions, GLPs were developed, which are oval-shaped polymeric elastic plates 0.25-0.30 mm thick, 9 mm long and 4.5 mm wide. The GLPs are usually inserted into the conjunctival cavity once a day.

In the conjunctival sac, GLPs quickly wet with lacrimal fluid, become soft, turn into gel, and gradually dissolve. The ability of GLPs to dissolve in the conjunctival sac is what distinguishes them from the previously used PVA-based films. Each film contained a certain amount of chloride and the prolongation properties of the selected polymers were evaluated using the technique of the so-called nickel test [81].

The results showed that the maximum concentration of nickel ions is reached within the first $3 \cdot 10^2 - 9.6 \cdot 10^2$ sec. Subsequent changes in the concentration of nickel ions depend on the properties of the polymer in which it resides,

When administered as an aqueous solution of nickel chloride, the concentration of nickel ions is practically undetectable after 30 min. The use of polymers increases the contact time of nickel ions with the eye tissues several times, with the shortest contact time being observed for PVP films and significantly longer, 2-3 times, for films made of PVS and especially PAA. This phenomenon is explained by the increase in the viscosity of the tear fluid when the polymer film dissolves in it; in the case of insoluble PVS films, by the slow diffusion of nickel cations from them. A comparison of the maintenance time of the maximum concentration of nickel ions for the PAA solution and films based on it prepared from polymers of different molecular weights showed that films prepared from polymers with a molecular weight of more than 300,000 contributed to a longer maintenance time of the maximum concentration of nickel cations in the conjunctival cavity.

Differences in the residence time of nickel ions in the conjunctival cavity suggested a possible dependence of this characteristic on the type of polymer and its molecular weight. This assumption was confirmed by testing a number of copolymers of acrylamide with other unsaturated compounds such as vinylpyrrolidone, ethyl acrylate, vinyl acetate, etc. The polymers were obtained by copolymerization of vinyl and acrylic monomers under free-radical initiation conditions [82,83]. By varying the molar ratios of the initial components, copolymers with different degrees of hydrophilicity were obtained. Comparative tests showed that the long-term maintenance of the maximum concentration of nickel ions is facilitated by the use of copolymers with an increased content of acrylamide units.

The slow diffusion property of nickel ions from PVA films was used to create so-called layer-by-layer films to improve the prolongation properties of the base. If the insoluble PVS film swells, causes discomfort in the eye and must be removed from the conjunctival sac, the introduction of 25-30% PVS in compositions with other, biosoluble polymers results in a two-layer film in which the insoluble layer is about 0.08 mm thick and cannot be felt by the eye. However, it has been shown that when the thickness of the insoluble part of the films is reduced, the residence time of the potassium ions introduced into such films does not exceed the contact time when using PAA and its copolymers /80/.

The need to improve the technological properties of the films requires either obtaining copolymers with reduced brittleness, or the introduction of plasticizers into the base composition. In [84] hydrophilic oligoethers were used as plasticizers, which, along with their main purpose, help to improve the prolongation properties of the films.

For convenient use of the films and to ensure their long-term storage in sterile conditions [85,86], there have been developed pencil dispensers that allow the extraction of the GLPs one by one without contaminating the remaining films and are supplied to the treatment network in a sterile form. The GLPs are sterilized by Υ-irradiation.

GLPs are recommended for the treatment of glaucoma [87], viral keratitis [88], adenoviral conjunctivitis [89], allergic eye diseases [88], to help with burns [90], mechanical trauma [91], postoperative treatment of patients, for keratoplasty, and for various preventive and preoperative eye treatments [92-94].

GLPs are especially convenient for glaucoma patients who have had to administer medications as drops several times a day for many years. Administering GLPs once a day greatly facilitates treatment and helps maintain normal intraocular pressure over a long period of time. Patients easily learn the way of self administration of GLP. The high efficacy of GLPs allows their wide application in the treatment of mass infectious eye diseases in countries with tropical climates.

Analysis of the literature data on the production and use of HLPs showed that almost all HLPs were obtained on the basis of synthetic polymers. The use of water-soluble cellulose derivatives for these purposes has not been given sufficient attention. There is only one work [95], where a film based on carboxymethylcellulose and methylcellulose was used for prolonging the action and effective use of various drugs in the treatment of eye diseases.

6.2 Physiologically active polymers based on water-soluble acetylcellulose (WRAC).

In the last 10-15 years a new field of physiologically active polymers has emerged at the junction of three sciences: biology, medicine and polymer chemistry. The expansion of the assortment of physiologically active polymers largely determines the successful solution of the problem of combating various malignant diseases.

Having started with the modest task of prolonging the action of known low molecular weight (HB) drugs, in recent years it has significantly outgrown its original framework and turned into one of the intensively developing and promising areas of polymer chemistry, whose foundations were laid by S.N. Ushakov.Physiologically active polymers are widely used in medical practice as dressing and suture materials, napkins, turundas, and also as injectables, orally, etc. Cellulose materials (medical gauze) are usually used as a starting product for obtaining dressings for antimicrobial materials due to their availability, good hygroscopicity, air permeability. Water-soluble high-molecular weight compounds, harmless to the body, are used as a carrier polymer to obtain physiologically active compounds administered inside the body.

Physiologically active polymers can be obtained both in insoluble form (dressings and suture materials) and in soluble form to prolong the action of drugs administered by injection.

To obtain water-soluble physiologically active polymers, antituberculosis, antitumor, antiviral, and blood-relieving drugs were used as drug substances, and BPAC, which has not been used for this purpose so far, was used as a carrier polymer.

Unlike the known cellulose ethers used as a polymeric base for drugs, such as methylcellulose, carboxymethylcellulose, etc., it is a complex ester. HPAC is a complex ester, due to which its acetyl groups are easily hydrolyzed without causing destruction of the polymer itself. This makes it possible, depending on the degree of substitution, to easily adjust (or control) the solubility of the physiologically active HPAC-based polymer. In addition, as noted above, it has been established that in the body the drugs are assimilated under the action of the KOA acetylation enzyme. On this basis, it can be assumed that the presence of acetyl groups in HPAC allows it to easily pass through the lipid membranes of the body cells and be harmlessly eliminated from the body.

To obtain physiologically active polymers, BPAC with a bound acetic acid content of 18% and a polymerization degree of 200 was used, because these parameters determined the best solubility of the polymer in Ringer's solution [96-97].

The aqueous solutions of HPACs have been noted to be physiologically indifferent, neutral in reaction, odorless and tasteless, and stable for a long time [98-99].

To attach drugs, aldehyde groups were introduced into the HPAC molecule by oxidation with iodine acid [100, 101, p.337]. After the introduction of aldehyde groups into HPACs, it is easy to attach tubazide, tetracycline, sarcolysin, and other drugs to it.

By varying the exposure time of iodic acid to HPAC, the content of aldehyde groups in it was regulated.

To confirm the formation of the aldehyde group in HPAC after oxidation with iodine acid, infrared spectra of control and oxidized samples were taken (Fig. 102). Since the absorption bands of the aldehyde groups coincide with the absorption bands of the acetyl groups, the IR spectra of the saponified HPAC samples were taken to prove the formation of aldehyde groups. Saponification of oxidized HPAC was performed under mild conditions with a 0.5N solution of sodium methylate in anhydrous methanol, ensuring complete detachment of acetyl groups [102].

It was found that in oxidized HPAC, an absorption band belonging to the aldehyde group appeared in the 1740 cm region^{-1} (Fig. 102) and whose intensity was higher the deeper the HPAC sample was subjected to oxidation. However, in an alkaline environment (during saponification), the carbonyl group can be oxidized to carboxyl. In [103], the authors argue that unambiguous assignment of the band in the 1740 cm region^{-1} to the vibration of carbonyl or carboxyl groups is impossible in oxidized cellulose, since both groups absorb in this region. However, the appearance of this new band in the spectrum of oxidized and saponified HPAC samples is reliable evidence of oxidative processes.

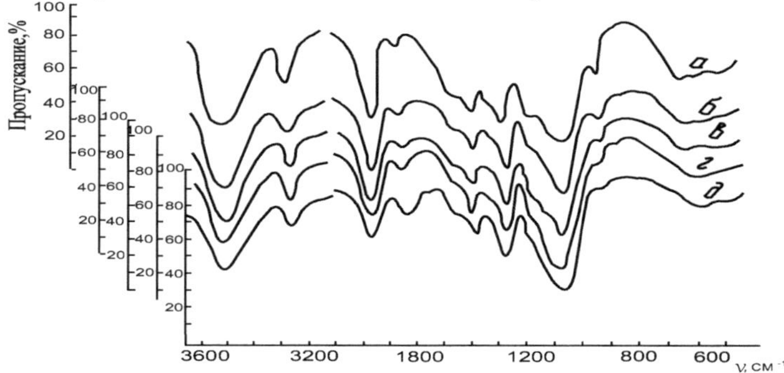

Fig.102. IR absorption spectra: a - initial HPAC
b-d-oxidized 30,60,120 and 180-60 s, respectively.

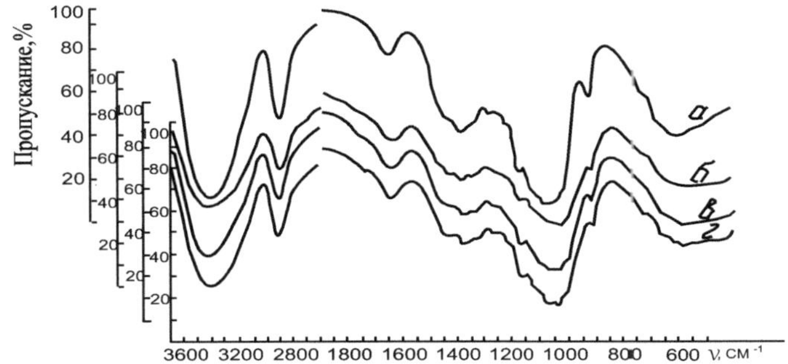

Fig.103. IR absorption spectra.

a - saponified initial HPAC;

b-g - VRAC oxidized (30,60,120)-60c and saponified.

The HPAC sample containing aldehyde groups was treated with an aqueous solution of tubazide, precipitated in acetone, and dried. The reaction was performed at 293^0 K-295^0 K in a neutral medium, with a bath modulus of 1:50. The product was purified by double resuspension in acetone.

The reaction of oxidized HPAC with tubazide can be represented as follows:

The change in the tubazide content in HPAC depending on the number of aldehyde groups is shown in Table 70.

Table 70.

Composition of products of interaction of oxidized HPAC with tubazide.

291

Oxidation time, $\tau \cdot 60$ c.	Number of alds groups per 100 ASGs	Number of connect ions. acetic acid,%	Nitrogen content,%	Tubaside content,%	Solubility in Ringer's solution
30	14,2	18,0	1,1	3,2	P
60	18,3	17,0	2,7	6,1	P
120	35,9	17,7	3,1	8,4	P
180	43,2	15,6	3,8	10,1	H

As can be seen from Table 70. the amount of aldehyde groups increases with increasing oxidation time and, accordingly, the amount of joined tubazide. However, an increase in the content of aldehyde groups above a certain limit (more than 40 per 100 AG3) is inexpedient because the tubazide-containing derivatives obtained on the basis of such preparations lost solubility in water and in Ringer's solution. Apparently, this is caused by a decrease in the amount of bound acetic acid (up to 15%) and by the formation of semi-acetal bonds within one molecule and between neighboring macromolecules.

It should be noted that no more than 50% of the aldehyde groups interact with tubazide. This circumstance is explained by the fact that some of the aldehyde groups at the second and third carbon atoms are in a semi-acetal bond with the primary hydroxyl group [102,103].

It should be noted that HPAC is a convenient analytical model to confirm this assumption because of its solubility in water and available polar solvents. For this purpose, the primary hydroxyl groups of HPAC were blocked by tritylation, after which the product was oxidized with iodic acid and tubazide was attached under the accepted conditions. The data obtained confirm that indeed the samples blocked by tritylation contain tubazide, about 2 times more than the non-tritylated: in the IR spectrum of tritylated HPAC appears a new band in the 1500-1600 cm region[-1] , relating to the phenyl radical (Fig. 104).

In the IR spectrum of oxidized HPAC treated with tubazide with increasing oxidation time, an increase in the absorption bands of 1680, 1600, 1560, 1500, 850,760 cm[-1] , which indicates the chemical interaction of oxidized HPAC. HPAC with tubazide to form the -C=N bond (1680 cm^{-1}) (Fig. 105).

Thus, the obtained physiologically active polymers based on HPAC and containing chemically bound tubazide (10%) were tested at the Uzbek Research Institute of Tuberculosis named after Sh. Alimov, where a marked prolonged action of tubazide was shown.

In oncology practice, the use of polymers is currently limited, although it is in the treatment of malignant tumors medical professionals have a shortage of new effective drugs.
Practice shows that a single injection of antitumor drugs, even in maximum tolerated doses, rarely produces a positive effect. On the one hand, it is explained by asynchronous division of cells that make up the tumor structure, on the other hand, by the ability of drugs to interact with the chemical components of the cell only in certain phases of the cell cycle. Taking into consideration that the duration of one cell division cycle varies between (20-30)-3600 s it should be considered that during this time a drug should contact the tumor for all its cells to be able to absorb the antitumor agent. However, some of the cells that are in an inactive state when conventional antitumor drugs are taken practically remain intact. In another part of cells, due to insufficient concentration of the drug in the blood, non-lethal damage to the cell may occur. That is why, as the time of "bombardment" of the tumor with a chemoagent increases, the probability of damage to the largest number of tumor cells increases. The therapeutic effect of chemotherapy in cancer patients is the more significant the more continuous is its effect on the tumor.

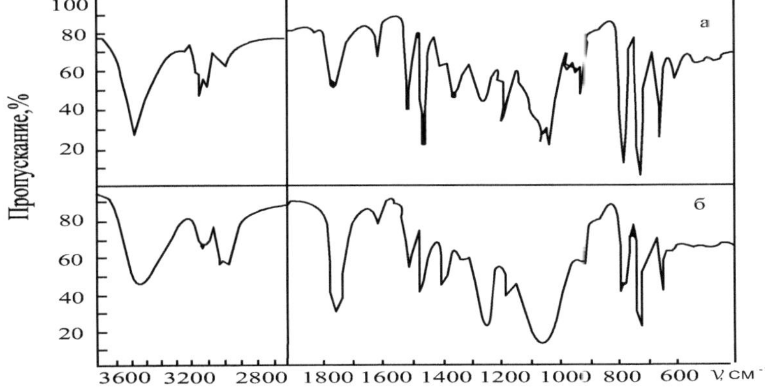

Fig.104. IR absorption spectra.

a - tritylated HPAC; b - tritylated and oxidized HPAC.

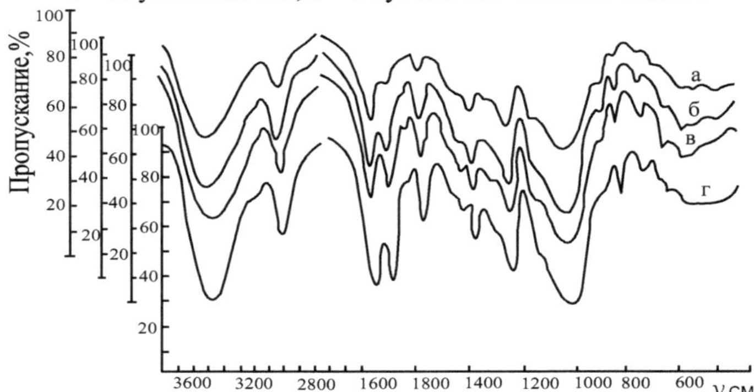

Fig.105. IR - absorption spectra of oxidized HPAC treated with tubazide.

(a. oxidation 30-60s; b.60-60s; c.120-60s; d 180-60s)

This effect in modern conditions is achieved most often by bringing chemotherapy to the tumors through the blood vessels (regional chemotherapy). However, in this case it is always necessary to take into account the leakage of chemotherapy drugs into the general circulation and excretion from the body, when sometimes up to 60% of chemotherapy drugs are lost. The only work in this

In this case, the antitumor drug Aurantine, which was administered to patients in a solution of polyvinylpyrrolidone, was clinically and experimentally studied [104]. At the same time, this antitumor antibiotic was determined in the blood serum for 6-7 days without losing its activity. In the works [105-109], antitumor drugs with prolonged action based on HPAT were synthesized.

For this purpose, the alkylating compound sarcolysin was selected

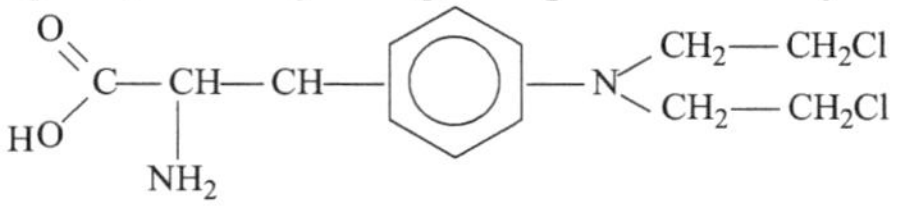

obtained by L.F. Larionov in 1956 in the Institute of Experimental Cancer Therapy of the Academy of Sciences of the former USSR and proved to be a powerful cytological substance with a broad spectrum of action. Its

294

universal action on a tumor, unfortunately, as well as the action of other antitumor drugs, is short-lived. According to the data of I.N. Romanova and A.N. Belousova (1960) the retention time of sarcolysin in the general bloodstream of an organism is on the average -2-3600 s [105].

To achieve this goal, sarcolysin was included in the structure of VRAC.

Sarcolysine was attached to HPACs by aldehyde groups introduced by exposure to iodic acid. VPAC samples containing aldehyde groups were treated with an aqueous solution of sarcolysin and precipitated in acetone. The reaction was performed at $293\text{-}295^0$ K in a neutral medium at a bath modulus of 1:50 and a sarcolysin concentration of 2%. The product was purified by double resuspension in acetone. Water-soluble products containing up to 12-13% chemically attached sarcolysin were obtained.

The addition of sarcolysin to the oxidized HPAC is confirmed by taking IR spectra, in which absorption bands appear in the 1680 cm region^{-1} related to azomethine (-C=N-) bonds, 1560 cm^{-1} corresponding to deformation vibrations of the H group, absorption bands 1600, 1500 cm^{-1} related to vibrations of double bonds of the benzene ring. With an increase in the oxidation time of HPAC to 2-3600 s, the number of aldehyde groups increases, due to which the interaction of sarcolysin with oxidized HPAC, its content in the product increases and accordingly there is also an increase in the absorption bands in the 1500, 1560, 1600 and 1680 cm^{-1} . The obtained antitumor drug sarcolysin prolonged action (SPD) was tested under experimental conditions to study its side effects and antitumor properties.Experimental studies to study the effects of BPAC and SPD on the body were conducted on 52 white mongrel rats. Animals were injected with 0.03 g of 1% SPD in the form of fiber and in the form of a solution.

This dose was chosen to be consistent (by weight) with the single dose used in the clinical setting. After surgical soft tissue dissection, the fiber-form SPD was immersed under the neck nodule muscle followed by suturing of the wound. LDS in the form of a solution was injected into the thickness of the tongue.

Animals were slaughtered 1 or 2 at different times from Zx 3600 s to 60 days from the beginning of the experiment.

The blood of the animals was examined every three days. After animals were slaughtered, the liver, spleen, kidneys, heart, lymph nodes, and soft tissues together with the direct injection of the drug were subjected to morphological study. A total of 154 blood smears and 306 morphological studies of internal organs were conducted. The studies showed that LDS administration was accompanied by only minor and short-term lymphopenia, which was not observed in the control group of animals that received only HPAI.

Morphological examination revealed that HPAC caused a weak local reaction in the form of non-small cell infiltration with the appearance of giant resorption cells at a later date (30 days). In the internal organs and lymph nodes there were reactive changes. These changes were reversible.

In contrast, LDS administration leads to moderately pronounced dystrophic changes in the internal organs with marked vascular abnormalities. In the lymphoid organs a picture of sharply pronounced reactive hyperplasia combined with hemosiderosis is revealed. In the late terms the morphological changes intensified, and the lymphoid tissue showed signs of hypoplasia. The changes found indicated a moderate toxicity of SPD.

Experimental studies on the effect of SPD on tumor were performed on 60 white mongrel rats, of which 40 were previously inoculated with sarcoma-45 tumor in the lateral surface.

The results of the study showed that in the first day the tumors in the animals receiving SPD lagged behind in their growth by an average of 32.5% compared to the tumors in control animals that did not receive SPD, The study of hemodynamics showed that in rats with tumors receiving SPD there was a development of leukocytosis due to a shift of the formula to the left. At the same time, pronounced neutrophilia and lymphopenia are observed.

As a result of malignant tumor growth there was animal mortality, which was more pronounced in the control group of animals that did not receive SPD. In this group, by 19 days from the beginning of the decline, the death rate was 83%, while in the animals of the first group it was 50%.

Based on the experimental data, it should be considered that LDS has a marked inhibitory effect on the development of malignant tumors and is superior to conventional sarcolysin in its therapeutic effect. However, the cytological properties of SPD are limited to 14 days. Probably, to obtain a more pronounced clinical effect, repeated administration of LDS is necessary, which should be confirmed by further experimental studies.

The use of such a drug is much more convenient for both patients and medical personnel. This circumstance is especially important in oncology practice.

In oncological practice for the treatment of ophthalmic diseases, along with sarcolysin, an effective anti-tumor drug, prospidine, is also widely used. However, the use in the form of intramuscular injection does not give the desired effect, since 75% of pure prospidine is removed from the body within 1 day.

In this regard, the use of prospidine directly in the tumor area is extremely important.

For this purpose, resorbable films containing prospidine were obtained, which were in direct contact with the eye lesion area, as well as in the form of fibers (tampon) after surgical intervention. Since, prospidine does not have functional groups that would allow chemical bonding to the HPAC molecule, it was introduced by the "inclusion" method.

The mechanism of prolonged action of prospidine introduced into the matrix of the polymeric carrier by this method is explained by the fact that the molecules of the drug substance slowly diffuse during operation, starting from inside the fiber outwards, thus providing gradualness and duration of the drug substance action.

To impart antimicrobial properties to fibers it is theoretically possible to use all pharmaceutical substances, however, to obtain antimicrobial drugs by the method of "inclusion" a number of requirements should be met: they should be soluble in the common solvent used to prepare spinning solutions, should be stable in the conditions of polymer molding and processing technology, stable to the action of temperature and components of spinning solution or settling bath accepted in this technology. In this regard, despite the apparent simplicity of the method, the successful works carried out in this direction are few.

The goal was also to obtain prospidine (dichlor N, N" gu (γ-chloro-β oxpropyl) N,N" dipyrotripiperazine) of prolonged action (PDP) in the form of a fibrous tampon and eye dosage films (ODFs) of standard size. Prospidine has the following chemical structure.

$$ClCH_2-CH-CH_2-N\bigcirc N\bigcirc N\bigcirc N-CH_2-CH-CH_2Cl$$
$$\quad\quad\quad OH \quad\quad\quad\quad\quad\quad\quad\quad\quad\quad\quad\quad\quad\quad\quad\quad OH$$

Prospidin has found wide application in oncological practice. Y.A. Ershova et al. found [110] that prospidine at a dose of 145 mg/kg causes inhibition of tumor growth by 58-64%.

Under the influence of the drug there is a decrease in the total number of intotically dividing tumor cells, a sharp morphological change and irreversible damage to the chromosome. Normal tissues (bone marrow and cornea) are little sensitive to the effects of prospidine. In spite of this, according to the data of N.S. Bogomolova et al. [111], the following was found typical for the drug: when administered by injection it disappears from the blood during the nearest 2-3600 s after injection, gets into the kidneys, lungs, pancreas at early times in relatively large amounts, and in relatively small doses in the liver, spleen, lymph nodes; irregularity of the drug administration and products of its transformation from different organs; fast excretion from the body (about 75% of the drug is excreted during the 1st day mainly with the urine).

In order to prolong the action of prospidine and to use it more effectively, fibers and films containing different amounts of prospidine were produced. The HPAC-based fibers were molded from aqueous spinning solutions with propidine additives. The concentration of the spinning solution was 10% (weight). The spinning solutions with propidine additives are stable and no signs of delamination were observed within 15 days. The fibers were molded in isopropyl alcohol (IPA) baths. The choice of IPS as a precipitation bath is due to the fact that IPS is a good precipitator for HPAC on the one hand, and on the other hand, the molded fibers are sterilized. Forming of fibers containing up to 15 per cent of prospidin proceeds satisfactorily, while further increase in the amount of prospidin leads to frequent breaks. The physical and mechanical properties of the fibers deteriorate somewhat with increasing amounts of the additive. This is apparently due to the fact that increasing the concentration of the additive leads to loosening and changes in the supramolecular structure, as well as in the intermolecular interaction due to the increase of the second component in the polymer matrix. As the fibers with prospidine were intended for use in the form of a tampon after surgical intervention, high indicators of physical and mechanical properties are not required. Based on the above studies, optimal molding conditions for molding water-soluble fibers with prospidine were selected with the following parameters: prospidine content 10%, spinning solution concentration 10% (weight), die drawing 100%, plasticizing drawing

25%, temperature 293-295⁰ K. The fibers molded under these conditions had satisfactory physical and mechanical properties ($\sigma - 6 \div 8cN/tex$, l=18+22%). Sterilization of fibers before use was carried out in alcohol.

The process of preparation of ocular drug films (ODLs) was as follows: the drug propidine was injected into an aqueous filtered solution of HPAC. The solution was carefully homogenized and applied in an even layer to the horizontal glass surface. The film was molded by slow evaporation of the solvent in room conditions. The dried films with a residual moisture content of 10-12%, was removed from the surface of the glass and transferred under a standard die, by means of which ophthalmic drug films of rounded form with certain dimensions (f = 9-0.4 mm) were obtained. The GLPs thus obtained were vacuum dried for 2-3-3600s to a moisture content of 3-5%. The obtained films contained 6 mg of prospidine, which were transferred to the Research Institute of Oncology and Radiology to study under experimental conditions on experimental animals (rabbits).

The results of biomedical tests showed that GPs based on BPAC and prospidine are an effective drug for the treatment of cancer eye diseases in ophthalmology and strongly increase the residence time of prospidine in the body (in the lesion focus). Blood analysis and histological studies of the organs of the experimental animals showed that no accumulation of BPAC in the body was observed.

Preparation of preparations with interferon-inducing and antiviral properties based on oxidized HPAC and oxybenzylamines.

Infectious diseases caused by viruses are spreading among people, animals, and plants due to the lack of effective treatment and prevention tools. The antiviral drugs used in medical practice do not solve the problem of combating viral diseases, since many antiviral drugs have mutagenic, antigenic, teratogenic, and toxic effects.

The discovery of interferon inducers, under the influence of which the body or cell culture produces its own interferons, was a new milestone in the fight against viral diseases. Intensive research in this direction led to the creation of interferon inducers "Megosin", "Poly-guacil", "Amexin", "Reafferon" and others, which are now successfully used in medical practice.

Interaction of oxidized water-soluble acetylcellulose {OVRAC) with low molecular weight compounds synthesized aromatic oxyamines derivatives [112,113] containing fragments of benzylamine (1), 2-oxybenzylamine (2), 4-oxybenzylamine (3) and 2.4- dioxybenzylamine (4) by reaction:

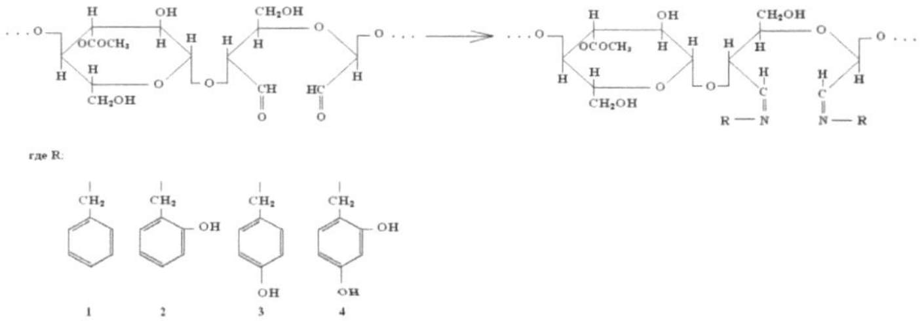

Aromatic oxyamines, in turn, were synthesized from the corresponding' oxyaldehydes according to the following schemes:

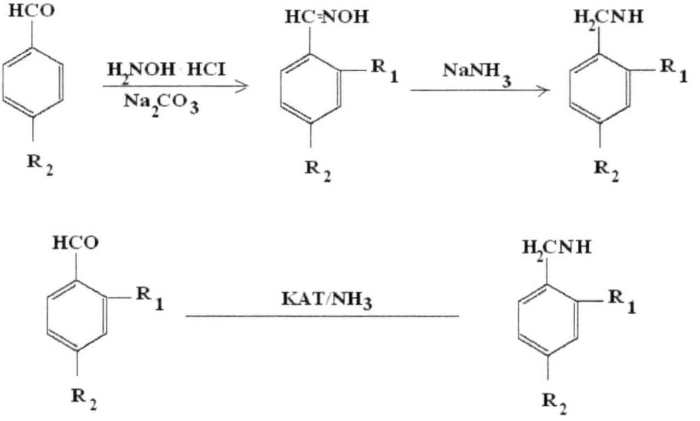

where:R R_{12} -H(benzylamine),
R_1 -OH, R_2 -H (2-oxybenzylamine),
R_1 -H, R_2 -OH(4-oxybenzylamine),
R_1 -R_2 -OH (2;4- dioxybenzylamine).

The synthesis of OVRAC derivatives with substituted aromatic oxiamines was carried out at temperatures of 40-60°C in an aqueous medium. The reaction products were isolated by precipitation in acetone and the precipitate was washed with acetone and alcohol sequentially to remove the unreacted low molecular weight substance. The product was purified by double resuspension in acetone. Infrared spectra of product 1 show absorption bands confirming the proposed structure: a broad and intense absorption band at 3300-3600 cm^{-1} related to valent vibrations of O-H and C-H bond, 1950-1120 cm^{-1} symmetric and asymmetric valent vibrations of C-O-C bond in glucopyranose, and valent vibrations of C=C bond of aromatic nucleus are observed at 1520-1560 cm^{-1} . Direct evidence for the formation of the | structure is the absorption at 1680 cm^{-1} related to the azomethine C=M bonds. The presence of an aromatic nucleus in the synthesized compounds is also confirmed by UV spectra where a "benzene" absorption band is observed at 265-270 mmc. The physicochemical characteristics of the synthesized azomethine derivatives of HPAC are given in Table 71.

Table 71.

Physico-chemical constants, IR-, UV-spectra of azomethine derivatives of HPAC.

№	R	Water solubility	[η]	IR, cm^{-1}		UV, mmc	
				V =C$_c$	V =N$_c$	λ$_{max}$	lg ε
1.	CH$_2$	P.	0,210	1540	1680	260	2,47
2.	CH$_2$	X.P.	0,2310	1525	1680	268	2.50
3.	CH$_2$ OH	P.	0,2400	1529	1684	270	2,40

| 4. | | P. | 0,2300 | 1534 | 1680 | 275 | 2,35 |

The study of interferon-inducing activity of the synthesized compounds showed that benzylamine, 2,4-dioxybenzylamine derivatives induced interferon with an activity of 20-40 IU/ml and had low antiviral activity. 2-oxybenzylamines had the highest interferon-inducing activity, named "SAVRACs" /117,118/. A comparative study of the cytoxicity of SAVRATs with known viral substances showed that the synthesized compound at a concentration of 2000 μg/ml had no noticeable effect on cell culture status, while gossypol at a concentration of 125 μg/ml strongly inhibited the proliferation of FEC cells (Table 72).

Table 72.

Cytotoxic effects of compounds in FEC cell culture

Drug cipher	Dose, μg/ml	Intensity of cytotoxic effects of drugs	
		24 hours	48 hours
Gossipol	125	++	+++
SAVRAC	125	-	-
-//-//-	250	-	-
-//-//-	500	-	-
-//-//-	1000	-	+
-//-//-	2000	+	++
-//-//-	4000	+++	++++

Note:
"+"-25% monosolic cell half-life.
"-++"-50% suppression of cell monosol,
"+++"-75th suppression of cell monosol,
"++++"-100%-nose suppression of monosolar cells,
-//-//-/- monolayer is fully preserved.

Antiviral and interferin-depressant drugs based on water-soluble aminoacetylcellulose have also been obtained [119-122].

In neutral media, the oxime-HVPAC synthase was first carried out by the action of hydroxylamine on HVPAC [114-118]. The oxime

HUVEC is a water-soluble compound, which is a white powder of the following structure;

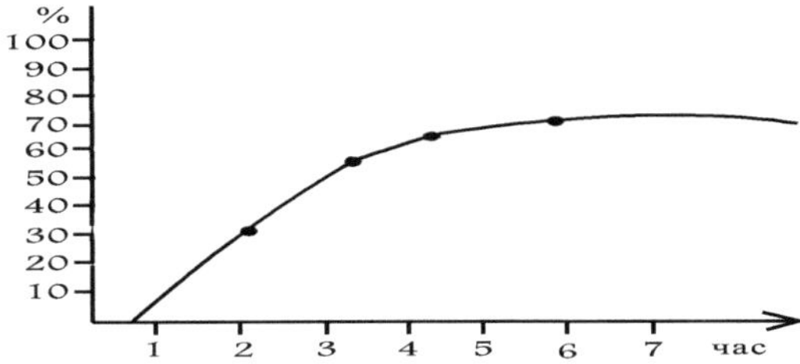

Further, water-soluble aminoacetyl cellulose /VaAC/, containing primary amino group in the elementary unit of cellulose molecule, was obtained by reduction of OVRAC oxime:

The effect of the reducing agent concentration in the range of 0.1-1.5% on the amount of newly formed amino groups in aminoacetylcellulose at the reduction reaction duration of 4 hours was studied. The dependence of the yield of water-soluble aminoacetylcellulose on the concentration of the latter leads to an increase in the yield of VAAC. Water solubility of VAAC worsens with the increase of amino groups, which seems to be associated with an increase in the amount of hydrogen bonds /"cross-linking"/ both within and between molecules of VAAC.

The dependence of the yield of VAAC on the reaction duration, temperature, and pH of the medium was also studied. The results of the study showed that the maximum yield of VAAC is achieved at a reaction duration of 3.5-4.0 hours (Fig. 106).

Fig.106. Dependence of VAAC yield on reaction time.

In many reactions, an increase in temperature leads to an increase in the reduction rate and an increase in the yield of by-products. It should be noted that increasing the temperature of the reaction medium to 50^0 C leads to a slight increase in the yield of VAAC. The optimum temperature for the synthesis of VAAC is the temperature of the reaction medium of 35-40 °C. To find out the possibility of reduction of OVRAC oxime in a weakly acidic medium the reaction was carried out at pH equal to 4-5. The data obtained indicate that the reduction reaction in weakly acidic media leads to an increase in the yield of the target product.

To determine the number of amino groups in VAAC samples, trinitrobenzosulfonic acid /TNBS/ was used, which interacts with amino groups to form a compound that absorbs light at a wavelength of 400 nm. The largest number of reduced groups was 80-85% of the total number of oxyimide groups in the aminoacetylcellulose oxime.

The chemical structure of water-soluble aminoacetylcellulose was proved using modern methods of physical and chemical analysis /IR, UV-spectra and elemental analysis/.

In the IR spectra of VAAC there are characteristic absorption bands confirming the structure of the synthesized compound: a broad absorption band at 3200 cm^{-1} -3400 cm^{-1} valent vibrations of the OH group, a wide band at 3050-3100 cm^{-1} corresponds to the valent vibrations of the NH$_2$ group, and at 1640 cm^{-1} observed strain vibration NH$_2$ group. A narrow band with an average absorption intensity at 1750 cm^{-1} , corresponds to the valence vibrations of the C=0 bond of the acetyl group.

By interaction of VAAC with benzaldehyde (VII), 2-oxy-(VIII), 4-oxy-(IX), 2,4-dioxy-(X), naphthaldehyde (XI), 2-oxy-(XII), 4-oxy-(XIII), 2,4-dioxynaphthaldehyde (XIV), aromatic oxyaldehyde derivatives with water soluble aminoacetylcellulose were synthesized:

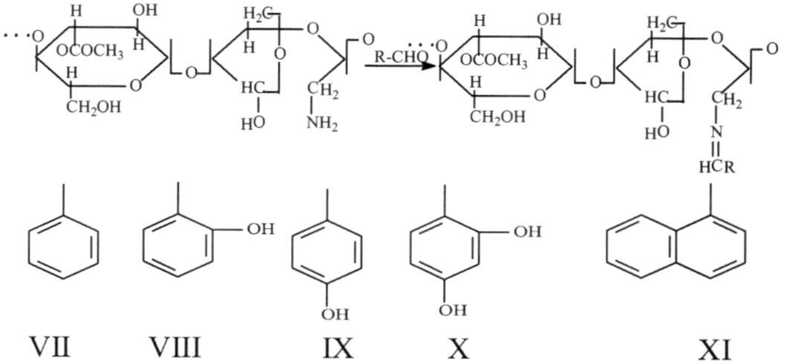

VII VIII IX X XI

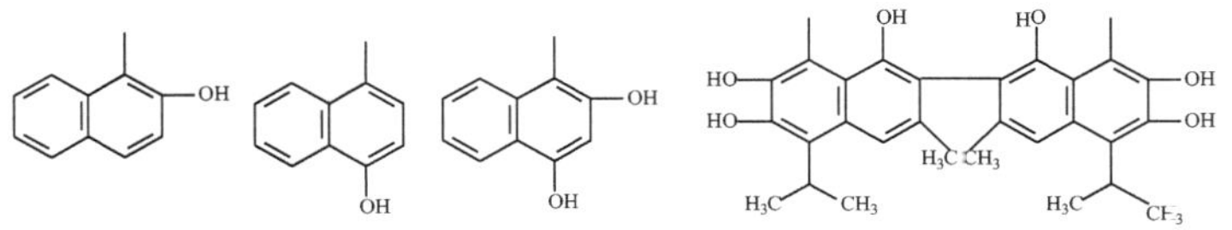

XII   XIII   XIV   XV

The physicochemical parameters of the synthesized compounds are given in Table 73. In the IR spectrum of the obtained compounds in the region of 3500-3450 cm^{-1} the valent vibrations of the O-H bond appear, and in the region of 1650 cm^{-1} the valent vibrations of the $C=N$ bond appear.

The condensation reaction of water-soluble aminoacetylcellulose with low molecular weight oxyaldehydes was carried out in aqueous medium, at room temperature, at slightly acidic pH values and with a reaction duration of 2.0-5.0 hours. After the reaction was completed, the product was isolated by precipitation in acetone. To remove the low molecular weight compound (unreacted aromatic aldehydes) the obtained substance was repeatedly washed with alcohol and acetone.

Chemical structure of the obtained compound is confirmed by infrared spectra data. Thus, a new absorption band at 1640 cm^{-1} corresponding to the azomethine bond /-H C=N-/ of the formed new compound appears in the product of condensation of VAAC WITH gossypol. This absorption band is absent in the spectra of the original WAAC and gossypol, as well as in the mechanical mixture. The condensation product of VAAC with gossypol "Homicel" is well soluble in water, insoluble in organic solvents.

Table 73.

Physical and chemical constants of derivatives of aldehydophenols, aldehydonaphthols and gossypol.

№	Reaction conditions			[η]	Solubility in water	Output %	IR, em⁻¹ $C=N$	UV, Mic.	
	pH	^0C	Time, h					Ma cλ	lgε
VII	7,0-7,5	25	3,5	0,85	p.	83	1570	262	2,42
VIII	6,5-7,0	85	2,5	0,41	p.	92	1676	266	2,35

IX	6,5-7,0	40	2,0	0,69	p.	86	1674	264	2,45
X	7,0-7,5	35	2,5	0,51	p.	80	1672	268	2,40
XI	6,5-7,0	20	5,0	0,61	t.r.	72	1675	272	2,22
XII	6,5-7,0	40	2,0	0,60	p.	74	1682	276	2,20
XIII	7,0-7,5	22	4,5	0,71	p.	70	1673	278	2,34
XIV	7,0-7,5	25	3,0	0,73	p.	76	1678	275	2,40
XV	6,5-7,0	40	2,5	2,5	p.	84	1680	280	2,48

Condensation of HAAC with gossypol was carried out at different values of ph and temperature of the reaction medium.

Table 74.

The yield of the product of condensation of gossypol with VAAC /CU/.

№	Number VAAC,g.	Number of gossip.g.	Time, h.	Temperature,^{0} C	pH	Output %
1.	0,5	0,05	6	30	4,5-5,0	55
2.	0,5	0,05	6	35	4,5-5,0	65
3.	0,5	0,05	6	40	4,5-5,0	84
4[x]	0,5	0,05	6	45	4,5-5,0	70
5[x]	0,5	0,05	6	50	4,5-5,0	80

x-increase in temperature leads to structural changes in the target product / color change and reduced solubility in water.

The antiviral effect of the synthesized substances was studied in the culture of Kul fibroblast cells /FEC/ against the vesicular stomatitis virus /VSV/ both when the cells were treated with the preparations and the virus simultaneously and when the cells were pretreated with the preparation before the introduction of the virus. The results obtained are shown in Figure 107.

VAC /VII-XV/ derivatives have an inhibitory effect on the development of BVS, moreover, the suppression of the virus activity is more pronounced when cells are pretreated with the compounds. In concentrations 3-4 times lower than cytotoxic dose it inhibits virus activity on the average by 1.5-3.0 lg TCD_{50} /ml, Antiviral activity of the studied compounds depends on their chemical structure. Thus, compound VII at a dose of 300 μg/ml inhibited virus activity by 1.0-1.2 lg TCD_{50} /ml, and at 500 μg/ml by 1.5 lg TCD_{50} /ml. Decreasing the concentration of the drug leads to disappearance of the antiviral activity. The appearance of the hydroxyl group in the phenyl part of the molecule leads to an increase in the antiviral activity. Compound XIII has relatively high antiviral activity among the studied compounds, which decreases the BVV infection activity by 100 and by 500-1000 times compared to the control when the cells are pretreated 24 hours before virus introduction at the dose of 125 μg/ml and at 250 μg/ml. The location of the hydroxyl group in the para position of the phenyl radical /IX/ leads to a decrease in

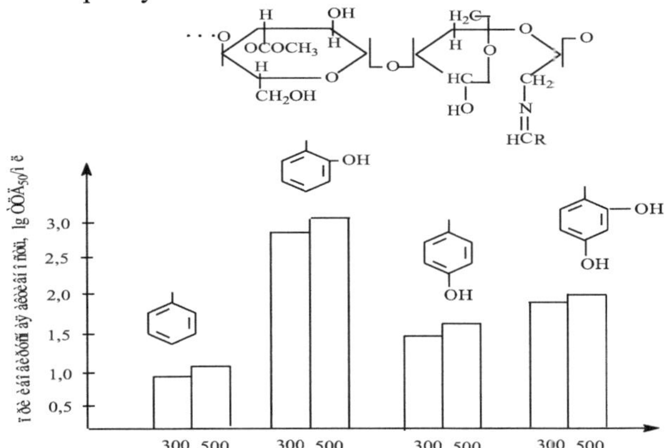

Figure 107: Antiviral activity of VAAC derivatives with oxyaldehydes.

antiviral activity of the drug. The appearance of two hydroxyl groups in the /X/ molecule leads to a slight increase in antiviral activity. The transition from the phenyl to the naphthyl system /XI/ is accompanied by an increase in cytotoxic activity. Thus, compound /XI/ has a toxic effect on FEC cell culture at a dose of 1200 μg/ml and has negligible antiviral activity. The data obtained testify to the relationship between the structure

307

and biological activity in the azomethine derivatives of HAAC: the location of the hydroxyl group in the ortho-position of the phenyl radical, leads to an increase in antiviral activity compared to its location in the other part of the molecule.

Antiviral activity and toxicity of XV were studied at concentrations of 62,125,250,500 µg/ml. Increasing the drug concentration up to 3000 µg/ml does not lead *to* cytotoxic effect on FEC cells. Compound XV has interferon-inducing activity in cell cultures. Thus, in the presence of DEAE-dextran polycation, it induces interferon formation with the activity of 1280 IU/ml. A correlation between the activity of the produced interferon and the concentration of the drug is observed.

Table 75.

Antiviral activity of gossypol derivative with VAAC against common viral infections.

Virus	Inhibition of virus multiplication	Animal protection in % in h.	
		24	4
Flu	2,6	50	35
Hepatitis	n.i.	60	40
Vesicular stomatitis	2,8	70	55

Note:n.i.-not investigated

The antiviral activity of XV was established during experimental infection on the model of influenza strain Aishi 68 H2/H2.The results show that the suppression of virus reproduction when using X V reaches 2.6 and 2.8 l g , i n d i c a t i n g high activity of the drug against this strain of influenza virus.

The activity of XV against mulberry mulberry caterpillar nuclear polyhedrosis virus was further studied. Determination of the preventive action of the preparation against the virus was carried out at concentrations of 0.05 and 0.025% by wetting a water solution of the preparation on a mulberry leaf. The test options were as follows: feeding the caterpillars once per age (3 in total) with the leaf treated with the preparation; feeding the caterpillars with the leaf treated with the preparation every 36 hours, taking into account their withdrawal period. Feeding with the leaf was started from the 3rd instar of caterpillars and continued until cocooning according to the above scheme. In the control variants, caterpillars were fed with a mulberry leaf moistened with pure water (1 time of age or 36 hours) and regular feeding of caterpillars. The preparations were tested during spring, summer and summer-autumn feeding periods on caterpillars of Tetra hybrid 3 and Ferganskaya-I hybrid. At the end of the fattening period, caterpillar viability /in%/, weight of one cocoon /g/, percentage of silk sheath and technological properties of cocoons were recorded. Experimental caterpillars were fed under the conditions accepted for feeding white-skinned breeds and mulberry silkworm hybrids (Table 76).

On the basis of testing the effectiveness of the drug, we can conclude that Gomycel at a concentration of 500 µg/ml increases the viability of caterpillars, improves cocoon weight, weight and percentage of the silk sheath of the cocoon. Technological properties of cocoons curled by experimental caterpillars are slightly higher compared with the control variants.

Table 76.

Viability of experimental caterpillars.

№	Contents of the option	Viability %	Statistical reliability
1.	Feeding caterpillars with a leaf moistened solution of CU preparation at a concentration of 125 µg/ml, water from the 1st day of the 3rd instar	63,0±1,3	0,966 0,683
2.	Same, 250 µg/ml	70,0±3,3	0,990
3.	Checklist moistened with water.Feeding according to the same scheme.	59,0±1,3	-
4.	Control of normal caterpillar feeding	62, ±1,4	-

Thus, the data obtained when testing the preparation Gomycel on spring feeding, allow us to conclude about the promise of the preparation as a means of increasing the viability of mulberry silkworm caterpillars.

Synthesis of hemostatic preparations based on HPAC.

Bleeding is one of the life-threatening complications of various diseases and conditions, especially in surgery, obstetric pathology, wounds, etc. [110,111]. In this connection, the problem of hemorrhage stopping has been and remains an urgent task of medicine.

At the same time, the arsenal of modern local hemostatic agents is relatively small. Currently used in medical practice hemostatic agents of local action, such as hemostatic sponge, hemostatic gauze, gauze tampons and napkins impregnated with 1% solution of feracryl, do not always meet the requirements of the practical doctor. These agents show moderate hemostatic activity, are not convenient for use and do not resorb.

There were obtained [112-114] hemostatic preparations in the form of polymeric form /fibers and film/ by "inclusion" of the hemostatic preparation "Lagoden" into the structure of HPAC.

Low-molecular-weight natural substances, such as lagohylline and lagohirzine, isolated from plants of the genus Lagohylus, have weak hemostatic activity due to insolubility in water.

In order to make them soluble and prolong their action, they were chemically attached to the ORACs.

In order to join the aldehyde group of VVRAC, a certain amount of it was dissolved in dimethylformamide so that the concentration was 5-6%. Lagoquiline or lagohirzine was added to the ORAC solution in a 4:1 ratio with the polymer. Anhydrous copper sulfate or phosphorus anhyllril was used as a water-releasing agent. The reaction was performed at temperatures 333, 345, and 355^0 K for 6 hours.

Table 77.

Dependence of the number of lagohylline and lagohirizine molecules bound to HUVECs on temperature and type of water-releasing agent.

°C	Water-releasing agents	Number of bound molecules		Reaction rate constant $K-10^{-3}$		Reaction activation energy, kJ/mol	
		Lagohilin	Lagohirzin	Lagohilin	Lagohirzin	Lagohilin	Lagohirzin
50	CuSO₄	50	36	1.9	1.2		
	P O₂₅	55	45	3.1	1.8		
60	CuSO₄	52	39	3.8	2.7		
	P O₂₅	57	48	6.4	4.1		

80	CuSO$_4$ P O$_{25}$	54 59	40 50	8.2 13.5	6.8 9.2	265 252	345 310
OVRAC contained 110 aldehyde groups per 1,000 AGZ.							

After completion of the reaction, the reaction mixture was filtered and precipitated in acetone followed by washing with precipitant.

The mechanism of the lagohylline (or lagohirzine) addition reaction can be depicted as follows:

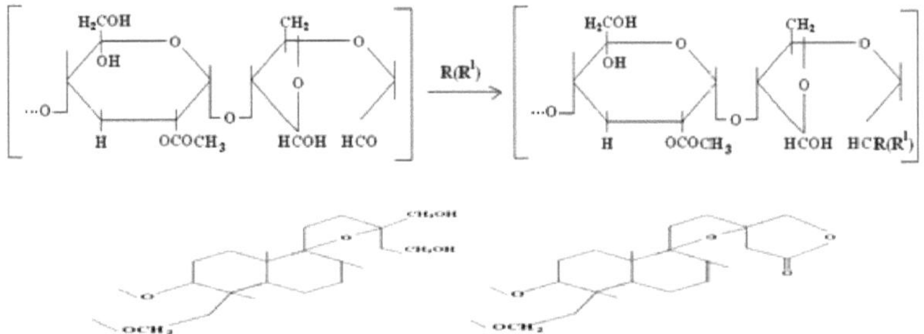

R-Lagohylline R -lagohyrisine[1]

The amounts of attached lagohylline (or lagochrysin) molecules with the macromolecule of OVRACs were determined by the difference in the number of aldehyde groups per 1000 anhydroglucoside link (AGZ) in OVRACs before and after the reaction. The data obtained are shown in Table 77.

As shown in Table 77, the number of immobilized molecules of lagohylline and lagohirizine increases with increasing medium temperature. Under the same conditions, lagohylline molecules react more actively with HUVECs as compared to lagohirzine. Under the conditions of these reactions, phosphoric anhydride is a more active water-releasing reagent than copper sulfate.

Since the reaction rate depends only on the concentrations of lagohylline or lagohyrsin, the reaction rate constants were calculated using the kinetic equation of the 1-order reaction:

$$K = \frac{2.3}{t} \lg \frac{C_o}{C}$$

The activation energies of the studied reactions were determined from the graph of the dependence of LgK on 1/t.

Figure 108 shows the kinetic curves of the esterification reaction of OVRAC with lagohylline. The nature of the kinetic curves of these reactions at different temperatures is identical.

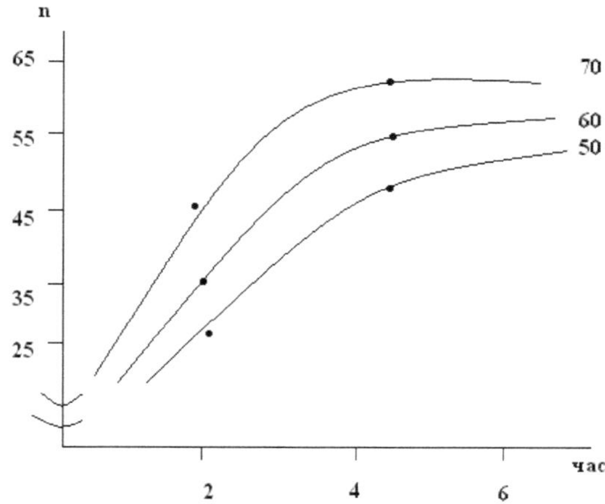

Figure 108. Dependence of the number of bound lagohylline molecules (n) on the reaction duration in the presence of $P_2 O_5$.

It turned out that the activation energies of both reactions in the presence of copper sulfate are greater than those of phosphorus-(V) oxide. This is apparently due to the more hygroscopic nature of phosphorus anhydride. The table also shows that irrespective of the type of water-releasing reagent, the activation energy of the interaction of lagohyrisine with OMVPAC is relatively higher compared to that of lagohylline.

A comparative hemostatic evaluation was performed for the obtained materials in which lagohylline and lagohirizine were chemically bound to OVRACs (Table 78).

Table 78.

Effect of the obtained preparations on parechymatous bleeding time in rats (average of 6 experiments in each case)

Substances used	M/sec, clotting time	%	p

Control (sheet, water)	258 ± 5.7	100	-
OVRAC	118 ± 8.7	41	0.01
OVRAC with lagahillin	52 ± 6.5	20	0.001
OVRAC with lagohirizine	43 ± 4.4	16	0.001

Analysis of the obtained data shows that OVRAC itself reduces the time of parenchymatous bleeding by 2 times in relation to the control. Preparations of IVRAC with lagohirzine and lagohirzine when applied to the wound surface of the liver instantly clot blood and reduce the time of parenchymatous bleeding by 5-6 times in relation to the control.

Also, HPAC-based fibers and films containing up to 5% lagoden in their structures stop bleeding and blood clotting much faster compared to the well-known hemostatic drugs (Table 79).

As can be seen from Table 79, polymeric materials based on HPAC and lagoden can stop parenchymal bleeding, which confirms their hemostatic properties and can be used in practical medicine as a resorptive styptic agent in the treatment of local bleeding.

Table 79.

Effect of water-soluble polymeric materials, hemostatic gauze and hemostatic sponge on parenchymal bleeding time in rats and rabbits (average of 10 experiments in each case).

Substances under study	clotting time		%		p	
	Rat	bunny	Rat	bunny	Rat	bunny
Control (medical gauze)	260 ± 10	240 ± 10	100	100	-	-
Fibers from VRAC	181 ± 20	168 ± 16	70	70	0,01	0,001
VRAC fiber with 5% moisture	139 5 ±	84 ± 10	52	35	0,001	0,001
VRAC film	271 ± 21	254 ± 30	106	105	0,05	0,05
VRAC film with 5% lagoden	120 6 ±	101 ± 10	46	42	0,001	0,001

| Hemostatic gauze | 156 ± 15 | 147 ± 14 | 60 | 57 | 0,01 | 0,01 |
| Hemostatic sponge | 169 ± 14 | 125 ± 15 | 61 | 52 | 0,001 | 0,001 |

Thus, physiologically active polymers based on water-soluble cellulose derivatives and low-molecular-weight drug substances are overexpressed materials used in various branches of medicine.

LITERATURE.

1. Ushakov S.N., Panarin V.O. Combination of penicillins with water-soluble polymers. DAN USSR, 1962, 125, pp. 1102-1104

2. Khomyakov K.P., Virnik A.D., Rogovin Z.A. Prolongation of action of drugs by using them in mixture with polymers or joining to polymers. Advances in Chemistry. 1964, т. 33, 29, c. 1051-1065.

Shostakovsky M.F., Sidelkovskaya F.P. Drugs based on polymers. Med. prom. USSR, 1961, № 3, pp. 6-13.

4. Kuznetsova N.P., Glikina M.V., Nemtsova N.N., Samsonov G.V. Formation and properties of water-soluble complexes of insulin with polyelectrolytes of different structure. Proceedings of the Second All-Union Symposium on Chemistry and Physicochemistry of Physiologically and Optically Active Polymeric Substances. Riga: Zinatne, 1971, pp. 113-117.

5. Weinstein V.A., Naumchik G.N. Interaction of polyvinylhirrolidone with the polyene antibiotic nystatin during co-deposition. The 3rd Symposium on Physiologically Active Synthetic Polymers and Macromolecular Models of Biopolymers. Riga: Zinatne, 1971, p. 17.

6. Yudovich E.A., Kamilov I.K., Ushansky Z.M. Influence of apricot gum on lengthening and strength of anesthetic action of solutions of novocaine, cocaine and Dicaine. Proceedings of the Tash. Pharm. Institute, 1962, vol. 3, pp. 289-299.

7. Wolf L.A., Meos A.I., Kirilenko Y.I., Kotetsky V.V. Polyvinyl alcohol fibers with special properties. JVCO IM. D.I. Mendeleev, 1966, vol. 1, 6, c. 654-656.

8. Plate N.A. Polymers in Medicine. Translated from English: World, 1969,-239 p.

9.Ushakov S.N. On obtaining films, filaments, poroplasts and thixotropic gels and aqueous complexes of polyvinyl alcohol and its copolymers. DAN USSR, 1960, v. 134, 3, p. 643-646.

10. Frome A.A., Nemenova N.M., Baranina M.A. Decotran transfusion reaction. Clin. medicine.1963, vol. 41, Ji 3, p. 96-1Ol.

11. Koziner V.B.Synthetic blood substitute polyvinylpyrrolidone. Pathology, Physiology and Experimental Therapy.1958. T. 2, No 6, p. 53-58.

12. Ushakov S.N. Synthetic polymers for medicinal purposes. L.: Medgiz, 1962, 42 p.

13. V., Konstantinova I.M., Meshchensky A.D., Privalova L.G., Selavda T.V. Study of kinetics of biodegradation of alloplastic material based on copolymer N vinylpyrrolidone and methyl methacrylate in model media and tissues of the body. Materials of the All-Union Symposium on Physiologically Active Synthetic Polymers and Other Polymers for Medical Applications. L., 1975, p. 5.

14. Vasilieva I.S., Suzdaleva V.V. Modern state of the problem of blood substitutes. Problems of Hemotology and Blood Transfusion. 1966, т. 11, 7, c.3-9.

15. Adel I.B., Dmitrev S.A., Collected: Ion Exchange and Its Application. Moscow: Academy of Sciences of the USSR, 1959, - 307 p.

16. Saldadze K.M., Pashkov A.B., Titov V.S. Ionic high-molecular compounds. Moscow: Goskhimizdat, 1960, - 360 pp.

17. Osborn G. Synthetic ion-exchangers. Moscow: Il, 1964, -506 p.

18. Helint A., Flanagan T., Selection of cation-exchange resins for therapeutic purposes. Ionic resins in medicine and biology. Moscow: IL, 1956, pp. 25-268.

19. Mac Chesney e., Nahod F., Tainter M. Cation exchange resins Ioncak therapeutic agent for sodium removal. resins in medicine and biology. Moscow: IL, 1956, pp. 269-28.

20. Yaschunsky V.G. Complexes and their use in medicine. DVho M. D.I. Mendeleev, 1965, vol. 10, 6, pp. 679-683.

21. Eighorn G. Chemistry of Coordination Compounds. Moscow: IL, 1960, 587 p.

22. Malek G., Hoffrman I., Herold M., Kolytz I. Antibiotics with directed penetration into the lymphatic system. Antibiotics, 1958, 1, pp. 45-50.

23. Malek G., Kolz I. On the use of lymphotron antibiotics in clinic. Antibiotics. 1958, № 4, c.34-37.

24. Ushakov S.N. On two phase gels of iodopolyvinyl alcohol. DAN USSR, 1961, vol. 139, 1, pp. 160-62.

25. Ushakov S.N. Results and prospects for the synthesis of blood substitute polymers with medicinal effect. Theses of reports of IX scientific conf. of IVS AS USSR, L., 1962, p.1-3.

26. Ushakov S.N., Kononova T.A. Synthesis of polyvinyl ethers of pelentanoic acid. DAN USSR, 1959, vol. 129, 6, p. 1309-IIIr.

27. A.S.I36551 (SSR.). Method for the preparation of pelentane esters. (Ushakov S.N., Kononova T.A.). Published in BI, 1961, J# 5.

28. Ushakov S.N., Trukhmanova L.B., Drozdova E.3, Markelova T.M. Synthesis of para-aminosalicylic ester of polyvinyl alcohol. DAN USSR, 1961, Vol. 141, No. 5, pp. 1117-1119.

29. Rabinovich I.M. On the prolongation of the action of chemotherapeutic drugs based on polymers. Materials of the Conference on Drug Therapy in Oncologic Clinic. L., 1964, pp. 133-34.

30. A.S. 159847 (SSSR.). Method for preparation of mixed penicillin hydrazides and copolymers of vinyl alcohol with unsaturated acids (Ushakov S.N., Panarin E.F.). Opubl. In B.I., 1964, No. 2.

31. Panarin V.F. Synthesis and study of water-soluble reactive and biologically active polymers acting on cell systems: Author's dissertation .... D. in Chem. - Л.: 1979.

32. Campbell T. A. Pharmacentlcal Aspects of a piamino salicylate Dialdehyde Starch Compound. J. Pharm. Sci., 1963, V.52, No. 1, pp. 76-78.

33. Rosenfeld V.L. Dextran, its characteristics and importance as a blood plasma substitute. Uspekhi biolog. chimii, 1959, vol. 3, p. 366-387.

34. Vasiliev A.E., Livshits A.B. Synthesis of N-aminoacyl derivatives of carboxymethyldextran. Proceedings of the Second All-Union Symposium on Chemistry and Physicochemistry of Physiologically and Optically Active Polymeric Substances. - Riga: Zinatne, 197, pp. 170-74.

35. Pat. 20310 (Japan). Coatings of tablets that dissolve in the intestine. (Tanabo Setsyaku I.K. - Republished in RZHim, 1967, no. 8, p. 872.

36. Prospectus "Cellulose azetat phthalat" Archlv der Pharmazlc, Dw. of Eastman Kodak Comp-Rochester /USA, 1968, N° 7, p. 147.

37. Gluzman M.H., Zaslavskaya R.G., Rubtsova V.N. Phthalylation of acetylcellulose of different acetylation levels.ZhPH, 1966,

38. Zatz J.Z. Knowles B., Monomolecular F1lm Properties of Some Cellulose Esters.

39. J.Pharmaci Sci., 1970, v. 59, no. 8, pp. 1188-1189.

40. Pat,3149038 (C1JA). Th1er Film Coceting for Tablets and the Like and Method of Coat1ng Sampson F. -onyó. in Official Gazette on Mat. pat. ved. United States, 1964, vol. 806, no. 3, p. 834

41. Pat. 3132075 (C11A) Solid Medicinal dosage forms coated with hydroxyethyl cellulose and hydrolised solyrene mallic anhydrid copolymer. published in Official Gazette of Mat. United States, 1964, vol. 802, 1, pp. 225-226.

42. Pat 3380998 (C1JA). Cellulose der1vat1ves and the1r appl1cat1on for coat1ng agents. / Kenji Naito, Osaka, Hioroaki Nomura, Nichinomija, Vasic Nogch, Minoo-published in Mat. Ved., United States, 1968, vol. 849, 11 5, pp. 167 - 468:

43. State Pharmacopoeia of the USSR. Hizdany, 1968, p. 667. Kryazev V.N., Harkova O.T., Yablokova L.V. Preparation of a2etone phthalylcellulose with different solution viscosity. Materials of the conference "State and prospects of development, manufacture and use of excipients for medicine production". Kharkov, 1982, p.1, p.

44. Zhigach K.F., Finkelstein M.Z., Timokhin I.M., Malinina A.I. Preparation of carboxymethylcellulose for blood replacement solutions. DAN of the USSR, 1958, Vol. 123, No. 3, p.471-474.

45. Pat. 3087853 (C1A). Water soluble Compositions consisted essentially of Lodine and Water soluble oxygen containing Polymer / Ho sman. W., Starke A. - Opubl. In Official Gazette of the Mat. Pag. ed. of the United States, 1963, Vol. 789, JE 5, p. 1348.

46. Heyg Allen. Polymer-Drug interaction stability of Agneons Gels Containing Neomycin Sulfate. J. Pharmac1 Sc1, 1971, v. 60, N° 9, pp. 1338-1341.

47. U.S. Patent 3495800. Form and bracket for cast1ng a s2ack of un1formed concrete slabs. / Donald J., Fisher. publ. in Official Gazette of Mat. of Patents, USA, 1970, vol. 871, no. 3, p. 830.

48. Shevchenko S.M., Bugrin N.A., Iskritzky G.V., German T.V., Litvinova V.A. On the use of microcrystalline cellulose in the manufacture of tablets. Theses of All-Union Scientific Conference "State

and prospects for the development, production and use of excipients for the manufacture of drugs", Kharkov, 1982, p.I. 50.

49. Pat. 2718667 (USA). Method of preparlng enteric cupsules/ Malm C.J., Gordon D.H., Rochester N.U. - Op. in Official Gazette of U.S. Pat. USA, 1955, v. 698, #3, p. 529.

50. A.S. 329699 (SSSR.). Method of obtaining microcapsules (Massimo op. Calanchi, Massimiliano Macherna, Martino Macchini).in B.I.,1972, no. 7,

51. Pat. 3557279 (C1A). Microencapsulation from an anil- inforflammatory drug / Morse Z.D. Princetion N.J. - Only J. In Official Gazette of Mat. of Pat. USA, 1971, vol. 882, 3, p. 1308.

52. Rosenberg G.Ya., Pokidova N.V., Koziner V.F. Problems of hematology and blood transfusion. 1953, 3, 57.

53. Sarimsakov A.A. Investigation of the chemistry of cellulose dialdehydes and its esters. Dissertation of Chemical Sciences, Tashkent, 1978. - 22 c.

54. Najimutdinov Sh., Sarimsakov A.A., Usmanov H.U. Chemical structure and reactions of dialdehyde cellulose. Cellulose Chem. Techn. 1975, 9, c. 617-639.

55. Physiologically active polymeric substances. Tashkent: Tashkent State University Press, 1976. - 176 c.

56. Some aspects of synthesis of polymers for medical purposes. Tashkent: Fan, 1978. - 20 c.

57. Virnik A.D. Chemical modification of cellulose and dextran to obtain biologically active materials and compounds. D. in Chemistry. Tashkent, 1969. - 335c.

58. Akhundova N.I. Inactivation of isonicotinic acid hydrazide preparations and its relationship with drug resistance of tuberculosis microbacteria. Autoref. dissertation of candidate of medical sciences. Baku, 1969. - 21 c.

59. Hilkin A.M., Schecter A.B. Collagen and its application in medicine. Moscow: Medicine, 1976, - 288 p.

60. Perepelkin K.E., Perepelkina M.d. Soluble fibers and films. L.: Chemistry, 1977. - 104 c.

61. Iirgenson B. Natural Organic Macromolecules. Moscow: Mir, 1966.- 554 p.

62.Nachinkin O.I. On the use of synthetic fibers based on polyvinyl alcohol for surgical purposes. Medical. industrial, USSR, 1965, p. 7-15.

63.Dyachik I., Kollar I., Pekareva I., Yangachik I., Yambrakh i. Preparation of fibers based on oxyethyl cellulose. Preprints of the 1st International Symposium on Chemical Fibers. Kalinin, 19777, vol. 2, c. 191-198.

64. Perepelkin L.P., Katalevsky E.E. Investigation of fiber forming process from low-substituted oxyethyl cellulose. Chem. Fibers, 1972, № 6, pp. 47-49.

65. Bezvernaya L.I., Tokarev L.V., Kalinina Y.G. Preparation of fibers from cellulose ethers. Moscow: NIITEchim, 1971, 31 p.

66. Geller B.E., Khamraev A.L. Influence of fiber formation conditions on physical and mechanical properties of fibers based on carboxymethylcellulose. Chem. Fibers 1972 $5, pp. 47-49.

67. Umarova L.N., Khamraev A.L., Geller B.e. Fibers based on carboxymethyl cellulose. Chemical Fibers, 1973, № 2, p. 64.

68. Khamraev A.L. Study of the process of obtaining carboxymethylcellulose fibers. Author's thesis ... Candidate of Technical Sciences. - M. - 1979. - 43 c.

69. Nadzhimutdinov Sh. Synthesis of functional polymers, their chemical transformations and physiologically active derivatives. Author's abstract of dissertation ... D. in Chemistry. Tashkent, 1980. - 60 c.

70. Nadzhimutdinov Sh., Sarymsakov AA, Nurmukhamedov Sh., Fayziev Sh.K., Nigmetov K.N. Fibers on the basis of medium-substituted carboxymethylcellulose. Materials of All-Union Conference on Chemistry and Physics of Cellulose. Tashkent, 1982, p.2, p. 49.

71. Rolev M.Y. Research in obtaining fibers on the basis of low-camel sulfate of cellulose. Author's thesis, Moscow, 1971, MTI.

72. A.S.235905 (SSSR). Method of obtaining cellulose-based modified loquin. (Gritskov I.V., Ivanov V.M., Gershovich F.S., Tonkopryad K.I., Napadaylo F.S., Shemshchur Z.A., Lashkov M.I.). publ. B.I., 1969, № 6, p.61.

73. Pat. 2687944 U.S.). Manufacture of Water-soluble textile and other materials. / - Jonson E.B., Webster J.P. -Published in Official Gazette of Mat. Pat. of the United States, 1954, Vol. 685, No. 3, p 1115.

74. Pat. 2545070 (USA). Cellulose Ether solutions and Method of spinning the some. Norman Luols Cox, Claymont. published in Official Gazette of Mat. Pat. ed. of the United States, 195, vol. 644, no. 2, p. 524.

75. Khamrayev A.L., Geller B.E., Burenin P.I. Medical fibers based on carboxymethylcellulose. Materials of the 1st All-Union Symposium "Polymers in Medicine". Tashkent, 1973, p. 47.

76. Volkov V.V., Krasnov A.M., Khramov G.L. Application of eye medication films in field conditions. Military Medical Journal, 1978, 2, pp. 43-46.

77. Burlakova T.N. Experimental data on penicillin permeability to intraocular fluids. Bulletin of Ophthalmology, 1948, No 27, pp. 1-5. И

78. Maychuk Y.F. Polyvinyl alcohol as a medicinal basis of antibiotics for ophthalmic practice. Journal of Ophthalmology, 1964, No. 5, pp. 350-354.

79. Erofeeva L.N. Study of eye films with penicillin group antimicrobials and antiviral prepac. rat florenal. Author's thesis. , 1974. 16 c.

80. F., Davydov A.B., Kondratieva T.S. Initial evaluation of prolongation properties of biosoluble polymers for ocular drug films. Chemico-Pharmaceutical Journal, 1974, vol. 8, no. 6, pp. 24-29.

81. Krichna V., Brow F.J. Polyvinyl alcohol as an ophthalmic vehicle. Amer. J. ophthalmol., 1964, vol. 57, N° 1, pp. 99-106.

82. Titsina I.F. The use of some polymers for prolongation of eye sciences. - M., 1970. - 27 c. Drugs. Autoref. diss. of medical sciences.

83. Maychuk Y.F., Davydov A.B., Khromov G.P. Biosoluble polymers as a basis for soluble films. Pharmacy, 1978, 1, p. 60-61.

84. Prospectus of the Ministry of Health of the USSR. Eye drug films. (GP). M., 1978. - 25 c.

85. G.L. Khromov, G.V. Shelukhanov, A.B. Davydov, K.P. Najestkin, Penal dispenser for eye medication films. Medical equipment. 1975, № 6, c. 44-45.

86. A.S. 402367 (SSR.). Device for storage and dispensing of solid dosage forms. (G.L. Khromov, G.V. Shelukhanov, A.B. Davydov, K.P. Najestkin, Y.F. Maychuk). Opubl. in B.I., 1976, #6.

87. Shulgina N.B., Lenkevich M.M., Aleksandrova Z.I. Sanation of the conjunctival cavity by means of drug films impregnated with medicinal substances. Bulletin of ophthalmology. 1975, № 5, c. 78-80.

88. Maychuk Y.F. Comparative evaluation of different eye dosage forms of dexamethasone. Viral and allergic eye diseases. Cheboksary-Moscow, 1973, pp. 105-107.

89. Peenkov M.A., Avrushchenko N.M., Khristenko L.A. Eye medicated films in the treatment of inflammatory diseases of the eyeball. Journal of Ophthalmology, 1978, No. 6, pp. 431-433.

90. Puchkovskaya N.A., Vaino-Yasenetsky V.V. Retrocorneal films, clinical and histological characteristics. Ophthalmology, 1973, 3, pp. 119-125.

91. Malakhonova NL, Murzin AA, Kanygina EL, Tulubenskaya NA, Khromov GL, Gerasimova GA Application of eye films with fibrinolysin in ophthalmic practice. Immunoglobulins and other blood products. L., 1976, p. 108-III.

92. Maychuk Y.F., Khromova G.L. Ophthalmic drug films with scanamycin. Vestn. of Ophthalmology, 1977, ¹ 6, p. 61-62.

93. Nikolaeva I.S., Bogdanova N.S., Pershin G.N., Kutchak S.N. Experimental study of ocular medicinal films with oxolin. Pharmacology and Toxicology, 1977, No. 3, pp. 346-348.

94. Shipanova A.I. Pharmanokinetic assessment of idoxuridine injected in ophthalmic films. Proceedings of the All-Union Congress of Ophthalmologists, 1979, pp. 133-134.

95. Smotkina N.V. Clinical observations on the efficacy of PVS, CMC, methylcellulose as prolongators of pilocartin, hemotropin and osopoloamin action. Materials of the 2nd All-Russian Congress of Ophthalmologists, Moscow, 1968, p.32.

96. Rakhmanberdiev G., Petropavlovsky G.A., Usmanov H.U. Water-soluble acetylcellulose, its properties and application possibilities. Cell. Chem. and Technol., 1978, Vol. 12, No. 2, pp. 153-176.

97. Rakhmanberdiev G., Mirnigmatova I.M., Gafurov T.G. Synthesis of physiologically active cellulose derivatives with some low molecular weight drugs. Uz. chem. journal, 1974, 1, p. 60-62.

98. Rakhmanberdiev G., Mirnigmatova Sh. M., Usmanov Kh. U. Synthesis of physiologically active polymer based on water-soluble acetylcellulose. Materials of symposium on physiologically active and other polymers for medical use. Leningrad, 1975, pp. 59-60.

99. Rakhmanberdiev G., Svetitsky P.V., Mirnigmatova Sh.M., Yadgarova N., Azizov I.A., Usmanov H.U., Muratkhojaev G, Mukhamejanov M. Positive decision on issue of author's certificate for application # 334903/05 from 13.04.1982 "Method. Obtaining of physiologically active polymer" Publication in the open press is prohibited.

321

100. Methods in Carbonhydrate Chemistry Academic Press.N.-Y., London, 1963, 3. p. 327.

101, Obolenskaya V.V., Schegoleta V.P., Akim G.M., Akim E.L., Kosovich N.L., Egolnova n.V. Practical works on the chemistry of wood and cellulose. Moscow, 1965, - 411 p.

102. Kotelnikova N.E., Petropavlovsky G.A. Comparative study of oxidation of microcrystalline and cotton cellulose by nadic acid. Cellulose Chem. Technol., 1974, T. 8, Já 3, 0. 203-214.

103. Rogovin Z.A., Yashunskaya A.G., Bogoslovsky B.L. Preparation of chemically colored toloknas. X111, 1950, vol. 23, 36, pp. 631-640.

104. Kochetkova V.A., Vorontsova L.E. Experimental and clinical study of prolongation of aurantin action with polyvinylpyrrolidone. Materials of conference on drug therapy in oncological clinic. L.,1964, pp. 86-87.

105. Romanova I.N., Belousova A.I. Action of alkyliruptic agents on ATP-ase activity of liver mitochondria. Biochemistry, 1972, no. 846-850.

106. Rakhmonberdiev G.R., Yusupova N.F., Mirkomilov T.M., Sidikov A.S. Synthesis of physiologically active polymers based on water-soluble acetylcellulose. "Prospects of creation of materials on the basis of raw materials Center Asia".T. 1997, P.29.

107. Rakhmonberdiev G.R., Yusupova N.F., Sagdieva Z.T., Myakova N.V. Properties of solutions of antitumor eye drug prolonged action. Kimye va pharmacia.t. ,1998 №1,c.6-9.

108. Rakhmanberdiev G.R., Yusupova N.F., Sidikov A.S., Muratova T.T.. A polysaccharide base for ocular drug cells. Chemistry of derivative compounds.,T,1998,p.188-160.

109. Pat.RUz. IHDR 9600501.1. of 28.05.1996 Antiociculatory eye preparation. Rakhmanberdiev G.R., Yusupova N.F., Muratova T.T., Muratkhojaev N.K.

110. A., Minakova S.M., Chernova V.A. On the antimitotic activity of prosimidine. Pharmacology and toxigology, 1973, M.P.79-81.

111. Bogonolova N.S. Suskova V.S., Chernyi V.A., Serebryakov N.G.. Blood content, distribution in organs and excretion from the organism of prospidine depending on the way of its application. Pharmacology and toxigology, 1976, No.6, pp.742-746.

112 .A.S. 1407000 (SSSR). Aslanov K.A., Ershov F.I., Rakhmonberliev G.R., Aulbekov S.A., Maulianov S.A, Lavrukhina L.A.

Poly-1,2-β-0,4-|3-6 acetyl-D-glucopyrancsyl]3-0-[1-oxy-2-(salicyliminoethyl]-D-erythrofuranose possessing interferon-inducing and antiviral properties against encephalomyocarditis vrus.

113. Amanova S.H., Mirzasv R.B., Maulyanov S.A., Rakhmonberdiev G.R., Aslanov H.A. Synthesis and structure of water-soluble aminacetylcellulose derivatives// Uzb Chem. 1994, № 1. C. 13-16.

114. Rakhmanberdiev G.R., Auslbekov S.A., Mukhamedjanov M., Maulyanov S.M. Antiviral agent based on water soluble acetylcelulose and oxyamylophenols. All Unions. scientific conference. Problems of modification of natural and synthetic fiber-abrasive polymers. Moscow,1991, section P. 3-14.

115. Rakhmanberdiev G.R., Maulianov S.M., Auslbekov S.A. Synthesis of oxidized water-soluble acetylcellulose derivatives. DAN Uz SSR, 1987, #10, p.36-37.

116. Rakhmanberdiev G.R., Amanova S.H., Mirzaev R.T., Madlyanov S.A., Aslanov H.A. Synthesis and pharmokinetics of 3A-SAVRAZ. DAN RUz, Tashkent, 1993,p.160.

117. SAVRAC. In Handbook of Antiviral Drugs. М.Медицина.1998,c.160.

118. Rakhmanberdiev G.R., Maulyakov S.A., Tukhsanov A.Kh. Aslanov Kh.A. Water solubility of acetylcellulose-basis for creation of antiviral substances and interferon inductors. Хим.природн.соед.1995,№1, c.165-167.

119. A.S.16956339 SSSR. Derivatives of aminoacetylcellulose and gossypol possessing interferon-inducing and antiviral properties. / Aslanov H.A., Rakhmanberdiev G.G., Ershov F.I., Auelbekcv SA, Tazulakhova E.B., Sayitkulov A.M., Achilova S., Amanova S.H., Maulyanov SA.

120. Amanova S.H., Aslanov H.A., Auelbekov S.A., Rakhmanberdiev G.G.. Synthesis of water-soluble aminoacetylcellulose.//DAN R.Uz.-#3. 1992-C.30-31.

121. Amanova S.H., Mirzaev R.T., Mamuyanov S.A., Rakhmanberdiev G.G., Aslanov H.A. Synthesis and structure of derivatives of water-soluble aminoacetylcellulose,//Uzb.chem.zhurnal.-1994.№I.-P.13-16.

122. Amanova S.H., Mirzaev R.T., Mukhamedjanov M., Rakhmanberdiev G.G., Aslanov H.A. Synthesis and biological activity of

derivatives of water-soluble acetylcellulose // All-Union conf. on problems of use of cellulose and its derivatives in medicine and microbiology.

123. Barkatan Y.S. Hemorrhagic diseases and syndromes. - Moscow: Melitsina, 1980, 335 p,

124. Petrovsky B.B., Chazov G.I., Anlresv S.V. Topical problems of hemostasis, molecular and biochemical parisological aspects. - Moscow: Nauka, 1980. 503 c.

125. Rakhmonberdiev G.R., Sidikov A., Yusupov R., Zainutdinov U. Hemostatic polymeric material. Patent RUz 1NDR 9b00549.1 from 04.06.96.

126. Yusupov R., Zainutlinov U., Rakhmonberliev G.R., Sidikov A. Synthesis and study of blood-stopping drugs // Proceedings of Tashkent Chemical Technology Institute, 1995. C. 109-112,

127. Rakhmonberdiev G.R., Sidikov A.S., Yusupov R.D., Zainugdinov U.N., Kazantseva D.S. Preparation of polymeric hemostatic materials based on "Lagoden" // Chemistry of natural compounds, 1997. 448 c. Industry.T..-1989.-p.59.

6.3. Physiologically active polymers based on cellulose sulfates Anticoagulant activity of sulfated polysaccharides.

The majority of biologically active substances with anticoagulant activity contain sulfuric acid esters in their structure and represent salts of sulfated polysaccharides [119]. One of the representatives of such substances is widely known heparin, isolated for pharmacological purposes from the liver and lungs of bovine cattle and used in medicine for the prevention of thrombotic disorders. Despite the wide clinical use of heparin, it has a number of side effects, *in this regard,* numerous studies on the creation of drugs with anticoagulant activity on the basis of sulfated polysaccharides are conducted. The anticoagulant activity of sulfated polysaccharides and the influence of the nature of their macromolecule and molecular parameters on the activity they exhibit have been studied.

The sodium salts of polysaccharide sulfates were chosen as the objects of the study: cellulose monosulfate (MSC) with $SP = 0.98$ and with the location of the sulfate group mainly at the 6th carbon atom, cellulose disulfate (DSC) with $SP = 2.0$ with the location of sulfate groups at the 2nd and 3rd carbon atoms, cellulose trisulfate (TC) with $SP = 2.84$ with the location of sulfate groups at the 2-,3-,6th carbon atoms; amylose sulfate (SA) with SP=2.43, guar gum sulfate (GMS) with SP=3.1 and polygalacturonic acid sulfate (PGS) with SP=1.06.

Study of anticoagulant activity of polysaccharide sulfate preparations was carried out in *in vitro* experiments. Studies were performed using a calibration curve for low anticoagulant $1.54 \cdot 10^{-4}$ $-7.7 \cdot 10^{-4}$ mg/ml (0.02-0.10 units) and for high $7.7 \cdot 10^{-4}$ - $4. \cdot 10^{-3}$ mg/ml (0.1-0.6 units). The concentration of anticoagulant was chosen so that the inactivation of the enzyme was linear. The analyzed substance and analog were added to platelet-poor citrate plasma. For this purpose, blood was centrifuged for 20 minutes at 2000 g. Blood for the study was obtained from a rabbit ear vein by drip.

Figure 29 shows the effect of sulfated polysaccharides on the clotting time of whole blood.

As can be seen from the data shown in Figure 109, blood clotting time increases with the addition of all drug samples. The greatest effect at this concentration of preparations was observed in heparin and samples of TCC, SPGC, CA, where the blood clotting time increased by 4 times or more in relation to the control to which 0.9% saline was added.

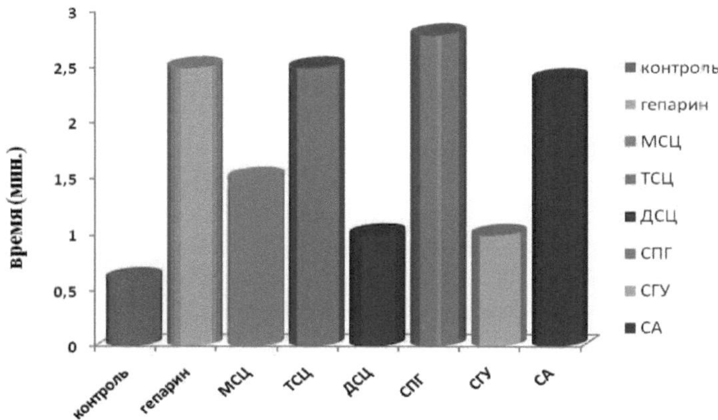

Fig.109 . Effect of sulfated polysaccharide samples on whole blood clotting time at a concentration of $4.6 \cdot 10^{-4}$ mg/mL on rabbit blood *in vitro*

From the studies of sulfated polysaccharides as direct anticoagulants, by tests of clotting time, ACTV, TV we can conclude that all the studied drugs have a direct anticoagulant effect. They significantly increased clotting time, ACTV and TV by two or more times. In the ACTV test, polymers with a linear structure were shown to be more than twice as active as those with a branched structure. The most active samples among cellulose sulfates were disulfate with CAN = 2.0 and with the arrangement of sulfate groups at 2-, 3-rd carbon atoms and trisulfate with CAN = 2.84 and their activity was close to that of heparin. For cellulose sulfates it was shown that it is the sulfate group located at the second AGE carbon atom which promotes the anticoagulant activity. Polygalacturonic acid sulfate samples with CP=1.1 and amylose samples with CP=2.43 also showed high activity. It is clearly shown that the value of anticoagulant activity does not depend on the anomeric bond configuration (α or β). At the same time, for polysaccharides with glycosidic bonds ($1\rightarrow4$) the presence or absence of sulfate at C-6 had no effect on their activity. Consequently, the presence of sulfate groups in polysaccharides can increase their specific and nonspecific binding to a wide range of biologically important proteins. Both the structure of the polysaccharide macromolecule and, possibly, its conformation in solution play an important role in these interactions.

Study of the antimicrobial activity of cellulose sulfates.

Various samples of Na-SC were examined for antimicrobial activity against various pathogenic microbes. Determination of microbial sensitivity to the preparations was performed by the paper disk method as follows:

The studied material - isolated pure culture was sown in a continuous "lawn" on the surface of the agar in Petri dishes. The bottom of the cup was divided into sectors according to the number of test preparations using a pencil on glass. Then sterile tweezers were used to put a paper disc soaked in the solution of a certain drug on each sector. The disks were placed at an equal distance from each other, from the center of the cup and its edges. The seeded cups were incubated in the thermostat for 18-24 hours. The results were determined by measuring the zone of delayed growth of the microbe. The presence of growth around the disk indicates the insensitivity of this microbe to the drug (Figure 30).

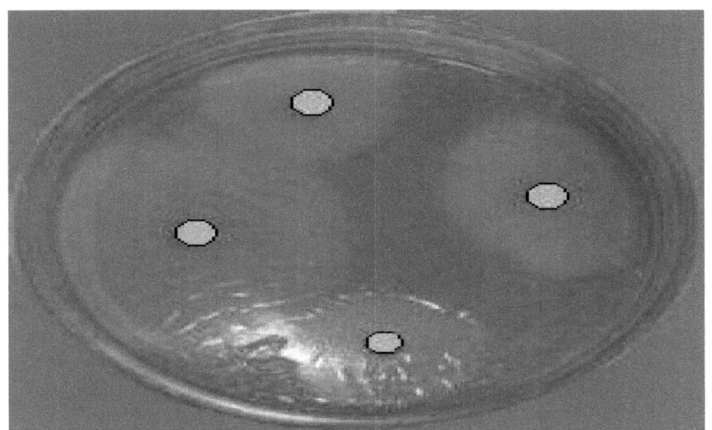

Fig.110. Determination of sensitivity of microbe Streptococcus group A to preparations *in vitro*

It was found that the preparation Na-SC has a powerful effect on microbes. It should be noted that both Gram-positive and Gram-negative microorganisms are sensitive to it (Table 80).

Table 80.

Information on the sensitivity of microorganisms to Na-SC under in vitro conditions.

№	Microorganisms	Concentration in %			
		10%	3%	2%	1%
1	Klebsiella	25,0±1,2	20,0±1,0	20,0±1,0	20,0±1,1
2	Eschirichii LP	20,0±1,0	15,0±0,7	10,0±0,5	7,0±0,5
3	Eshirihi LN	22,0±1,1	15,0±0,8	12,0±0,7	12,0±0,5
4	Candida fungi	-	-	-	-
5	Streptococcus group A	25,0±1,2	20,0±0,9	20,0±0,8	15,0±0,7
6	Group D Streptococcus	15,0±0,7	10,0±0,5	10,0±0,4	8,0±0,3

7	Staphylococcus aureus	25,0±1, 2	18,0±1, 0	15,0±0, 7	15,0±0, 5
8	Staphylococcus aureus hemolyticus	25,0±1, 2	20,0±1, 1	20,0±1, 1	10,0±0, 5
9	Staphylococcus epidermus.	15,0±0, 7	15,0±0, 7	13,0±0, 4	10,0±0, 3
10	Pseudomonas bacillus	6,0±0,2	-	-	-
11	Proteus	14,0±0, 5	8,0±0,3	8,0±0,4	

From this we can assume that this drug can be widely used to treat purulent inflammatory diseases and escherichiosis; it is especially effective for clabsiellosis.

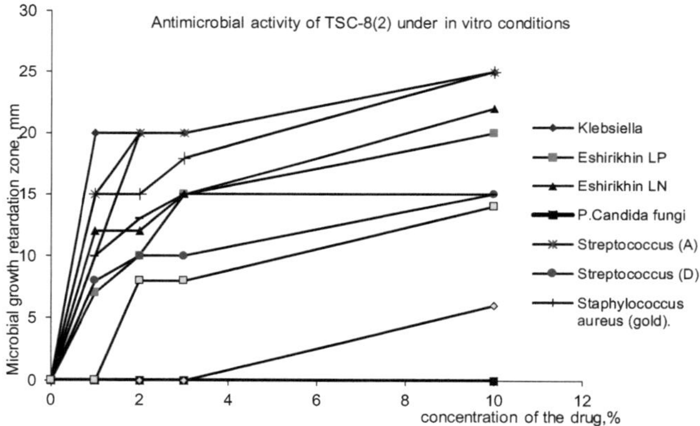

Fig.111. Antibacterial effect of Na-CC under *in vitro* conditions

Based on the molecular parameters of the investigated Na-CC samples, the "structure-properties" relationship was studied (Figure 112).

The graph shows that preparations with an SP of up to 2.0 and almost the same molecular weight have a negligible effect on microbial growth. In the C3 interval from 2.0 to 2.2, an increase in activity can be observed. Starting from SP 2.2, the activity changes insignificantly. These results indicate that a substitution of at least two hydroxyl groups in the AGEs is required for the antimicrobial activity of SC.

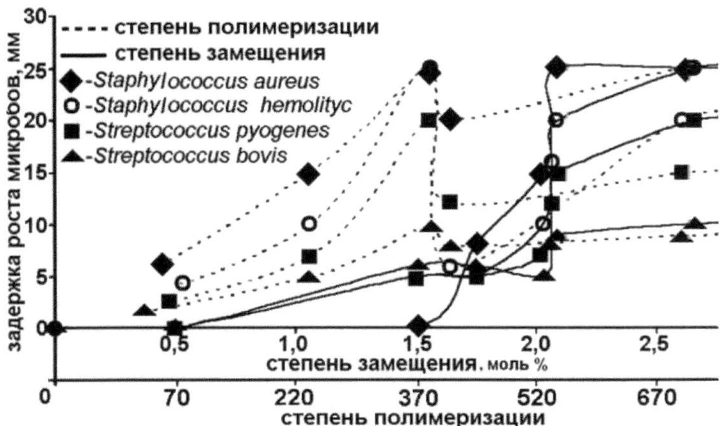

Fig.112. Dependencies of microbial growth retention on NW and SP of SO samples.

The effect of the SP value of SCs on their biological activity was studied. It is interesting that in this case we can see a similar change in biological activity. With an increase in SP of samples up to 370 - 400 a gradual increase in biological activity is observed, with a further increase in SP the biological activity practically does not change or changes insignificantly.

On the basis of microbiological tests of SC with different molecular parameters, it was suggested that the samples with weak action appear to act only on the cell wall of the microbe, disrupting the synthesis of peptidoglycan and thus having a bacteriostatic effect.

SOs, which have a pronounced effect on microbes, seem to act in the direction of disruption of the functions of the cell membrane or ribosome of the microbial cell, which manifests as the death of the microbe and, thus, have a bactericidal effect.

Based on the results of the study of the sensitivity of microbes to CC derivatives, we can conclude that the molecular parameters of CC play an important role in determining their activity. The pathogenic microbes studied show almost no sensitivity to cellulose derivatives with CP less than 1.0. This is due to the fact that at this SP the hydroxyl group 6 of the carbon atom is mainly replaced, and although, as computer simulations show (Fig. 6), the sulfate group is located on the surface of the macromolecule, their amount seems to be insufficient to bind to the surface wall of the microbe. With an increase in NW greater than 2.0, SC begins to have an effect on the microbe. This can be explained by the fact that when the SP is greater than 1, the substitution of hydroxyl groups 2 and 3 of the carbon atom begins, which can be seen from the computer simulation of SC. In this case, the amount of sulfate groups, to a certain extent, allows interaction with the surface wall of the microbe. At a CH of about 3.0, almost all hydroxyl groups are replaced and, most importantly, the hydroxyl groups of the 2nd and 3rd carbon atoms are replaced almost completely, as can be seen from the computer simulation of SC.

Computer models were created using the complex program ChemOffice-2002, indicating the location of SO_3 Na-groups on the macromolecule with increasing Na-CC NW. Below are models of sulfuric acid cellulose ethers of 16 glucoside units with different CC.

№	Molecular parameters of computer models of Na-CC	Macromolecule surface view	
		Front view	Side view
1	6-O substituted Na-mono-sulfate cellulose		
2	6,2-O substituted Na-di-sulfate cellulose		
3	6,3-O substituted Na-di-sulfate cellulose		
4	2,3-O substituted Na-di-sulfate cellulose		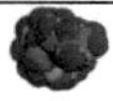
5	6,2,3-O substituted Na-tri-sulfate cellulose		

Fig. 113. Computer model of SC derivatives with different SPs (dark balls indicate sulfate groups).

It can be seen from the figures that NW SC, starting from 2.0, the biologically active group, SO_3 Na, almost covers (begins to cover) the surface of the macromolecule and therefore their ability to act with other specific groups (with specific groups of microbial envelope proteins) increases.

To conclude on the possible mechanism of antimicrobial activity, we can say that the exact mechanism of interaction between SOs and the microorganism is still unknown, but various explanations for their antimicrobial activity can be assumed.

Under normal conditions (~Ph 7), the sulfate groups have a negative charge, which is usually responsible for the demonstrated activity. It is assumed that the polyanionic nature of SC binds to the cell membrane through electrostatic interaction with the positively charged cell membrane of the microbe.

Thus, the antibacterial activity of sulfated polysaccharide samples was studied for the first time. It was found that the sensitivity of pathogenic bacteria depends on their species and the molecular characteristics of the modified polysaccharides. The introduction of the sulfate group into the polysaccharide molecule provides the manifestation of a wide range of antimicrobial activity.

For the purpose of molecular modeling of the interaction of polysaccharide sulfates with pathogen glycoproteins, we used the AutoDock modeling program for docking calculations .

Docking calculations were performed on the glycoproteins of various pathogens, including HIV-1 (3dno, the molecular structure of HIV-1 gp120 trimer in its attached state with CD4, a viral model like protein), E. coli (O176 O-antigen of E. coli). Also used in docking modeling were the Lamarckian genetic algorithm (LGA) and Solis & Wets local search method . The initial position, orientation and torsion of the ligand molecules were set chaotically. Each docking experiment was obtained from 10 different runs, which were stopped after a maximum of 250000 energy calculations. The size of the aggregate was set as 150. In the searches, the transfer step was applied as 0.2 Å, the quaternion and torsion step values were set to 5.

Infection of body cells by a pathogen occurs in several stages. The main stage is the interaction (fusion) of the pathogen with the body cell, which involves receptors (proteoglycans localized on the cell membrane) and domains (glycoproteins) located in the pathogen envelope. Spatial complementarity of the interacting structures forms the basis of such interactions. The key-lock analogy is usually used to describe it. This interaction is the initial and undoubtedly necessary stage of any infectious process.

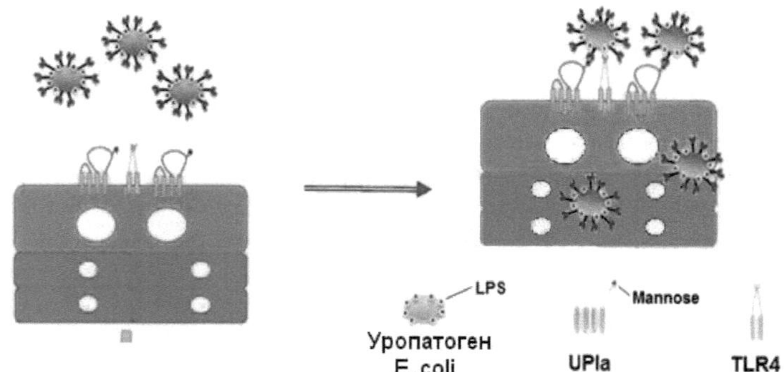

Fig.114 . Interaction between innate host defense and
***Ureapathogenic E.coli* in the bladder ducts.**

Computer simulations establish free energy (ΔG)-3.58 and 3.54 kilocalories/molecular masses, inhibition constant (K_i) 3.42 mm and electrostatic energy-0.04 kilocalories/molecular masses.

At

Fig. shows the conformations of the compounds and their respective minimum potential energies.

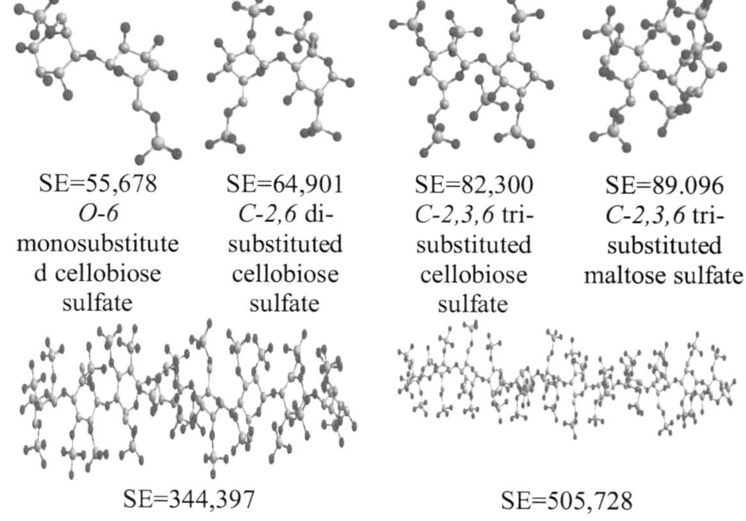

C-2,3,6 trisubstituted
polyglucose sulfate (SP=8)

C-2,3,6 tri-substituted
polyglucose sulfate (SP=12)

Fig. 115. Computer models of sulfate derivatives of polyglucans after their minimization of potential energies.

The results of the computer simulation studies showed that as the degree of esterification of free hydroxyl groups of SC increased, their interaction with the protein parts of the leukemia virus epitope increased.

Antiviral activity of sulfated cellulose derivatives against tobamoviruses.

The results of the study show the manifestation of high antiviral activity of cellulose sulfate derivatives against tobamoviruses. Sufficient concentration for 90±5% inhibition of viral infection of samples with SP above 1.5 was 0.5-1.0 mg/ml, samples with SP below 1.5 was 5.0-7.5 mg/ml. The number of sulfate groups in the polysaccharide chain and chain length were found to be the main factors in manifesting the biological activity of sulfated polysaccharides. Samples with CP above 1.50 and SP above 337 at a concentration of 1.0 mg/ml showed high activity (Fig. 27).

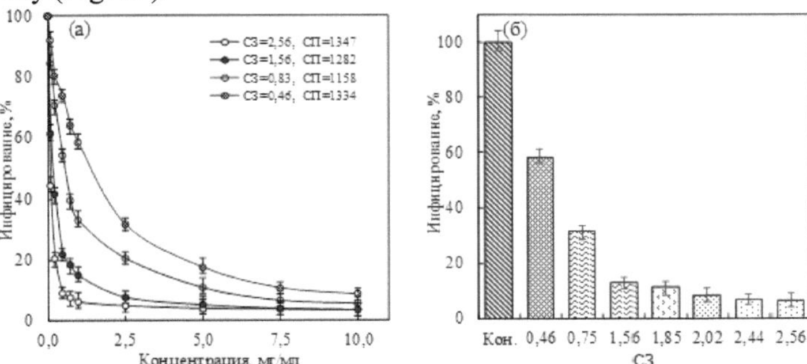

Fig.116. Effect of concentration (a) and NW (b) on the antiviral activity of cellulose sulfate derivatives against BTM (concentration (a) 0.10-10.0 mg/ml, concentration (b) 1.0 mg/ml, $P \leq 0.05$).

The results obtained show the direct interaction and destruction of the protein envelope of BTM under the action of cellulose sulfate derivatives in solution. Under the action of sulfate derivatives, when BTM virions are located at a mutually close distance, aggregation is observed, and the balance of mutual attraction of the protein envelope subunits is

333

destroyed. As a result, thinning, breaking, bending, destruction, and sticking together of some parts of the viral bacilli in the protein envelope are observed (Fig. 37). As the value of sulfate derivatives NW increases, their inhibitory effect on the protein envelope of BTMs, which undergoes destruction more deeply, increases.

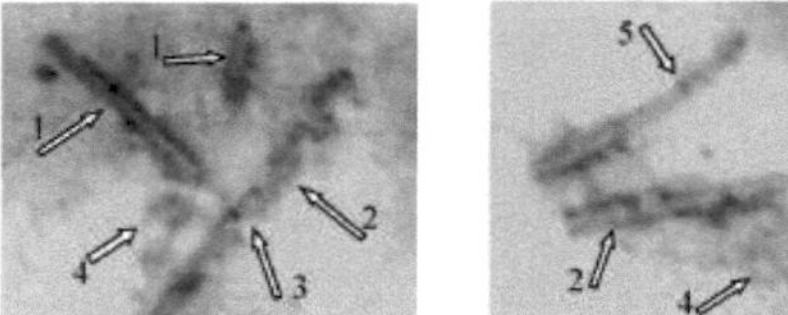

Fig.117. Changes in the protein sheath of BTM under the action of cellulose sulfate derivatives (Na-CC CP=1.56, 2.02; (1)-broken virion; (2)-bending of protein sheath; (3)-broken protein sheath; (4)-20S proteins; (5)-thinning of protein sheath; (6)-deflection of virion).

The results show that sulfated oligosaccharides based on cellulose sulfate derivatives exhibit high elicitor activity against plant tobamoviruses. The elicitor activity increases with increasing concentration and SP, as well as with decreasing SP (Fig. 29). High activity (75.2-88.2%) of cellulose sulfatoligosaccharides (MM=7000-15100 Da) with SP 1.54-1.64, SP 22-46, at 1.0 mg/ml concentration was found.

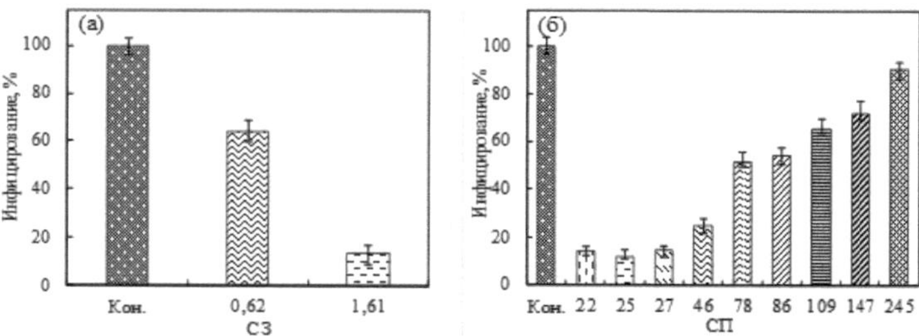

Fig.118. Dependence of elicitor activity of sulfated oligosaccharides against BTM of *N.glutinosa* plant on SZ and SP (sulfated oligosaccharide 1.0 mg/ml, SZ=1.54, SP=22, MM=7000 Da, *P≤0.05*).

After treatment with sulfated oligosaccharides, the preservation of high activity (85.9-90.9%) for 24-240 hours was shown. It was also found that the effect of cellulose sulfatoligosaccharides reduced the diameter of necrotic spots formed by BTM in plants. After treatment with sulfated oligosaccharides, the increase in elicitor activity against BTM and the preservation time corresponded to the activity of phenylalanine ammonia lyase (PAL) in plants, the increase in phenolic compounds and their preservation time (Fig. 30).

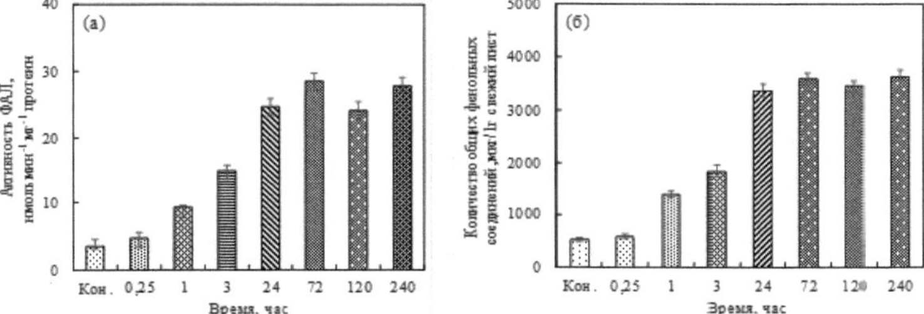

Fig. 119. FAL activity of *N.glutinosa* plant after treatment with sulfated oligosaccharides (a) and change in total phenolic compounds (b) (sulfated oligosaccharide CZ=1.54, SP=22, MM=7000 Da, $P \leq 0.05$).

6.4. Physiologically active polymers based on carboxymethylcellulose.

Today there is no area of practical medicine where polymers have not been used, and their reasonable application would not contribute to improving the effectiveness of therapeutic effects or to solving problems that are still unresolvable. In this aspect, great merit belongs to the schools of academicians N.A. Plate and V.A. Kabanov, their students who have been and are still engaged in the creation of a new generation of polymeric forms of structural products and medicinal preparations for medical purposes.

The first medicines were created mainly on the basis of compounds of plant and animal origin, by trial and error, whose composition was not precisely determined. These medicinal compounds mostly had a complex and not precisely determined composition.

With the development of organic chemistry, individual synthetic drugs of known structure began to be introduced into medical practice, which over time almost completely displaced the drugs based on natural compounds.

With the development of the pharmaceutical industry and the introduction of a large number of drugs of synthetic origin into medical practice, resistance and decreased sensitivity of the body to these drugs, and an increase in their negative, side effects on healthy organs of patients began to occur more often in patients.

Therefore, in recent years, intensive research has been conducted to create a new generation of drugs from plant and animal raw materials with both low and high molecular weight structures.

The development of polymer chemistry contributed to the widespread introduction of soft and hard structural polymeric materials (artificial organs, films, coatings, fillers, adhesives, etc.) into medical practice.

In the pharmaceutical industry, water-soluble, biodegradable polymers are widely used as carriers of biologically active compounds (BAS), proliferators, fillers, thickeners and emulsifiers in the production of pharmaceutical polymers (DP).

At the current stage of development of polymer science and the pharmaceutical industry, the creation of a new generation of polymeric forms of original drugs and medical devices based on natural and synthetic polymers is of particular importance.

In recent years, with the development of nanotechnology in this direction, large-scale research on the creation of nanostructured drugs and medical devices on a polymeric basis is being conducted.

Analysis of the state of development of chemistry and physicochemistry, including nanochemistry and nanotechnology, towards the creation of pharmaceutical, original drugs shows that polymeric, nanostructured biologically active compounds, in contrast to traditional low-molecular weight drugs, have a number of features. Before proceeding to the peculiarities of polymeric forms of drug polymers, it is necessary to consider their main distinguishing features from traditional low-molecular-weight drugs [1].

The polymeric form of a drug product generally consists of the active ingredient (drug substance) and the dosage form, which provides the drug to the body. The effectiveness of a drug is determined not only by the content, structure and properties of the drug, but also to a large extent by its dosage form.

Polymeric forms of drugs can be conditionally divided into two groups that differ in the principles that determine their biological activity.

The first group of polymeric forms of drugs includes compounds whose biological activity is due to their polymeric structure, i.e. their molecular weight (MM) structure, molecular weight distribution, the nature and content of functional groups and physical and chemical properties of the macromolecule as a whole.

Low-molecular-weight analogs of these polymers in most cases do not have the biological activity characteristic of these types of polymers.

The mechanism of action of this type of polymers is not related to their disintegration into low molecular weight biologically active fragments but to the properties of macromolecules, in particular to cooperative polymer-polymer reactions between body biopolymers and polymeric forms of the biologically active compound. This mechanism of action is not characteristic of low-molecular-weight drugs. This type of drug polymers (LP) with intrinsic biological activity. This type of LP with "intrinsic" biological activity is diverse, just like low-molecular-weight drugs. LPs with "intrinsic" biological activity can be conditionally divided into four large groups **(Fig. 1)** [2].

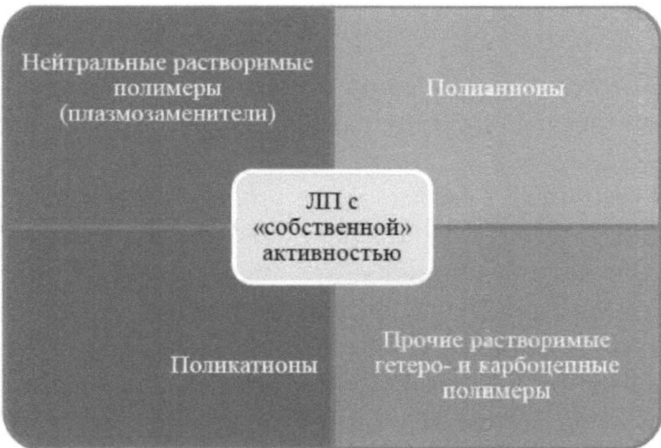

Fig.120. Polymers with intrinsic biological activity.

The second group of polymeric forms of drugs includes compounds whose structure includes polymers that are relatively inert in relation to the body, otherwise known as "carrier polymers" (PN) and low-molecular-weight or high-molecular-weight drugs.

Such polymers can be conventionally referred to the group of polymers of "grafting" type because in the overwhelming cases low- and high-molecular weight drug compounds (active principles) are chemically attached (grafted) to the PU by different types of chemical bonds: covalent, ionic, coordinational, etc. In polymers of the second group in most cases the biological activity is due to the grafted fragment.

The biological activity of this group of LPs varies due to the basic principles of molecular construction of the carrier polymer and low-molecular- or high-molecular-weight LPs.

1. A soluble, inert to the body biodegradable or biosoluble carrier polymer that can be eliminated from the body naturally;
2. The type of bond between the carrier polymer and the biologically active compound.

A. - a low molecular weight biologically active compound;
B. - a high molecular weight biologically active compound (biopolymer).

According to the composition, structure, properties and principles that determine their biological activity polymers are conditionally divided into two groups:

1) polymers - carriers 2) polymers with intrinsic biological activity. The first group of LPs are polymers that are inert with respect to body organs or have low toxicity, are mostly water-soluble, biodegradable or are excreted from the body by natural pathways.

The presence of polymers of the first group in the structure of drugs contributes to a significant improvement in the physicochemical and some biological properties of low-molecular-weight BAS.

LPs of the second group include polymers that have their own biological activity. Their activity is related to their structure, molecular parameters, the nature of functional groups, etc. The mechanism of action of LPs of this group is practically not related to their degradation in the body, but is caused by the properties of macromolecules and cooperative reactions of LP macromolecules with the biopolymers of the body.

Based on the analysis of the results of medical and biological studies of the above two groups of polymers, we concluded that the choice of polymers soluble in water, polymers subject to biodegradation in the body without forming toxic low molecular weight fragments or biologically inert polymers that can be excreted completely naturally in the body is preferred when creating a new generation of drugs. Such polymers are mostly natural compounds with hetero-chain structure.

With the right choice of the carrier polymer, a decrease in the toxicity of low-molecular-weight BAS and a prolongation of their duration and concentration in blood plasma in the free and bound state are achieved.

It is known that the permeability of polymers through the body's biomembranes is much lower than that of low-molecular-weight BAS, so there is a slow and wide distribution of PL in the body and the metabolism of low-molecular-weight BAS in the circulation slows down, which allows through the proper choice of the carrier polymer to control the pharmacokinetics of PL.

Despite the rather long history of creation of low-molecular-weight drugs and to date, no general principles of creating their prolonged forms with selective action have been developed. In contrast, the basic principles of molecular design of polymer-drug substance systems have been developed for LPs for quite a long time.

However, this does not mean that a set of desired properties can be accurately assigned to any drug substance.

In the molecular design of drugs consisting of a drug substance (drug) and a carrier polymer classified in the first group of polymers, depending on the final goals, it is advisable to start the research with the selection of a drug and a carrier polymer.

When creating drugs, not all drugs can be components of polymeric dosage forms. It is of interest to create LPs using known drugs when the polymeric form of the drug substance is intended:

- for the treatment of chronic diseases that require frequent administration of medications, for a sufficiently long period of time;

- for drugs with low therapeutic latitude, the inclusion of which in the structure of the carrier polymer will increase their selectivity;

- for drugs with properties of bioregulators (hormones, vitamins, etc.), which must be administered to the body constantly and in small doses;

- for topical medications, the distribution of which throughout the body is undesirable because of the possibility of side effects;

- for unstable drugs, such as enzymes, biopolymers, which are not stable in the free state and have high toxicity at high doses.

The possibility of attachment of surfactant to the carrier polymer is determined by the presence of functional groups capable of interacting with surfactant under sufficiently mild conditions with the formation of chemical bonds of various nature or molecular complexes. If necessary, reactive functional groups capable of reacting with the active drug centers can be included in the structure of the carrier polymer.

An ideal carrier polymer that meets all of these requirements does not yet exist and is unlikely to be created.

According to the mechanism of action, LPs can be divided into four groups:

1.   Drugs acting outside cells (antibiotics, enzyme inhibitors, anticoagulants and their neutralizers, etc.)

2.   LPs acting on the cell surface. This group of LPs interacts with cell membrane receptors (hormones, catechol amines, etc.)

3.   intracellular drugs (antibacterial, antitumor, etc.)

4.   LPs that act selectively on one or more cell types.

The above approaches, requirements for the conditional groups for the LPs, LPs, and polymer carriers are of a general nature, and when choosing a strategy for the synthesis of LPs, it is necessary to proceed from the specific tasks set.

Another group of LPs are water-soluble but biodegradable compounds containing LPs. The terms "microparticles", "nanoparticles", "microspheres" are used to characterize this group of LPs. It is known that some water-insoluble heterochain carrier polymers that can be attached to drugs are capable of biodegradation with the formation of water-soluble fragments under mild conditions in the body. These include some polysaccharides and their derivatives (starch, cellulose derivatives, chitosan, sephadex, agarose, streptokinase, albumin, etc.).

Finely dispersed LPs - "nanoparticles" and "nanocapsules" sized 20-80 nm, containing LPs are capable of producing stable colloidal solutions.

In the case of biodegradability, of great interest are "nanoparticles" containing the appropriate drugs aimed at suppressing tumor cells, since these cells have an increased tendency to endocytosis. According to A.E. Vasiliev, nanoparticles containing therapeutic agents should contain three types of components: a drug that affects the cell, agents that recognize cells of the right kind, and penetration factors that ensure the entry of the drug inside the cell.

The possibility of selective transport of drugs to "target" organs, selectivity, the ability to control the movement of nanoparticles in vitro under the influence of external factors opens up great opportunities in the direction of creating a new generation of promising, original drugs.

The second group of LPs includes polymers with intrinsic activity. In medical practice, LPs with intrinsic biological activity have been widely used for a long time. This group of LPs can be conditionally divided into four groups: neutral polymers, polyanions, polycations, and other polymers (Fig. 1).

Based on the above, it is established that the creation of a new generation of original LPs belongs to the developing field of polymer chemistry and drug technology. The successes in this field, first of all, seem to be related to the purposefulness of the ongoing research, from the correct choice of approaches to their synthesis, elucidation of the mechanisms of action of drugs and establishment of the relationship between the composition, the fine molecular structure and the effectiveness of their pharmacological action.

Among the most interesting directions in the field of creation of PLs that have recently emerged are the creation, establishment of the composition, structure and pharmacological properties of nanostructured PLs and nanoparticles included in the structure of polymers.

These nano-LPs are of great interest in the creation of new drugs to target their transport to organs and target cells.

From the above presented composition, structure, properties, principles of their creation and biological effects nanostructured polymeric therapeutic systems are of great interest for practical medicine. Research in this direction is developing and the practical possibilities provided by LPs are in many cases unique and cannot be achieved using low-molecular-weight drugs.

The third group of medical polymers includes structural polymers. Currently, in practical medicine, structural polymeric materials are widely used, the creation of which is inconceivable without the use of polymers. These include hard and soft structural products such as artificial endoprostheses, vessels, heart valves, cavity fillers, as well as surgical threads, dressing materials, wound coverings, hemo- and enterosorbents, hemostatic materials, and others.

This group of polymers has special requirements that differ from those for medicinal polymers.

Polymeric forms of drugs with intrinsic activity.

The composition, structure, and properties of LPs with "intrinsic" biological activity are varied, as are the properties of low-molecular-weight drugs.

This group of LPs with "intrinsic" biological activity can also include inert polymers intended for the creation of structural polymeric products with long-lasting use, by making artificial locomotor and other (artificial blood vessels, heart valves and organs), which are classified as a separate group and are not considered in this communication. Also, the results of studies of biopolymers, which are directly used in medicines in their pure form as enzymes, hormones, heparin, etc., are not given in this report.

As noted above, soluble neutral polymers were assigned to the group of LPs with "intrinsic" activity. This group of polymers is the most studied and widely used in medical practice. This group includes blood substitutes and plasma substitutes with detoxifying action. They are designed to maintain an acceptable volume of circulating blood and maintain the necessary osmotic pressure over time, which ensures the recovery of blood loss. Then the polymer should be eliminated from the body naturally without accumulation in vital organs. Given the need to introduce a sufficiently large amount of these polymers into the body, it is necessary to strictly control their toxicity. LPs of this type should be practically nontoxic and nonantigenic.

The most widely used antishock blood substitute since the middle of the last century is "clinical" dextran (polyglucin, macrodex, etc.). It is obtained by partial acid gyrolysis of a high-molecular-weight bactericidal polysaccharide dextran.

Dextran by chemical origin is partially branched $1,6$-α-polyglucosine, whose side chains are attached to the main chain, through $1,2$-, $1,3$- and $1,4$- bonds [7]. Dextran-polyglucosine with $M\omega =50,000\pm10,000$ is used in clinical practice. The number of $1,6$-α-bonds in the main chain is 93%, the remaining 7% are $1,3$- and $1,4$- side chains. At the same time, there are no side $1,2$-bonds in polyglucosin.

The therapeutic effect of dextran is determined by its colloidal-osmotic properties and solution viscosity and permeability through capillary vessels.

As a result of many years of research, it was found that dextran with $M\omega> 200$ thousand is toxic. Its fractions with $M\omega \leq 50$ thousand are eliminated through the kidneys, the less $M\omega$ the faster it is eliminated from the body. Dextran in the body also undergoes enzymatic biodegradation with a decrease in Mω, which facilitates its elimination through the kidneys. The final breakdown product of dextran is glucose, which is utilized by the body.

6-O-(2-oxyethyl) starch and poly-α, β-(N-oxyethylaspartamide), and poly-N-(2-oxypropyl)-methacrylamide were considered competitors of dextran as a blood substitute. The latter, as a carbo-chain polymer, has not been subjected to biodegradation and has not found wide application.

Until recently, such blood substitutes with detoxifying properties as low-molecular-weight polyvinylpyrrolidone (hemodez and its corresponding foreign analogues), polyvinyl alcohol (polydez), which is a carbon-chain polymer and cannot biodegrade in the body and dextran (repolyglucin) with Mω = 35±5 thousand were widely used in medical practice. these blood substitutes are capable of forming complexes with both low and high molecular weight compounds in the body due to hydrogen bonds, hydrophobic effects, complex-forming and other non-covalent interactions. The upper limit of the molecular weight of the above detoxicants should be such that the resulting complexes are easily excreted from the body via the kidneys. Because of partial deposition in the body polyvinylpyrrolidone with Mω = 70-100 thousand has not been used in medical practice as an antishock blood substitute for a long time.

Polymeric iodine complexes were widely used in medical practice, where the latter is located in the polymer structure not as an I_2 molecule, but as an I^+ ion, which, unlike molecular iodine, has no toxicity. Polymers containing I in their structure$^+$ had a high antiseptic effect. Similar complexes of iodine with polyvinylpyrrolidone were also known and used in medical practice as antiseptics.

Polymeric forms of drugs with intrinsic activity also include polycations, whose biological activity is very diverse and is associated mainly with their polyelectrolyte nature.

It is known that proteins, nucleic acids, and a number of polysaccharides as biopolymers of the living organism belong to polyelectrolytes.

When injected into the body mainly heterochain biosoluble polymers with oppositely charged functional groups with respect to the body biopolymers, cooperative interactions occur with the formation of polycomplexes with sufficient strength.

Ionenes were among the first polycations whose biological activity was studied in order to find out whether they can be used as drugs in practical medicine.

Ionenes are heterochain polymers that contain quaternary nitrogen atoms at certain distances from each other in the main polymer chain and have the following general structure:

$$\left[-(CH_2)_n-\overset{R}{\underset{R}{N^+}}-(CH_2)_m-\overset{R}{\underset{R}{N^+}}- \right]_P \quad X^-\ X^-$$

Where pi m is the length of the hydrocarbon chain;
R is an alkyl or aryl radical;
X^- is an anti-quaternary nitrogen ion;
P is the degree of polymerization.

Ionenes are usually obtained by the Menschutkin reaction, either by polycondensation of dialkylaminoalkyl halides, or by cationic polymerization with cycle opening followed by alkylation.

Ionene polymers have high bactericidal activity, which depends on the ionene structure.

The presumed mechanism of the bactericidal action of ionens is explained by their adsorption to bacterial cell walls.

When ionens penetrate the bacterial cell, polyelectrolyte complexes with DNA can be formed. The strength of the complex is in direct dependence on the ionen structure. It was found that the ganglion-blocking activity of the ionens is in direct dependence on the number of methylene groups n andm in their structure.

For a long time the direct interaction of ionens with biopolymers of the organism heparin has been proved as follows. During the interaction of sulfate and sulfamine fragments of heparin with ionenes strong polyelectrolyte complexes are formed that do not have anticoagulative properties [8].

Taking this mechanism into account, a polymeric preparation polybren was created that can neutralize heparin injected into the bloodstream during surgeries with the use of a heart-lung machine.

It was found that the activity and toxicity of the ionenes depend on their structure and the density of their charge. High anti-heparin activity of ionenes where n=2, m=5 and x=Br (protamine sulfate) was shown in relation to ionenes where n=3, m=6 and x=Br (polybren). It was found that protamine sulfate is less toxic with respect to polybren and has higher anti-heparin activity.

Ionenes are not the only polymers capable of neutralizing heparin. Other polyionic polymers, such as poly-(tret-amines) containing additional amide and carboxyl groups, have similar properties [9].

The natural aminopolysaccharide chitosan containing secondary amino groups can also be conventionally referred to the group of polymeric ionens.

A chitosan derivative, sulfohitosan (sulfoparin), containing sulfate and sulfamine functional groups, with anticoagulative properties similar to those of heparin, has been synthesized [10].

Biosoluble and biodegradable polyanions also belong to the polymeric forms of drugs with "intrinsic" biological activity.

At present, the biological activity of a number of polyanions has been established; in particular, their antitumor, antiviral, immunomodulatory, interferon-inducing, and other biological activities have been noted.

The biological activities of polyanions can be explained by the fact that most human body biopolymers belong to polyanions. In this regard, it can be assumed that the biological activity of exogenous polyanions is related to competitive mechanisms rather than to the formation of polycomplexes, as in the case of polycations. A number of reviews [11-17] and a large number of original studies performed in recent years [18-20] have been devoted to biologically active polyanions. In this communication we have made an attempt to determine the basic principles determining the biological activity of polyanions and the prospects for their application in the context of our own studies.

It is known that polyanions containing sulfogroups, such as polyvinyl sulfonate, polyvinyl sulfate and dextran sulfate, sulfate chitosan and other biosoluble polymers are to some extent analogs of the sulfo-derivative polysaccharide, heparin.

They have anticoagulant and antitumor activity [21], are interferon inducers. Höchst produced a sodium salt of polyethylene sulfonate, the drug Pergalen, as an anticoagulant, which was the first synthetic polyelectrolyte used in clinical practice; however, due to its high toxicity, its production was discontinued.

Heterochain polyanions containing carboxylic groups, such as carboxymethylcellulose, carboxylcellulose, carboxymethyl dextran, alginic acid and their sodium salts are less toxic than polysulfates.

Carbo-chain polycarboxylates, such as polyacrylic acid and polymethacrylic acid, copolymer of ethylene with maleic anhydride are relatively toxic and not susceptible to biodegradation due to the carbo-chain nature of the polymer main chain.

It was found that the main criterion for high biological activity of polyanions is a high density of carboxyl groups on the main chain of the polymer and carboxyl groups should be isotactic or syndiotactic and have MM≥ 1.0 thousand.

The polyanions described above are carbon-chain polymers and are not capable of biodegradation in the body. Biodegradable polyanions include carboxymethylcellulose and its periodate-oxidized modification by unsubstituted hydroxyl groups in the C position$_2$ and C_3 periodate-oxidized amylose, starch, dextran, microcrystalline cellulose, and alginic acid. These polymers have a high density of carboxyl groups and contain "weak acetal bonds. They exhibit weak antiviral activity, can weakly inhibit tumor growth, and have fairly low toxicity.

For a sufficiently long circulation of these polymers in the body, provided that they are eliminated from the body by biodegradation, the optimal values of their MM should be in the range of 30-50 thousand.

Based on the results of studies of both heterochain and carb-chain polyanions, we can conclude that the relevant task is to "separate" and establish the optimal activity and toxicity for each type of polyanions. Toxicity of the polyanions increases at MM ≥ 50 k, when their antitumor activity is preserved even at MM < 10 k (at these values of MM the polyanions are low-toxic). It is also shown that the antiviral activity of polyanions is relatively high at MM> 30 kM.

The mechanism of antiviral action of polyanions is apparently associated with the activation of macrophages and inhibition of viral replication in the early stages of viral infection.

Considering the above, apparently the most rational clinical application of polyanions lies in the combination or chemical attachment by different types of chemical interactions with low-molecular-weight antiviral drugs and antiviral vaccines, in which polyanions can act as immunoadjuvants. The immunoadjuvant activity of polyanions is mostly related to their negative charge. For example, the immunoadjuvant activity of the polyanion dextran sulfate is quite high when the original dextran shows no activity.

Interferonogenic activity of polyanionic polymers is known.

Interferon is a protein produced by the body as the fastest response to a viral infection and stimulates the body's nonspecific resistance. Often the production of interferon does not always keep up with the multiplication of the virus in the body. Therefore, induction of interferon in the body by means of interferon inducers is one of the means to control viral infections.

Among polymers, the most active interferon inducer is the synthetic polynucleotides poly(I) and poly(C), which have a large therapeutic scope.

Polyanions exhibit similar activity to varying degrees.

Based on the results of our research together with the company "Niarmedikpharma" a polymeric form of interferon inducer based on polyanion - carboxymethylcellulose and antiviral polyphenol gosipol called "Kagocel" and "CelAgrip" (development of scientists of IHFP AS RUz) have been introduced into medical practice as a preventive and therapeutic agent for viral influenza and acute respiratory viral infections.

The last group of polymeric forms of drugs with "intrinsic" biological activity includes other polymers, which are difficult to group according to their chemical structure. The mechanism of their action with the body is often not fully established and unclear.

One of the conventional groups of such polymers includes moly-N-oxides of tertiary amines containing N-oxide functional groups in the polymer chain or in the side groups.

Poly-2 vinylpyridine-N-oxide was found to have an antifibrotic (antisilicosis) effect.

When administered intravenously or by inhalation, it inhibits the developing silicosis.

The activity of this preparation depends on its MM = 30-150 k and on the degree of transformation of tretamine groups into N-oxide groups and configuration of substituents at the main polymer chain. An assumption was made that the antisilicose activity of poly-N-oxide is related to the adsorption of the weakly basic polymer on cells with weakly acidic SiO_2 due to the cooperative interaction. Other poly-N-oxides such as poly-N-allylpiperidine-N-oxide, poly-N-dimethylaminostyrol-N-oxide, etc. act similarly to poly-2-vinylpiperidine-N-oxide.

All poly-N-oxides are water soluble, non-toxic and have a pronounced antifibrotic activity. In addition to poly-N-oxides, a large number of biologically active polymers with "intrinsic" activity is known. For example, the aminopolysaccharide chitosan has pronounced

347

bactericidal activity. By chemical modification and regulation of the degree of deacetylation a number of chitosan derivatives with different biological activity have been obtained [22-24]. Also known is poly-O-butyl alcohol, vinylin, which is used in medicine for the treatment of wounds, burns, frostbite, ulcers, in gastritis and colitis due to its bacteriostatic, enveloping and anti-inflammatory action. As noted above, polymeric forms of polyphenols with MM=6.5-15 thousand have antiviral activity [25].

Molecular construction of drug polymers. In molecular construction of drugs, the main elements are a biosoluble or biodegradable, relatively inert with respect to the organism, carrier polymer and a low molecular weight or high molecular weight drug substance.

The creation of drugs of this type provides prolonged action of the drug substance, selectivity of its action in relation to the organs of the body, positive effects by changing the hydrophilic-hydrophobic balance, reducing the side effect and long-term preservation in the organism of therapeutic concentration of the drug substance, a significant reduction in toxicity and increasing the specific action of the drug substance.

When creating polymeric forms of drugs by molecular design, there is no need to incorporate drugs into the structure of the polymer when a rapid onset of action on the body is needed, when the drug is used to treat a particular disease in a short period of time or it is administered to the body once.

The positive effect achieved by inclusion of the drug substance in the structure of the inert polymer includes long-term circulation of the drug substance in the body or in the bloodstream due to its gradual release over time from the structure of the carrier polymer. In this case, the therapeutic concentration of the drug in the body is maintained, which ensures reduction of the frequency of its administration into the body and its amount.

A schematic representation of the action in the body of drugs and their polymeric forms is shown in Fig. 121.

As can be seen from the figure, the change in the concentration of a low molecular weight drug in the body over time, in the intervals of its repeated administration has a "scissor" structure. In this case, the concentration of the drug substance over time above the therapeutic dose can adversely affect healthy organs.

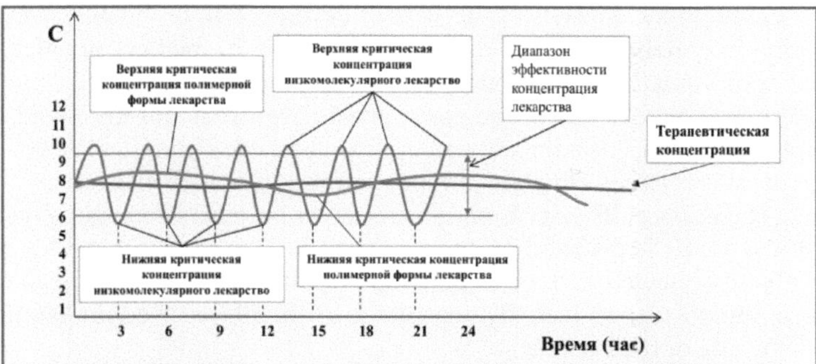

Figure 121. Changes in the concentration of a drug substance depending on the number of times it is injected into the body.

When the concentration of a drug decreases, in time, below the therapeutic concentration leads to a loss of effectiveness of this drug on the diseased organ of the body, the dose can be negative.

When the polymeric form of this drug is injected into the body, its therapeutic concentration is preserved for a fairly long time, which helps to reduce the frequency of its administration in the body, which helps to save this drug and increase the effectiveness of treatment of the disease and reduce its negative effect on healthy organs.

These advantages of polymeric forms of drugs are in direct correlation with the type of chemical interaction (type of chemical bonding) between the drug and the carrier polymer.

The advantages of polymeric forms of drugs include the selectivity of their action in relation to diseased organs, reducing its side effects, toxicity and increasing the specific action of drugs.

Another important form of LV is the regulation of the hydrophilic-hydrophobic balance between the LV and the carrier polymer.

By changing the hydrophilic-hydrophobic balance, it is possible to convert drugs insoluble in water to a soluble state by selecting polymeric carriers with high hydrophilicity.

Using this principle, water-soluble polymeric derivatives of the hydrophobic, water-insoluble polyphenol, gossypol, were obtained through its chemical attachment to macromolecules of the hydrophilic polymer dialdehyde carboxymethylcellulose [26].

Based on the analysis of the state of development in the direction of creating polymeric forms of drugs and their use in medical practice, a number of limitations were found, which include:

1. Adsorption of LP in the gastrointestinal tract is not great and its effect when this LP is administered orally is possible only after the active ingredient is released from the polymer in the stomach or intestine;

2. The permeability of capillary barriers for polymers depending on its MM and MMR is different in different organs. Therefore, when they are parenterally injected into the body, the permeability of the LP through the biomembranes determines the separation of the place of administration from the site of their action.

3. The isolation of polymers from the body is a difficult problem compared to low-molecular-weight drugs.

In particular, the permeability of polymers through the renal biomembranes depends on the MM and the charge of the polymer, the excretion (decomposition) of polymers with bile is slow.

Given the above, it is preferable to use polymers as carriers that are easily metabolized with cleavage of the main chain, thus avoiding their deposition in organs and tissues of the body.

If undesirable effects occur such as overdose, allergies, etc. LPs are difficult to excrete from the body quickly. In view of the above, it is necessary to check the tolerability of LPs before use and accurately establish the required dosage of the drug.

Molecular design of drug polymers requires step-by-step evaluation, selection and control of the following steps: selection of the carrier polymer and evaluation of its biological inertness, acute their chronic toxicity, biosolubility and biodegradability and, if necessary, the possibility of its functionalization, type of bond between the carrier polymer and the low molecular weight drug, selection of the type of the attached low molecular weight drug and the need for its prolonged action in the body, possible ways of attachment of drugs to carrier polymer and selection of the type of chemical

In the molecular design of LPs, not all LPs can be included in the structure of polymers.

The creation of polymeric forms of drugs is relevant and of practical importance:

1. For low-molecular-weight drugs used with sufficiently frequent administration for a long time;

2. For drugs with high toxicity, the use of which is limited.

3. For drugs poorly soluble in water, to obtain their water-soluble modifications.

4. To achieve selective delivery of drugs to "target organs" without their negative effect on healthy organs.

5. To prolong the time of action of drugs in the body.

6. For drugs not stable in storage forms of drugs for the treatment of diseases that require rapid exposure to their diseased organs, for drugs for a single introduction into the body.

The carrier polymer selected as the matrix-base for the creation of new PL should be non-toxic, organotropic and easily biodegradable, since it determines the physicochemical and partially biological properties of PL.

The carrier polymer must meet the following requirements when creating LPs:

1. Must be biosoluble and biodegradable.

2. The MM and MMR of the carrier polymer determines the duration of its circulation in the bloodstream. To have the necessary MM when it is necessary to get the LP inside the cells, by endocytosis, at the same time to excrete them through the kidneys, the MM of the carrier polymer must be low enough.

This contradiction can be resolved by selecting biodegradable carrier polymers.

The carrier polymer must contain reactive functional groups or be easily functionalized. Reactions between the functional groups of the carrier polymer and the low-molecular-weight drug should be easy and unambiguous.

The carrier polymer must be organized, nontoxic, shared with blood, and not antigenic.

The main and widely used hetero-chain carrier polymers in the creation of polymeric forms of drugs include: dextran, starch, carboxymethyl-, methyl-, oxyethylcellulose, alginic acid, chitosan, pectin, gelatin, serum albumin, globulins, antibodies, polyamino acids, and among the carbo-chain carrier polymers - polyvinylpyrrolidone, polyvinyl alcohol, polyethylene polyamine, substituted acrylates and acrylamides, homo- and copolymers of acrylic and methacrylic acids, etc.

In the molecular design of polymeric drug forms, the type of chemical bonding between the carrier polymer and the low-molecular weight compound is of great importance. The nature of the bond between the drug and the carrier polymer is a determining factor in the creation of the target application of the drug. Depending on the mechanism of action, its localization in the body, the chemical bond between the drug and the carrier polymer is of great importance, since the hydrolytic stability determines the mechanism of biological activity of the drug.

Depending on the orientation and localization of LPs, they can be divided into three groups:

1. LPs acting outside the cells. Enzyme inhibitors, anticoagulants and their neutralizers, antibiotics acting on extracellular bacteria, parasites, etc. can be referred to this type of drugs.

From the listed LPs, the LPs should be separated gradually, preserving the original structure, maintaining a minimum therapeutic concentration in the blood, intercellular and other body fluids for a long time. For prolonged circulation in the bloodstream, LP should have a sufficiently high Mm and at the same time endocytosis of LP should be minimal. The rate of cleavage of PL from the carrier polymer in the bloodstream or other body fluids should be such that the bulk of the PL has time to be detached, but its concentration in the body should be commensurate with the rate of excretion or metabolism of the PL. Otherwise, toxic effects and cumulation of bound drugs occur. These negative effects on the body of drugs are related to the nature and type of bonds between the carrier polymer and drugs and can be overcome by selecting the types of chemical bonds. Thus, by selecting the appropriate type of bond between the carrier polymer and the drug, its steric and charge environment, it is possible to regulate the rate of drug release from the carrier polymer and influence the activity and duration of drug action on the organism.

2. LPs acting on the cell surface. This type includes LPs whose activity is due to their action on the receptors on the surface of the target cell membrane. In this case, there is no need for LP penetration inside the cell.

In this type of LPs, it is desirable to preserve their activity also in polymeric form, i.e., their activity on the cell surface should be preserved without hydrolytic detachment of LPs and LPs should directly affect the cell receptors.

Other variants of the interaction of PL on the cell surface include PL bound by cell-specific polymers. This type of polymers is not active with respect to the cell surface, and from it the LV is detached on the cell surface by its enzymes or by short-lived enzymes secreted by the cell. This mechanism of action of PL on cells is not fully understood.

3. intracellular LPs. This group includes LPs capable of penetrating into cells through the cell membrane, such as antibacterial and anticancer compounds. The mechanism of penetration of low-molecular-weight drugs and LPs differs significantly. In particular, while LPs enter cells by passive diffusion or active transport, the mechanism of LP penetration into cells is endocytosis, which distinguishes their therapeutic effects.

The endocytosis mechanism of PL in the cell has a number of advantages, in contrast to the penetration of low-molecular-weight drugs into cells. By means of endocytosis of LPs, it is possible to penetrate into cells of LPs that are not able to penetrate through cell membranes. The greatest effect is achieved with the penetration into cells of drugs affecting lysosomes. In this case, the lysosomal enzymes break down the LP into the LV and the carrier polymer, which is also subjected to destruction.

The type of bonding between the polymer-carrier and the LP and the structure of the LP determines the method of its penetration into the cell, the localization of its site of action, and the mechanism of action of the LP on the organism. The binding of a drug to the carrier polymer can be performed by various types of chemical bonds, such as covalent, ionic, coordinational, cooperative, etc. Among these bonds, covalent and ionic bonds are of great importance. Ionic binding of PL of polymeric nature, such as proteins to carrier polymers, i.e., polyelectrolyte complexes, can lead to a gradual release of the active principle (protein) from the carrier polymer, since PL is in "defect" loops and is released from the carrier polymer in the process of restructuring the complexes in the body. In most cases of ionic binding of low-molecular-weight LV to polyelectrolyte carrier polymers, the bond is not strong enough and the LV is rapidly destroyed by changes in pH and ionic strength of the medium.

The most acceptable type of bonding between the LV and the carrier polymer is covalent bonding of various types. Covalent bonds can be divided into four types according to their stability:

1. labile bonds gradually hydrolyzed without the participation of enzymes, in consequence of changes in pH and ionic strength in the body;

2. relatively labile bonds, which are subject to slow hydrolysis without the participation of enzymes, although they break down quickly depending on steric and charge effects.

3. relatively stable bonds, which are hydrolyzed only enzymatically at an appreciable rate.

4. stable bonds, which in most cases are neither in vitro nor in vivo. Such bonds include amine, azo- and ester bonds. They can be subjected to disintegration by other mechanisms under the influence of neighboring groups, but not by the hydrolytic mechanism. The kinetic parameters of LV and LP excretion are variable, the course of the process is complex and not complete.

Chemical attachment of the LP to the carrier polymer can be performed directly and by incorporating reactive groups into the LP and the carrier polymer. Each of these approaches has its own advantages and disadvantages, and the choice of a PL strategy depends on the specific requirements for the PL.

Specific transport to the LP organs is the main factor determining their selectivity.

Three levels of LP selectivity are known to act on the surface or inside cells [27,28]. At the first level, the carrier polymer has a negative effect on non-target cells when the LP enters the body.

At the second, higher level, the concentration of PL around the "target cell" is quite high compared to other cells, in other words, the concentration of PL around non-target cells is lower or at the level of low-molecular-weight drugs introduced into the body.

The third highest level of selectivity is when the LP acts exclusively on the target organ cells.

For LPs acting on the cell surface, the first two levels of selectivity are acceptable, resulting from the hydrophilic-lipophilic effect of the LP in its distribution between plasma, lymph and intercellular fluid.

In this case, due to the low permeability of LPs through the capillary and other barriers of the body, the drugs attached to polymers are transformed from generally active substances to locally active or limitedly active ones.

In the case of intracellular LPs, their selectivity is achieved in addition to the first two levels by changing the way they penetrate the cell.

Cells capable of increased endocytosis absorb more PL than other cells. For example, polyanions are susceptible to increased endocytosis and are of great interest as a good carrier for drugs with intracellular effects and a second level of selectivity is easily achieved.

The third level of selectivity is achieved when LPs recognize certain types of cells (target cells) through specific effects on them. Recognition is achieved when LPs are exposed to the surface of the cell membrane. In this case, the LP remains on the cell surface or penetrates into the cell, through endocytosis. The ability of the LP to penetrate the cell is determined by the specific distribution of the carrier polymer on the side of the cell surface elements of the desired species, as in the case of affinity chromatography, but in the case of cells the process proceeds in vivo.

Carrier polymers with cationic functional groups are mostly sorbed on the cell surface and their selectivity is determined by the negative charge of the cell surface. Taking this into account, it is necessary to take into account the chemical and immunological properties of the target organs to select the necessary ligands for attachment to the carrier polymer and the necessary drugs when designing the LPs.

Based on the above, we can conclude that the problem of LP specificity and penetration into the target cells has not been fully studied and there is fragmentary information on the penetration of LP into tumor target cells.

All of the above types of LPs are water-soluble compounds.

However, the drugs can also be attached to water-soluble carriers that are able to interact with receptors on the surface of insoluble carrier polymers. In this case, the drug is attached to water-soluble, fine-dispersed, biocompatible carrier polymers. Such systems are of interest for diagnostic and research purposes.

The terms microparticle, microsphere, nanoparticle are often used for such insoluble carrier polymers.

It is known that some heterochain polymeric carriers, insoluble in water, to which the LP is attached, under mild conditions can decompose with release of water-insoluble fragments. In particular, the insoluble sorbent sephadex to which the LP is attached is capable of unlimited swelling due to gradual degradation and transition to a soluble state. This group of LPs can also include cross-linked carrier polymers, to which the LPs are attached, which in the body swell and gradually degrade with the transition of fragments into the soluble state.

Another type of microparticles includes drug carriers or microparticles of drugs themselves that are insoluble in water and are able to penetrate into cells in the insoluble state or are sorbed on their surface.

Such drugs include compounds with particle size in the nanometer range with a relatively narrow particle size distribution.

Such biologically active LPs include two types of compounds:

1. Nanoscale carrier polymers containing LVs as spherical nanocapsules or nanoparticles;

2. Nanopolymer systems or nanostructured polymers containing nanoparticles of drugs, including nanoparticles of metals with biological activity.

The above types of LPs can be made selective by incorporating ferromagnetic nanoparticles into their structure, which allows in-vivo control in the body, nanoparticles by means of a magnetic field.

Creation of LPs with micro- and nanoscale particles contributes to an increase in the target transport of drugs to the target organs.

This effect is achieved through selective adsorption or absorption of nanoparticles, by certain types of cells whose surface has a biospecific affinity for the surface of polymeric nanoparticles, which have an increased tendency to endocytosis.

In general, this type of LP must contain three types of components to perform its functions:

- The drugs that act selectively on certain cells;
- agents that recognize only certain cells through biospecific effects;
- permeation factors that ensure the entry of drugs into the cells.

In view of the above, we have synthesized a number of biologically active LPs with specific properties.

Based on a water-soluble, functionalized carrier polymer, an antiviral polymeric preparation with interferon-inducing properties was synthesized. In this case, a hydrophobic, water-insoluble natural polyfinol - gossypol - was chosen as a drug.

Due to a change in hydrophilic-lipophilic properties in the process of addition of gossypol to the carrier polymer, the obtained LP is well soluble in water. The obtained LP had direct antiviral activity due to the attached gossypol and had interferon-inducing activity due to the polyanionic structure of the carrier polymer.

This dosage form of the drug was introduced into medical practice as a prophylactic and therapeutic agent for viral influenza and acute respiratory viral infections.

Given the high antiviral activity of these drugs we developed polymer-polymeric nanocomplexes based on the polymeric form of the substance "CelAgrip" and a polymeric substrate of polyanionic Na-carboxymethylcellulose. Varying the conditions of synthesis of polymer-polymer compositions and conditions of formation of films on their basis, transparent biosoluble and biodegradable antiviral films were obtained, where the polymeric form of the substance "CelAgrip" in the matrix is evenly distributed with a particle size of 20-35 nm. The obtained biodegradable films showed high antiviral activity against ophthalmoherpes. Thus, antiviral biosoluble nanostructured eye films "GlazAvir" for the prevention and treatment of eye diseases of viral etiology were synthesized for the first time.

Another area of research in the creation of new LPs was the formation of metal nanoparticles with bactericidal and bacteriostatic properties in the structure of a polyanionic carrier polymer, Na-carboxymethylcellulose.

Silver, which has bactericidal and bacteriostatic properties, was chosen as a biologically active metal.

The synthesis of silver nanoparticles in the carrier polymer structure was carried out by photo-irradiation of AgCMC solutions in an excess of Ag+ ions. Solutions, hydrogels, films and LP powders containing silver nanoparticles of spherical and rod-shaped shape sized 5-250 nm were obtained by varying the ratio of the reacting components and their concentrations and reaction parameters. Stabilization of silver nanoparticles in the structure of the polymer matrix is explained by the envelopment of silver particles by macromolecules of polyanion capable of mutual repulsion. The obtained LP exhibited high biological activity against a wide range of bacteria and fungi.

To date, we have achieved certain successes in the creation of new, original polymeric forms of drugs and medical devices.

In particular, a polymeric form of interferon inducer with direct antiviral activity with the following structure was synthesized by nucleophilic substitution of NaCMC avoiding oxidation product with natural polyphenol - gossypeol:

Where gossypol is attached to the polymer chain by a semi-acetal bond. The synthesized drug was subjected to medical-biological and clinical trials at the Gamaleya Research Institute of the Russian Academy of Medical Sciences. This drug is registered in Russia and since 2007 its oral form under the name Kagocel has been established at Niarmedic Pharma.

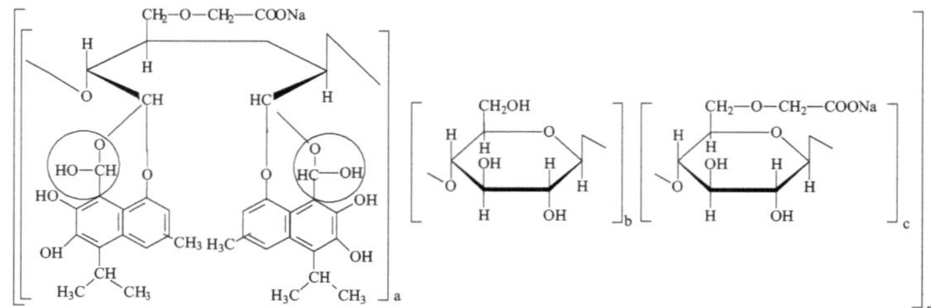

Fig.122. Structural formula of Cagocel.

Kagocel is intended for the prevention and treatment of viral influenza, acute respiratory viral infections and herpes in adults and children.

Today Kagocel is registered in Russia, Ukraine, Belarus, Armenia, Azerbaijan and Moldova.

We continued the research in this direction and synthesized an analogue of the drug Kagocel with the following structural formula.

A drug called CelAgrip has passed the full stage of pre-clinical and clinical trials in Uzbekistan and is approved for use as a prophylactic and therapeutic agent in the treatment of viral influenza, acute respiratory infections and herpes.

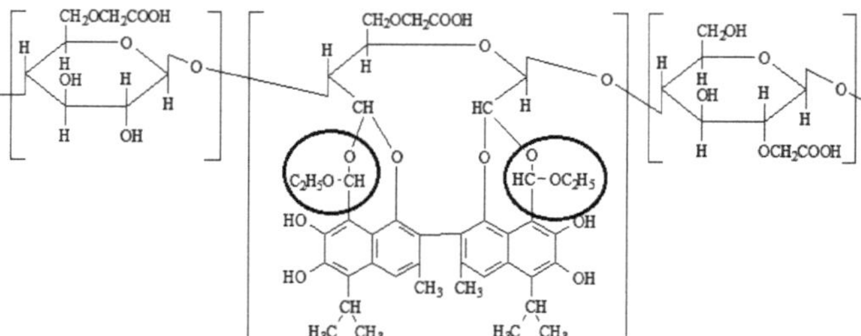

Fig.123. Structural formula of CelAgrip.

Since 2010, the institute's pilot production facility has been manufacturing the substance of CelAgrip, and its dosage form of 0.1 g tablets containing 12 mg of the substance is manufactured at Radix LLC.

Due to the use of only local raw materials in the production of the drug, the cost of one blister pack #10 of "CelAgrip" is $1.2. The cost of a blister pack of CelAgrip is $1.2 versus the cost of a similar pack of Kagocel, which costs $8-$10. US DOLLARS.

It is known that among diseases of the organs of vision diseases of viral etiology account for a large proportion. Today, a number of drugs in the form of eye drops are widely used for the treatment of eye diseases of viral etiology.

Among these drugs, the most effective is Ophthalmoferon. To achieve therapeutic efficacy, this drug is injected into the eyeball every hour during the day, and every two hours at night, which causes some discomfort to patients.

Considering the above, we first developed biosoluble eye films based on water-soluble NaCMC and the drug substance CelAgrip

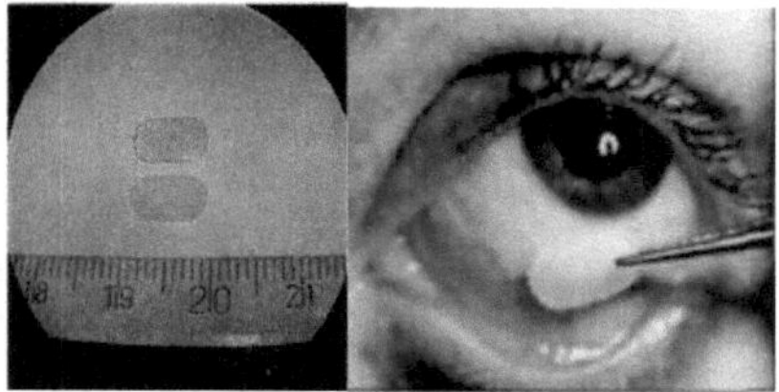

Fig.124. Glazavir antiviral eye medication films.

EyeAvir" eye films are films of purified NaCMC of a certain NW and SP to regulate the rate of their swelling and dissolution containing up to 20% of the substance CELAgrip. Atomic force microscope studies of the GLP structure showed that the CelAgrip substance in the polymeric matrix is evenly distributed and the particles have a rounded shape with a dimension of 30±5 nm.

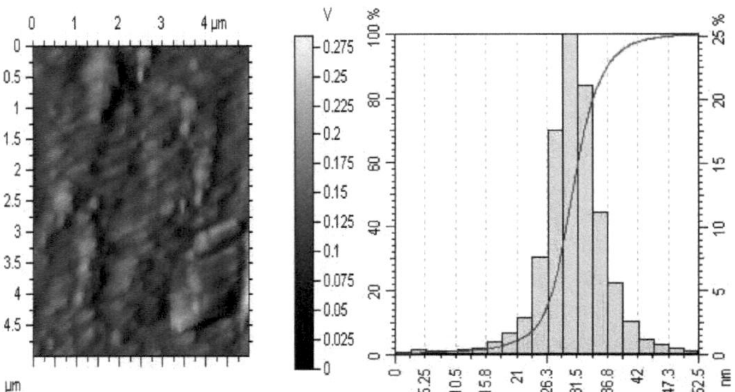

Fig.125. Electron micrographs of HLPs and homogeneity of substance nanoparticles distribution in the film structure.

The oval films, measuring 3-4.5 mm in width and 7-9 mm in length and 20~100 µm in thickness, were injected into the eyeball once a day, which helped maintain therapeutic concentration of the active ingredient at the required level. The films do not interfere with vision due to the fact that CelAgrip substance is approved as an antiviral agent at the clinic of the Institute of Immunology of the Academy of Sciences of the RUz by agreement of the patients, limited clinical trials of GlazAvir were conducted and positive results were obtained.

Given the spread of tuberculosis in recent years in many countries on the one hand and the emergence of high resistance to the known anti-tuberculosis drugs of Mycobacterium tuberculosis, we conducted studies of a polymeric, prolonged form of anti-TB drug that has no resistance "Celazon".

Celazon is a water-soluble powder containing 20-25% of the polymer bound to the macromolecule through the azomethyl bonding of isonicatinic acid hydrazide.

Medical and biological tests have established that to maintain the therapeutic concentration of the anti-TB drug in blood HINC is administered orally three times a day. When Celazon is administered once a day, the therapeutic concentration of HINC in the blood is maintained for 48-72 hours. Studies have shown that Celazone has no resistance when administered in the body for a long time.

In recent years in many countries of the world there has been a boom towards the creation of biomedical drugs and medical devices based on silver compounds and nanoparticles. This is due to the relatively harmlessness and lack of resistance in silver preparations.

The main problem in the creation of medical and biological drugs based on silver nanoparticles is the instability of these drugs during their storage. This is due to the fact that reducing the size of silver particles to nanoscale values increases their surface energy, contributes to their agglomeration with their subsequent precipitation. And the biological activity of silver varies in a wide range depending on the size and shape of silver nanoparticles, which has been unequivocally proved before us by the literature data.

Based on the results of our research, we found a way to stabilize silver nanoparticles in the solution of ionogenic natural polymers during their photo-irradiation with UV rays.

Based on the results of spectroscopic studies of nanocomposites of silver nanoparticles and NaCMC we made the assumption that the negative carboxylate anion of CMC is a trap for the positively charged silver cation and the reaction mechanism and sequence according to the Moth-Henry principle can be considered as an electron-stimulated atomic process.

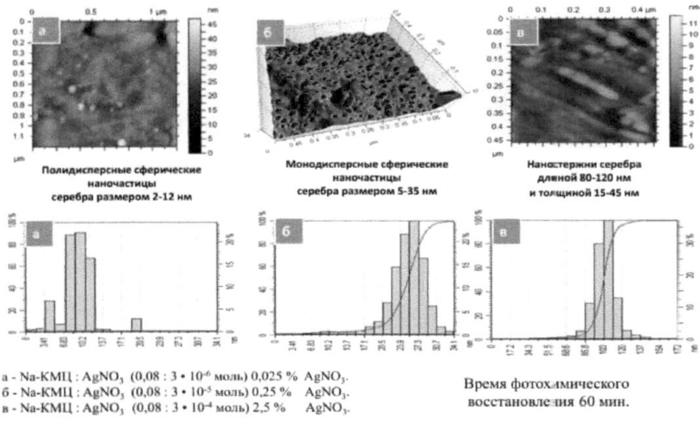

a - Na-КМЦ : AgNO$_3$ (0,08 : 3 • 10^{-6} моль) 0,025 % AgNO$_3$.
б - Na-КМЦ : AgNO$_3$ (0,08 : 3 • 10^{-5} моль) 0,25 % AgNO$_3$.
в - Na-КМЦ : AgNO$_3$ (0,08 : 3 • 10^{-4} моль) 2,5 % AgNO$_3$.

Время фотохимического восстановления 60 мин.

Fig.126. AFM images and histograms of silver nanoparticles distribution in CMC gels.

As can be seen from the figure at low concentrations of silver ions formed nanoparticles of spherical shape size 2-30 nm, and with increasing concentrations - rod-like structure size - the length of 230-700 nm and width of 40-80 nm.

Based on the results of the research, the drug "Aspayk" was created for practical surgeons to prevent the process of adhesions in the internal organs in the postoperative period.

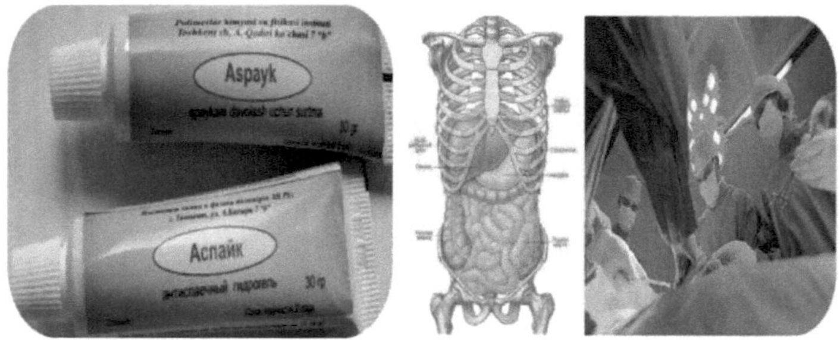

Fig.127. Aspike gel for prevention of adhesions in patients during surgical intervention.

Aspike creates a temporary barrier between the organs where the serous membrane is traumatized or disrupted during surgery.

For Aspayk, preclinical trials have been completed and the drug is in the clinical trial phase.

On the basis of bactericidal nanocomposition of NaCMC and silver nanoparticles we created a hydrogel "Baksergel" intended for the treatment of bedsores, porphytic ulcers and burns.

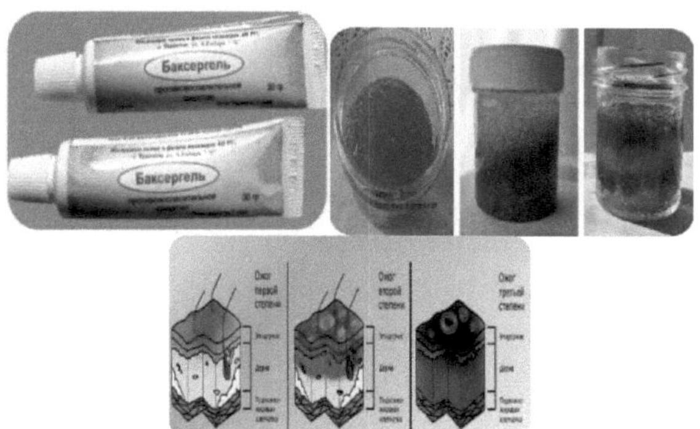

Fig. 128. "Baxergel" - hydrogel with silver nanoparticles for treatment of burns and cervical pathologies.

Due to its high bactericidal activity, the drug promotes rapid healing of wounds and burns, eliminating the formation of pus.

When applying Baxergel to a wound, it is not necessary to remove it because of its ability to resorb. The wound is not traumatized during its treatment.

Preclinical trials have been completed on Baxergel and it is in the clinical trial phase.

Based on the bactericidal nanocomposition of NaCMC and silver nanoparticles, a technology for the production of bactericidal biodegradable films for the treatment of wounds and burns was developed.

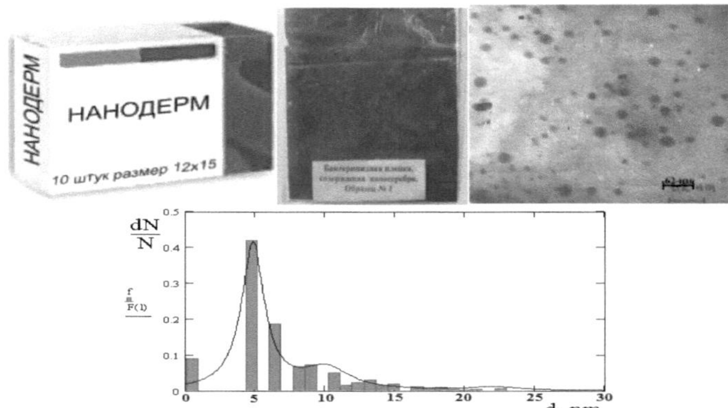

Fig.129. "Nanoderm" is a bactericidal, biodegradable film with silver nanoparticles.

Bactericidal film Nanoderm size 12-15÷18-24 cm and a thickness of 40-120 microns after packing is subject to self-sterilization in 48 hours. It is highly bactericidal, does not traumatize the wound if necessary, it is easily removed from the surface of the wound by washing out and does not traumatize the wound.

When applied to burn wounds, it significantly reduces evaporation of fluid from the wound surface and performs part of the function of the skin.

For Nanoderm drug, biomedical trials are nearing completion.

Based on the nanocomposition of NaCMC and silver nanoparticles the technology was developed and pilot production of bactericidal solutions for the treatment of cotton articles and materials was mastered. The content in the solution of purified NaCMC determined by NW and SP is 1.0-2.0%, and the content of silver nanoparticles is 0.0648±0.01%.

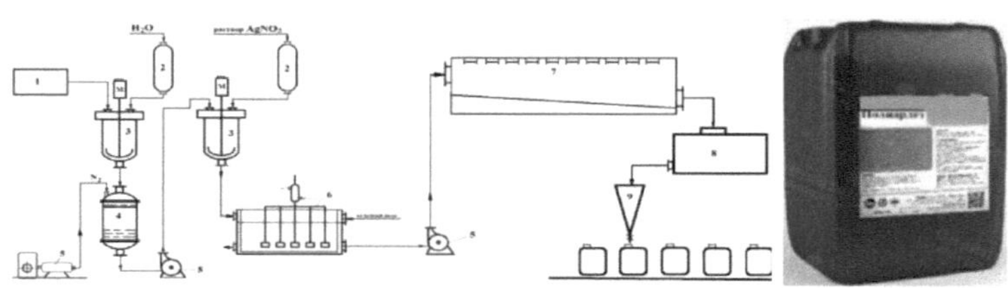

Fig.130. Polyardez polymeric disinfectant solution with silver nanoparticles.

On the basis of LLC "IP Mirus Textile Group" on the basis of a license agreement established production of bactericidal socks.

Manufactured socks have antifungal effect, reduce the level of perspiration of the feet and prevent unpleasant odor.

Unlike similar socks "Socks 47" produced in Russia and "Silver Socks" produced in Iran, our socks have high bactericidal activity and most importantly can withstand 10-15 washings with known detergents without reducing the bactericidal activity.

Thus, using nanotechnology techniques, it is possible to create a new generation of nanostructured drugs and medical devices with unique properties.

However, at present, the long-range effects of nanoparticles in the body are not fully elucidated. To find an answer to this question it is necessary to conduct in-depth basic research, which requires a large expenditure of financial resources.

Based on the above studies, we can conclude that the creation of new PLs in this way belongs to the developing field of chemistry, biology and pharmacology of high-molecular weight compounds. The basic principles of creating new LPs considered in this communication should be taken into account in the molecular design of polymeric forms of biologically active compounds.

The successes in the field of LP synthesis will apparently be related to the experimental elucidation of the mechanisms of LP action and the choice of targeted ways and approaches to their synthesis.

Successes in the direction of creating new drugs with targeted action on the body will be associated with: a correct and scientifically based choice of polymer carriers, drugs, the establishment of a fine molecular structure of drugs and the effect of their biological activity, the implementation of targeted transport of drugs to the target organs and mechanisms of penetration of drugs and drugs into cells.

In the creation of new LPs, the covalent bonding between the LP and the carrier polymer is of great importance.

In contrast to LPs, in the case of LPs the concept of structure is much broader. In this case, in addition to the structure of the LP and the carrier polymer, compositional and structural heterogeneity play a significant role, since the interactions of the LP with the body macromolecules are cooperative, i.e., the polymer effect is manifested, with the regularity of structure to achieve the stability of the resulting poly-complexes between the LP and the body being a determining factor.

Structural heterogeneity of LPs can often lead to various biological effects, including the manifestation of toxicity.

The stereochemistry and the MMR of the polymer play an important role in the manifestation of the biological activity of LPs. In the case of polyanions, it was found that biological activity and toxicity are a function of the MMR. With this in mind, when creating LPs, it is necessary to isolate the narrowest fractions of the MMR and study their activity.

Based on the above, we can conclude that LPs and the practical possibilities of their creation in many cases are unique and cannot be achieved with low-molecular-weight LPs.

LITERATURE

1.	Shtilman M.I. Biomaterials are an important area of biomedical technologies // Bulletin of Russian State Medical University, 2016.

2.	Vasiliev A.E. Medicinal polymers. Results of Science and Technology, 1981. vol.-16. p. 3-119.

3.	Shtilman M.I. Polymers for medical and biological purposes. Moscow: Academkniga; 2006. 400 c.

4.	MilushevaR Yu, Inoyatova FH, Batyrbekov AA, Rashidova Sh. "Hepatoprotective, anticoagulant and immuno-stimulating properties of chitosan BOMBYX MORI derivatives", Advances in chitin science, 2011, XI, 180-184

5.	Yunusov KE, Rashidova SH, Sarymsakov AA, Atakhanov AA Medico-biological polymers and prospects for their creation. Pharmaceutical Journal. Tashkent, 2011. - № 3. - C. 27-35.

6.	Kasymova G.Z., Sabirova R.A., Milusheva R.Yu The effect of different forms of chitosan on the redox processes in the liver in the metabolic syndrome Scientific and practical journal "Physician-graduate" № 3.3 (52), 2012 S.432-437 AcadSci book "Russia.

7.	Sidebotham R.L. "Adv. Carb. Chem.", 1974, 30, 371 - 444.

8.	Rembaum A. "Applied Polymer Symp.", 1973, no. 22, 229 - 317.

9.   Ferutte P. "IeFarmaco, Ed. Sci.", 1977, 32, no. 3, 220 - 236

10.S.Sh.Rashidova, R.Yu.Milusheva, Advances in Chitin Science 10th International Conference of the European Chitin Society, Volume X1 St.-Petersburg, 2011, pp.230-235.

11.Regelson W. In: "Water-Soluble Polymers". Ed. by N. M. Bikales. N. Y. - L., Plenum Press, 1973, p. 161-177.

12.Butler G.B. "J. Polymer Sci., Polymer Symp. 1975, № 50, 160-180.

13.Regelson W., Morahan P., Kaplan A. In: "Polyelectrolytes and their applications". Eds. A. Rembaum, E. Sefeny. 1975, 131-144.

14.Breslov D. S. "Pure Applied Chem.", 1976, 46, no. 2-4, 103-113.

15.Ottenbrite R. M., Regelson W., Kaplan A., Carchman R., Morahan P., Munson A.. In: Polymeric Drugs. Eds. L. G. Donaruma, O. Vogl. N. Y. - San Francisco-L., Acad. Press, 1978, p. 263-303.

16.Levy H. B., Ibid, p. 305-329.

17.Regelson W. "J. Polymer Sci., Polymer Symp.", 1979, no. 66, 483-538.

18.Sarymsakov AA, Junusov HE, Atakhanov AA, Rashidova SH, "Medical and biological polymers and prospects for their creation" Pharmaceutical Journal, 2011.

19.Yunusova H.E., Atakhanova AA, N.A. Ashurova, A.A. Sarymsakov,. Rashidova S.S. "Physicochemical studies of cotton cellulose and its derivatives containing silver nanoparticles" Chemistry of Natural Compounds: Volume 47, Issue 3 (2011), Page 415-418

20.Vokhidova N.R., Sattarov M. E., Kareva N. D., and Rashidova S. Sh. Fungicide Features of the Nanosystems of Silkworm (Bombyxmori) Chitosan with Copper Ions //Microbiology, 2014, Vol. 83, No. 6, pp. 751-753. © Pleiades Publishing, Ltd., 2014.

21.Regelson W., Holland J. F. "Nature," 1958, 181, no. 4601, 46-47.

22.Kudyshkin V.O., Futoryanskaya A.M., Milusheva R.Yu., Kareva N.D., Rashidova S.S. Synthesis of grafted copolymers of Bombyxmori chitosan and N-vinylcaprolactam. // DANRU, № 2, 2010 P.60-63.

Rashidova S.Sh., Milusheva R.Yu., Yugay S.M. "NanomaterialsBasedonChitosanBombyxmori", Advancesinchitin Science, 2011. XI. 230-235.

24.R.Yu. Milusheva, W. Song, S.Sh. Rashidova // Bionanocomposites Based on Chitosan Bombyx Mori and SepiolitePangel S9, Advances in Chitin Science, 2011. XI, 132-133.

24. A.A. Sarymsakov, Y.B. Lee, S.Sh. Rashidova. Ocular drug films for the treatment of viral eye diseases. Tinbo Korean Gazette, 2014.

26. Ershov F.I., Sarymsakov A.A., Nesterenko V.G., Rashidova S.S. et al. Patent of the Russian Federation № 2270708, 2003; № 2002755, 1993; № 21478, 2000.

27.Ringsdorf H. In: "PolymericDrug Delivery Systems". Ed by Kostelnik B. N. Y., Gordon Breach, 1978, p. 197.

28.Trouet A. In: "PolymericDrug Delivery Systems". Ed by Kostelnik B. N. Y., Gordon Breach, pp. 157-173.

Sarymsakov A.A., Rashidova S.S., Aripova T.U. "New in prevention and treatment of influenza and ARVI", Tashkent 2010, p. -110.

Vokhidova N.R., Nurgaliyev I.N., Ashurov N.S., Yugay S.M., Rashidova S.Sh. Obtaining and stabilization mechanisms of cobalt nanoparticles based on chitosan bombyx mori 2015 CIS-Korea conference on science and technology, 2015. P. 270-273

31. Yunusov H.E., Atakhanov A.A., Ashurova N.A., Sarymsakov A.A.,. Rashidova S.Sh. Physico-chemical studies of cotton cellulose derivatives containing silver nanoparticles. // Journal of "Chemistry of natural compounds" Uzbekistan 2011 № 3. Pages 370-373.

32.Vokhidova N.R., Pirniyazov K.K., Rashidova S.Sh. Formation and stabilization of metal nanoparticles in polymeric matrices. Review article. Proceedings of the Institute of Chemistry and Physics of polymers, Academy of Sciences of Uzbekistan. Tashkent, 2011. - C.128-145.

33.Yunusov HE, Atakhanov AA, AA Sarymsakov, S. Rashidova, SV Lobanova KV, Tashpulatov J. Silver nanoparticles formed on the polymeric matrices and their bactericidal properties. Pharmaceutical Journal, [1] 1, 2010, pp. 55-59.

B.L. Oksengendler, N. Turaeva, S.E. Maksimov, F.G. Dzhurabekova, Features of radiation defect formation in nanocrystals embedded in a solid-state matrix, JETF, vol.138, issue 1 (7), 1-6(2010).

I want morebooks!

Buy your books fast and straightforward online - at one of world's fastest growing online book stores! Environmentally sound due to Print-on-Demand technologies.

Buy your books online at
www.morebooks.shop

Kaufen Sie Ihre Bücher schnell und unkompliziert online – auf einer der am schnellsten wachsenden Buchhandelsplattformen weltweit! Dank Print-On-Demand umwelt- und ressourcenschonend produziert.

Bücher schneller online kaufen
www.morebooks.shop

Printed by Books on Demand GmbH, Norderstedt / Germany